ATOMIC SPECTRA

H. G. Kuhn, F.R.S.

ATOMIC SPECTRA

1962

ACADEMIC PRESS

NEW YORK

FIRST PUBLISHED 1962

Published throughout the world except in
the United States by
LONGMANS, GREEN & CO. LTD.

Published in the United States by
ACADEMIC PRESS INC.
111 Fifth Avenue, New York 3, New York

Printed in Great Britain by
J. W. Arrowsmith Ltd. Bristol

To

James Franck

Acknowledgments

We are indebted to the following for permission to reproduce material previously included in their publications:
The Nova Acta Regiae Societatis Scientarum Upsaliensis for material based on a table by B. Edlén; The Physical Review for material based on tables by C. L. Pekeris and H. E. White & A. Y. Eliason; Springer-Verlag for material from *Handbuch der Physik* by H. Bethe.

Preface

When Messrs. Longmans asked me, many years ago, to write a book on atomic spectra, I intended to prepare an English version of the book *Atomspektren* which I had written as part of the *Hand-u. Jahrbuch d. Chem. Physik* (Akad. Verl. Ges. 1934). The war, and afterwards various preoccupations, delayed the work so much that the original plan had to be altered. This is a new and essentially different book, taking account of later developments and changed requirements. Some figures and parts of the text of the German book could, however, be used, and the generous permission to do this freely was given by the Akademische Verlags Gesellschaft and is herewith gratefully acknowledged.

There appears to be a need for an up-to-date book on atomic spectra, treating the subject in an introductory manner, yet more thoroughly than general textbooks on modern physics are able to.

The approach is that of the Physicist, not the Mathematician, starting from observed facts and classical concepts, and intentionally stressing the correspondence between classical and quantum Physics. Mathematical concepts are defined and explained as far as they are essential for an understanding of the basic Physics. For complex mathematical techniques the reader is referred to the original literature and to texts such as *Theory of Atomic Spectra* by Condon and Shortley (Cambr. Univ. Press 1935) and *Quantum Theory of Atomic Structure* by J. C. Slater (McGraw-Hill 1960). The theoretical introduction in chapter II is not intended as a textbook of quantum mechanics but tries to collect formulae and methods in the form required in later chapters.

It is hoped that the book will be useful for more advanced undergraduate—and graduate—work, and also to research workers in the fields of Physics, Chemistry and Astronomy. In working for a degree, the student may omit chapter V and some more specialised paragraphs in other chapters without finding the text incoherent. For the benefit of the research worker, many recent developments which could not be fully described have been mentioned with references to the literature.

To aim at completeness in references was obviously impossible in a book of this kind, and the author wishes to apologise for the necessarily somewhat arbitrary selection of references. Apart from especially important papers, those containing full references to other work have been quoted preferentially. The failure to do justice to Russian literature is regretfully admitted; it is entirely due to technical difficulties.

I am profoundly indebted to Dr. G. W. Series who read through almost the entire manuscript and suggested some most valuable improvements, and also to Dr. D. M. Brink who read some sections and gave some valuable advice. I also wish to express my thanks to Prof. S. J. Foster, Prof. R. Gebauer, Dr. G. Herzberg, Dr. F. S. Tomkins and Dr. M. Fred for making prints of spectrograms available for reproduction, and to Mr. C. W. Band for making some excellent photographic reproductions. Mr. D. N. Stacey and Mr. J. M. Vaughan gave valuable help in proof-reading. I specially wish to acknowledge my indebtedness to the publishers whose understanding, helpfulness and efficiency made the work so much easier for the author.

H.G.K.

Balliol College, Oxford.
April 1961.

CONTENTS

IV. PERIODIC TABLE AND X-RAY SPECTRA

V. COMPLEX SPECTRA

VII. WIDTH AND SHAPE OF SPECTRAL LINES

APPENDIX

PLATES

CHAPTER

I. Introduction

1. The spectroscopic method

The light emitted by atomic gases in a discharge tube is due to the motions of electrons in the atom, and by studying this light we can gain information on the electronic structure which determines the physical and chemical properties of the atom. In principle, we could investigate the light in two different ways; we could set up a very sensitive oscilloscopic apparatus to record the electric vector **E** of the radiation field or one of its components, say E_x, as a function of the time t. Alternatively, we could send the light into a spectroscope which effectively carries out a Fourier analysis of the function $E_x(t)$. The "spectrum" can be regarded as a function $A(\nu)$, a record of the amplitude A as a function of the frequency ν.

The way in which the characteristic features of mechanical motion can be impressed on a light wave and then deduced from the spectrum may be illustrated by the following experiment which could be carried out in actual fact. Light from a highly monochromatic source, of frequency ν_0, is passed through a device which modulates its amplitude E at the frequency δ. It may be imagined as a uniformly rotating disk whose optical density varies sinusoidally as a function of the angle of azimuth or, more realistically, as an electro-optical Kerr cell operated by a radio-frequency generator. The amplitude of the emergent light is then given by

$$\begin{aligned} E(t) &= (a_0+a_1 \cos 2\pi\delta t) \sin 2\pi\nu_0 t \\ &= a_0 \sin 2\pi\nu_0 t+\tfrac{1}{2}a_1 \sin 2\pi(\nu_0+\delta)t+\tfrac{1}{2}a_1 \sin(\nu_0-\delta)t. \end{aligned} \tag{1}$$

If the light enters a spectroscope of extremely high resolving power, three spectral lines of frequencies ν_0, $\nu_0+\delta$ and $\nu_0-\delta$ and intensities $a_0{}^2$, $(a_1/2)^2$ and $(a_1/2)^2$ will be observed. Eq. (1) can be regarded as a simple example of a Fourier expansion of the function $E(t)$. The statement that the amplitude function $A(\nu)$ has finite values only for the three values ν_0, $\nu_0+\delta$ and $\nu_0-\delta$ of the variable ν and assumes the values a_0, $a_1/2$ and $a_1/2$ specifies the function $E(t)$ completely, except for the phases which are usually unimportant.

If the motion happened to be of a more complicated type, but still periodical, lines of frequencies $\nu_0\pm 2\delta$, $\nu_0\pm 3\delta$, . . would be observed, and a knowledge of their amplitudes and phases would enable us to derive the function $E(t)$.

If the motion is not strictly periodical, the spectroscope will no longer show perfectly "sharp" spectral lines. If the frequency or transparency of the Kerr cell drifts slowly or if it changes abruptly after long intervals of constancy, the spectrum will be continuous with pronounced peaks at approximately the frequencies of the spectral lines observed for strictly periodic motion.

Of the two kinds of information mentioned, the second one which consists in the knowledge of the Fourier amplitudes $A(\nu)$ or their squares $I(\nu)$, is not only easier to obtain for light emitted by atoms, but also proves to be of a specially useful and convenient form: atomic spectra are found to consist, to a high degree of approximation, mostly of spectral lines. The Fourier analysis performed by the spectroscope shows essentially a "discrete" spectrum, a number of definite values of ν for which the intensity $I(\nu)$ is finite while practically vanishing for other values of ν.

In any mechanical system, a discrete spectrum of frequencies is characteristic of periodic motions which are confined to a definite part of space, such as planetary motions in contrast to cometary motions; they are conveniently referred to as *bounded* motions.

It was the principal aim of the early spectroscopists to find an interpretation of the spectra of atoms in terms of a *model*, by applying the classical laws of mechanics and electrodynamics to the motion of particles having suitably chosen properties. Even in this general form, the problem led to difficulties of a fundamental nature: the integral ratios of frequencies (higher harmonics) which every classical model would yield, were not found in atomic spectra. But further difficulties arose when classical laws were applied to the more definite model which Lord Rutherford and others had evolved on the basis of well-established experimental facts such as the scattering of α-particles by atoms.

The atom is now known to consist of a nucleus, of positive electric charge $Z e$, and of Z electrons, each of which has the charge $-e = -4{\cdot}8029 \times 10^{10}$ esu and the mass $m = 9{\cdot}1085 \times 10^{-28}$ gm. The atomic number Z assumes all integral values from hydrogen, $Z = 1$, to the heaviest element known at present, $Z = 102$. Even for the lightest elements, the nucleus is so heavy, compared with the electrons, that it can be regarded as a fixed centre of attraction, except for the description of some very fine details of spectra. All the main properties of each atom are therefore determined by the single parameter Z.

Goudsmit and Uhlenbeck[1] later extended the model by replacing the *point electron* by the *spinning electron*, endowed with an intrinsic angular momentum as well as with mass and charge.

Under the influence of the electrostatic attraction by the nucleus, together with the mutual repulsion, the electrons would be expected to carry out complicated, planetary motions. But, as one can most easily see for the hydrogen atom with one electron, the gradual loss of energy by radiation would cause the electron to approach the nucleus more and more closely, emitting a continually increasing frequency, and finally to fall into the nucleus.

The laws of classical mechanics thus proved to be incapable of explaining either the structure of spectra or the stability of atoms, and new and more general laws had to be evolved. The empirically found relationships in atomic spectra, more than any other facts, showed the way to these new laws of quantum physics. The "old" quantum theory of Bohr and Sommerfeld which formed the first step in this development is still frequently used, mainly because of its close relation to classical mechanics. For rigorous, quantitative work, it has been superseded by the more complete and self-consistent methods of quantum mechanics. It appears that quantum mechanics in its present form is capable of describing the structure of spectra in their essential features. Numerically, however, the computations are extremely complicated and have not yet been carried very far. It is likely that semi-empirical methods, combining spectroscopic data with theoretical concepts, will continue to play a great part in the study of atoms.

A new phase in atomic spectroscopy started with the discovery of hyperfine structures and isotope shifts of spectral lines. These structures are caused by properties of nuclei other than their charges, properties which are characteristic not only of the different elements but of their different isotopes. Such properties are the mechanical angular momentum, the magnetic moment and the distribution of charge and magnetism in the nucleus. Hyperfine structure research has thus become a valuable tool in the study of nuclei.

Radio-frequency and micro-wave resonance methods have given very accurate values of hyperfine structures and fine structures of ground states and metastable states of atoms, and in the form of double resonance techniques also for excited states. Results obtained by these methods of spectroscopy, taking this word in a wider sense, will frequently be referred to in this book, though its contents will mainly be confined to optical spectra, a term which is meant to include infra-red and ultra-violet and even X-ray spectra.

2. Spectral series and the combination principle

As the accuracy of wavelength measurements increased, mainly owing to the perfection of diffraction gratings, it became more and more evident that atomic frequencies do not occur in the integral ratios which classical mechanics predicts for any model of the atom. Systematic, empirical search for regularities in the frequencies of spectral lines eventually provided the foundations on which quantum theory could develop.

Liveing and Dewar[2] (1879) found that spectral lines could often be arranged in "series"; within each series, the frequency differences and the intensities decrease regularly towards increasing frequencies. The series of lines in an atomic spectrum takes the place of the integral multiples of frequencies (harmonics) of the classical theory.

The starting point for further advance was Balmer's discovery that the series of lines in the visible spectrum of hydrogen could be accurately expressed by a simple formula. It was extended by Rydberg[3] and Ritz[4], and in this generalised form,

$$\tilde{\nu} = R/n^2 - R/n'^2 \qquad (n = 1, 2, \ldots, n' = 2, 3, \ldots) \tag{2}$$

it describes the complete atomic spectrum of hydrogen, from the far ultra-violet to the infra-red. The constant R is known as Rydberg's constant; $\tilde{\nu}$ is the wave number per cm.

Rydberg also found that series of lines in other spectra can be represented by differences of expressions $T_n = R/(n+\alpha)^2$ where α is a constant for any given series and R and n have the same meaning as before. But this latter formula is only an approximation, and other, similar formulae can be used.

The fact, however, that the wave-number can be expressed as the difference of two "terms" represents an accurate and far more general relation. On the basis of the work of Hartley, Rydberg and others, it was first clearly formulated by Ritz and is known as the *Rydberg–Ritz combination principle*. It states that wave-numbers of spectral lines can be expressed by differences of "terms" in such a way that other differences of these terms are again equal to wave-numbers of lines in the same spectrum.

As an example, one may find that the wave-numbers of two spectral lines can be written as differences of four terms

$$\tilde{\nu}_a = T_2 - T_3, \qquad \tilde{\nu}_d = T_1 - T_4$$

in such a way that other differences of these four terms,

$$\tilde{\nu}_b = T_2 - T_4, \qquad \tilde{\nu}_c = T_1 - T_3$$

also appear as wave-numbers of lines in the spectrum. This is equivalent to the statement that exact relations exist between the wave-numbers $\tilde{\nu}_a$, $\tilde{\nu}_b$, $\tilde{\nu}_c$ and $\tilde{\nu}_d$, namely

$$\tilde{\nu}_b - \tilde{\nu}_a = \tilde{\nu}_d - \tilde{\nu}_c,$$

$$\tilde{\nu}_c - \tilde{\nu}_a = \tilde{\nu}_d - \tilde{\nu}_b.$$

Such constant differences are in fact found in all spectra, and many have been established with very great accuracy.

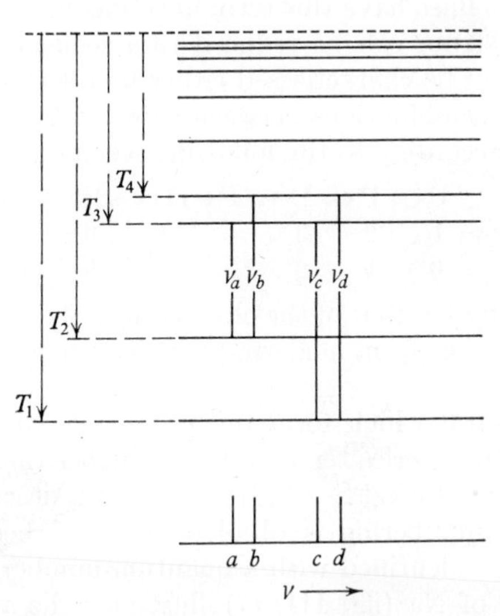

FIG. I, 1. Term diagram and combination principle.

It is customary to represent term values graphically as vertical distances of levels from a conveniently chosen reference level, and the wave-number of a spectral line as level difference, indicated by a vertical line joining the two levels. Sometimes, these vertical lines are laterally displaced in such a way that their horizontal distances equal the wave-number distances of the lines. The spectrum can then be drawn below the term diagram as shown in fig. I, 1. But it is often more convenient to indicate the possible combinations of terms by sloping lines whose lengths and slopes have no direct significance. This method has been employed in the classical collection of term diagrams of simple spectra, by Grotrian(5).

The combination principle allows one to convert the experimental values of wave-numbers of lines into term values, a process known as *term analysis*. The term values contain essentially the same information as the wave-numbers of the lines, but in a form which is more compact and more directly connected with the theory. A series of lines arises from the combination of one term with a series of terms. In simple spectra, all term series converge to a common limit which can be used for defining the zero point of the scale of term values (fig. I, 1). This is, e.g. implied in empirical expressions for term values of the type $R/(n+\alpha)^2$ which vanish in the limit $n \to \infty$. The value of any term is then equal to the wave-number of the limit of any series whose lines have this term in common.

A simple spectrum can be reduced to a sequence of term series each of which can be characterised either by a letter or a number k or l so that any combinations between values of k or l differing by 1 are "allowed", according to the following scheme:

$$
\begin{array}{rcccccc}
 & S \leftrightarrow & P \leftrightarrow & D \leftrightarrow & F \leftrightarrow & G \leftrightarrow & H -- \\
k = & 1 & 2 & 3 & 4 & 5 & 6 \\
l = & 0 & 1 & 2 & 3 & 4 & 5
\end{array}
$$

The numbering by k is that of the old quantum theory. The numbering by l, which arises from modern quantum mechanics, will be used throughout.

In alkali spectra, which form the prototypes of simple spectra, the lowest term of a series tends to be the higher the larger l. In the older literature, the lowest S-, P-, D- . . term was described as 1S, 2P, 3D . ., but this numbering is obsolete and has been replaced by one in which n is identified with a quantum number of an electron.

The example of Na (fig. III, 17) illustrates the arrangements of terms and the nomenclature. If we call n_0 the n-value of the common term of a series of lines, the following nomenclature is often used:

Principal series:	$n_0\mathrm{S} - n\mathrm{P}$
Sharp or *second subsidiary* series:	$n_0\mathrm{P} - n\mathrm{S}$
Diffuse or *first subsidiary* series:	$n_0\mathrm{P} - n\mathrm{D}$
Fundamental or Bergmann series:	$n_0'\mathrm{D} - n\mathrm{F}$.

In these symbols, n is to be regarded as a mere index number, not as a factor. It can assume successive integral values $\geqq n_0$; the second term of any of these expressions is also called *running term*.

These definitions and fig. III, 17 show that diffuse and sharp series have a common limit $n_0\mathrm{P}$ and that the difference of the wave-numbers of the limit of the principal series and the common limit of diffuse and sharp series is equal to the wave-number of the first line of the principal series $n_0\mathrm{S} - n_0\mathrm{P}$. The discovery of this fact in alkali spectra, the so-called *Rydberg–Schuster law*, was one of the important steps in the analysis of series spectra. A corresponding relation exists between the limits of diffuse and Bergmann series and the first line of the diffuse series and is known as *Runge's law*. In this purely formal way the work of the early spectroscopists has brought a great deal of order into the apparent chaos of spectra.

An interpretation of this wealth of data and numerical relations was first achieved by the quantum theory of Bohr and Sommerfeld. By postulating the frequency relation and interpreting the term values as energy values of stationary states of motion, Bohr attributed a new physical reality to the combination principle. The classical experiments of Franck and Hertz[(6)] proved the correctness of these postulates.

The number n of the Balmer formula (1) was now identified with a quantum number n associated with the motion of the electron. For the three degrees of freedom of the motion of a point particle, three quantum numbers appear in the theory, but in hydrogen-like atoms, in the absence of external fields, two of these do not affect the energy or term value, a fact known as *degeneracy*.

In atoms with N electrons the number of degrees of freedom, and thus of possible quantum numbers, is N times greater. The occurrence of comparatively simple spectra with regular series of terms shows that approximate treatments are often possible, ascribing the spectrum to the motion of one electron, in a field of force due to the presence of the other electrons. The coarse structure of the alkali spectra could thus be explained by a single particle model, but the doublet structure of these spectra found no room in a theory ascribing only three degrees of freedom to the electron. The difficulty was resolved by the hypothesis of the *spinning electron*, due to Goudsmit and Uhlenbeck, adding a fourth degree of freedom of a peculiar kind and thus a fourth quantum number to the description of the motion of an electron.

Pauli's principle finally led to a most striking interpretation of the periodic properties of atoms.

A given atomic species possesses several spectra, according to its state of ionisation. The ions which have only one electron, like He^+, Li^{++} . ., have hydrogen-like spectra. Their wave-numbers are given by (2) if R is replaced by Z^2R, where Z is the nuclear charge number. This is the simplest case of the general rule stating that spectra of atoms and ions with the same number of electrons, so-called *iso-electronic* spectra, are similar. The likeness is, however, not nearly as quantitative for atoms with more than one electron.

A frequently used nomenclature may be explained by the example of aluminium:

arc spectrum: spectrum of the neutral atom, symbol Al I,
first spark spectrum; spectrum of the singly ionised atom, symbol Al II,
second spark spectrum; spectrum of the doubly ionised atom, symbol Al III.

3. Spectroscopic data

The primary results of optical spectroscopy consist mainly in the values of the wavelengths of spectral lines. The unit of wavelength is the *Angstrom* unit (A) which was originally defined as 10^{-8} cm. But interferometric comparison of wavelengths with one another proved to be more accurate than comparison with the standard metre, and a new international Angstrom unit (I.A) was subsequently defined as the 6438·4696th part of the wavelength of the cadmium red line emitted under specified conditions of a discharge and measured in dry air at 18°C and a pressure of 760 mmHg. Values of wavelengths in tables (λ) usually refer to air under these normal conditions except for very short wavelengths measured in vacuum spectrographs. Wave-numbers $\tilde{\nu}$ (or σ) usually refer to vacuum and are always given in units of cm^{-1} for which the term *Kayser* (K) has recently been proposed. Thus:

$$\tilde{\nu} = \frac{10^8}{n\lambda}.$$

The accepted value for the refractive index n of air at standard conditions, based on the most recent measurements, is given by[7]

$$(n-1)\times 10^7 = 643{\cdot}28+294981(146-\tilde{\nu}^2)^{-1}+2554(41-\tilde{\nu}^2)^{-1}.$$

The frequency is connected with the wave-number by $\nu = c\tilde{\nu}$. The most recent measurements of the velocity of light have given the value $c = (2{\cdot}997924 \pm 0{\cdot}000008)\times 10^{10}$ cm sec^{-1}.

A commission on wavelength standards and spectrum tables set up by the International Astronomical Union has recommended that a line of the isotope 86 of Krypton be adopted as a new primary standard, so that the Angstrom unit would be re-defined by the relations

$$\lambda_{\text{vac}} = 6057{\cdot}80211\text{A},$$
$$\lambda_{\text{stand. air}} = 6056{\cdot}12521\text{A}.$$

This definintion has been formally adopted by the General Conference on Weights and Measures in 1960. The report of the commission also gives valuable information on standards in the visible and ultra-violet region.

In X-ray spectroscopy, a one thousand times smaller unit, the X-unit, is used, but an empirical Siegbahn X-unit has been defined by the relation 1A = 1002·06 X-U.

In radio-frequency techniques, frequencies are measured rather than wave-numbers, and the accurate knowledge of the value of c is important for comparison with optical data. With the present value of c, one finds

$$1/1000\,\text{cm}^{-1} = 0{\cdot}0333535\,\text{Mc/s}.$$

The most recent wavelength tables are those by Harrison[8]; they are more complete than the older tables of Kayser[9]. The accuracy of the values in such tables varies widely and is often considerably less than the number of decimals would indicate. Most values were measured by means of gratings and their accuracy varies between 10^{-2} and 10^{-3} A. Interferometric comparisons with the primary cadmium standard are far more accurate, and a number of krypton- and neon-lines thus measured have been adopted as secondary standards by the International Astronomical Union[10]. Wavelength standards in the vacuum ultra-violet are less accurately known. They are sometimes derived from wavelengths in the visible and near ultra-violet with the use of the combination principle[11–14].

A table of term values, including their configuration assignment and their Landé g-values, is now available.[15] With the exception of the rare earths which will be covered in a volume to be issued later, these tables are virtually comprehensive, covering all published and much unpublished material up to a specified date. It has superseded the tables of atomic energy states by Bacher and Goudsmit[16] which contains data up to 1938. References to later work on term analysis have been compiled by Shenstone[17] and Meggers[18,19] and by Brix and Kopfermann[20].

Differences of wave-numbers of adjacent lines are sometimes known much more accurately than their absolute values, as the result of interferometric measurements, and are often of greater theoretical interest. Such differences are usually quoted in wave-numbers rather than wavelengths, and for small differences, the relation

$$\Delta\tilde{\nu}/\tilde{\nu} = -\Delta\lambda/\lambda$$

can be used for conversion. The unit $1/1000\,\text{cm}^{-1}$ is convenient for use in high resolution spectroscopy and is sometimes called *Milli-Kayser* (mk).

As mentioned before, term values of simple spectra are often measured from the series limit, and increasing term values correspond to decreasing energy. Owing to the difficulty of establishing series limits accurately in more complex spectra, their term values are generally tabulated on a scale in which the ground state defines the zero-point and term values increase with increasing energy. When spectral lines are described by differences of term symbols such as 2S − 3P, the first term of the difference *always* indicates the level of lower energy, according to the definition of term values in simple spectra.

While wavelengths of spectral lines can be regarded as independent of discharge conditions, unless these are quite extreme or unless the very highest accuracy is required (see VII.E), the intensities depend on the conditions of excitation in the most marked and complicated way. The factors characteristic of the atoms themselves, the line *strengths* or the *transition probabilities*, are therefore known for comparatively few lines only, though they are of considerable interest, especially in astrophysics. As a rough indication of orders of magnitude, wavelength tables usually quote estimated intensities, on some arbitrary scale, and specify the type of excitation (arc, spark, glow discharge).

Term values are sometimes expressed in *electron volts* (ev), according to the relation

$$eV = -hTc.$$

The value of V in volts is the ionisation potential of the atom in the state described by the term value T, if the latter is defined as in simple spectra. The conversion relation is

$$1\,\text{ev} = 8066\,\text{cm}^{-1}.$$

The wavelength corresponding to 1 ev is 12400 A.

4. Survey of the simple spectra

The optical spectra of the elements of the first column of the periodic table and the strongest lines in the spectra of the second and third column can be conveniently described as a special class called *simple spectra*. The same applies to the iso-electronic ions which are included when the terms *alkali-like*, *alkaline earth-like* and *earth-like* spectra are used. Also many of the strongest lines of the elements of the sub-groups form simple spectra; e.g., Cu, Ag and

Au have lines forming alkali-like spectra, Zn, Cd and Hg show alkaline earth-like spectra, and Ga, In and Tl earth-like spectra.

The alkali-like spectra and their term structure will be discussed in detail in chapter III. It was largely the study of these spectra which led to the combination principle and to the concepts which made an understanding of complex spectra possible. In place of the single series of terms of the hydrogen-like atoms, alkali atoms show a sequence of term series, namely one S-series, one P-series, etc., and therefore one of each of the types of line series listed on p. 6.

We can largely understand the structure of the alkali-like spectra by applying the methods of quantum theory to the motion of one electron, the "valency electron", in the field of a central force due to the essentially rigid "core" formed by the nucleus and the remaining electrons. The fact that this force obeys a law different from the inverse square law accounts for the term structure being different from that of the hydrogen-like atoms.

Each term of the alkali-like spectra, except the S-terms, is found to be double. The *doublet* structure can be explained by the magnetic interaction of the electron spin with the orbital motion.

The spectra of helium and of the di-valent alkaline earth elements show two types of lines, *singlets* and *triplets*. The analysis of the spectrum leads to two term systems, one in which all terms are single and one in which all terms are triple except the S-terms which are single. Apart from the multiplet structure, each of these systems resembles that of an alkali atom, with one S series, one P series, etc.; but the lowest S-term of the *triplet* system is missing. Combinations between singlet and triplet terms are "forbidden", i.e. lead only to faint lines.

According to chemical experience, alkaline earth elements have two valency electrons, but in the terms which give rise to these "simple spectra" one electron only is excited at any time. The excitation energy, however, must be imagined as rapidly and periodically passing from one to the other electron in a kind of resonance process which causes the existence of two instead of one system of terms. The magnetic interaction of the spins of the two electrons with the orbital motion produces the triplet structure.

The spectra of the earth-like atoms are very similar to alkali spectra. They show a system of doublet terms, with one S-series, one P-series, etc., and the S-terms are again simple. But the lowest S-term is missing, so that the ground term of the atom is a P-term.

It is found that the *simple* spectra in the first three columns of the periodic table can be explained with the use of the quantum numbers of one electron only, except for the spin quantum number. This is possible, partly because only one electron is excited but also because only one electron has a non-vanishing orbital angular momentum. This definition of "simple" spectra is somewhat arbitrary but will be used in this book for convenience.

5. The complex spectra

The elements of the second and third column—and their iso-electronic ions—show some lines which do not fall under the category of simple spectra. They have generally rather higher excitation potentials and are caused by the excitation of more than one electron or by the excitation of one of the more tightly bound electrons. These and the spectra of most other elements may be described as *complex* spectra. They are sometimes referred to as *multiplet* spectra, because the grouping of lines in multiplets is more conspicuous than the arrangement in series. Only few members of any one series are generally known and the limits to which the term values converge are not always the same for different types of terms.

An important aid in the analysis of complex spectra is the study of the Zeeman effect. Also hyperfine structure investigations can sometimes be of help in the term analysis.

The structure of complex spectra can best be described in connection with their theoretical interpretation. The purely empirical survey in this chapter may be restricted to the statement of a simple rule governing the occurrence of multiplets in the different columns of the periodic table. This can best be expressed in the form of table I, 1,

TABLE 1

1	2	3	4	5	6	7	8
	singlets		singlets		singlets		singlets
doublets		doublets		doublets		doublets	
	triplets		triplets		triplets		triplets
		quartets		quartets		quartets	
			quintets		quintets		quintets
				sextets		sextets	
					septets		septets
						octets	

where the number at the top is that of the column and at the same time the number of electrons outside closed shells. The table applies mainly to the short periods, only the alternation of even and odd multiplicities for odd and even numbers of electrons holds rigorously throughout.

The high multiplicities in the right half of the table are mostly confined to the sub-groups. On the whole there is a distinct tendency towards lower multiplicities from the middle to the right in the table, in the terms of lower energy.

The same relations hold for ions with the same number of electrons. The close similarity between the spectra of atoms and of ions forming an iso-electronic sequence has been mentioned before. As in one-electron spectra, the frequencies increase rapidly with the stage of ionisation, and the spectra move into the vacuum ultra-violet. In spite of this, the spectra of the ions are often more completely known than those of the neutral atoms, and the comparison between iso-electronic spectra has contributed considerably to their interpretation.

CHAPTER

II. Theoretical Methods

A. CLASSICAL MECHANICS OF THE ATOM

1. Multiply periodic systems

Some acquaintance with the fundamentals of the classical treatment of systems of several degrees of freedom is necessary for an understanding of the quantum theory of atoms, and the following paragraphs intend to summarise briefly the main concepts and results. We shall be mostly concerned with the type of motion described as *bounded* or planetary or *multiply periodic*, in which the energy is conserved and in which also the particles remain in a finite part of space. In atoms, the loss of energy by radiation is so slow compared with the periods of the electronic motions that the atom behaves in most respects as a *conservative* system. Only in the discussion of the width of spectral lines will this assumption have to be abandoned.

For a system of one degree of freedom, such as a linear oscillator with a given potential function, the full description of the state of motion requires two parameters, such as the coordinate and velocity at time $t = 0$. But with a suitable choice of one of these parameters as a constant of motion, e.g. the energy, the other merely determines the phase. The motion is singly periodic so that the coordinate or any other variable can be expressed as a single Fourier series, i.e. by the superposition of one simple harmonic motion and its higher harmonics. This can be conveniently written in complex form:

$$x(t) = \sum_{n=-\infty}^{+\infty} A_n \exp(2\pi i n \nu_0 t) \qquad (n = 0, 1, 2, \ldots) \tag{1}$$

where n is an integral number. The Fourier coefficients can be expressed by Euler's formulae, as integrals extended over the whole period τ:

$$A_n = \frac{1}{2\pi} \int^{t+\tau} x(t) \exp(-2\pi i n \nu_0 t)\, \mathrm{d}t. \tag{2}$$

The possibility of using harmonic functions for expansions, and the validity of (2) are based on the property of *orthogonality* of these functions: the integral, taken over the full period, of the product of any one function with the complex conjugate of any other function of the same set is zero, e.g.

$$\int_{t}^{t+\tau} \exp(2\pi i n_1 \nu_0 t) \exp(-2\pi i n_2 \nu_0 t)\, dt = 0 \quad \text{for} \quad n_1 \neq n_2. \tag{3}$$

A system of f degrees of freedom has f periods, and the Fourier analysis of any variable contains f fundamental frequencies, with higher harmonics and combination frequencies:

$$x(t) = \sum A_{m,n}\,.\,.\, \exp[2\pi i(m\nu_1 t + n\nu_2 t + \,.\,.)]. \tag{4}$$

The state of motion is now determined by $2f$ parameters. They can again be chosen so that f of them are general constants of motion, such as energy, constant momenta, etc., and the remaining f are phase constants. The classical problem is regarded as solved when any coordinate can be written as a function of time and of these $2f$ constants.

If the energy of the system is not constant, as the result of radiation or other energy losses, the motion is no longer strictly periodic and the expansion of $x(t)$ leads to a Fourier integral:

$$\begin{aligned} x(t) &= \frac{1}{\sqrt{(2\pi)}} \int_{-\infty}^{+\infty} A(\nu) \exp(2\pi i \nu t)\, d\nu \\ A(\nu) &= \frac{1}{\sqrt{(2\pi)}} \int_{-\infty}^{+\infty} x(t) \exp(-2\pi i \nu t)\, dt. \end{aligned} \tag{5}$$

The *spectrum* $A(\nu)$ is now continuous; but if the rate of change of energy is small, the motion is nearly periodic and the function $A(\nu)$ shows pronounced maxima which can be considered as spectral lines of finite width.

2. The Hamiltonian and the action function

In Cartesian coordinates, the state of motion of a system of f degrees of freedom, e.g. of $f/3$ point particles, at a given time t is described completely by the f values x_k of the coordinates and the f

velocities $\dot{x}_k$ or momenta $m_k\dot{x}_k$ at that time. The main step towards the solution of a mechanical problem consists in the *separation of variables.* Instead of the arbitrarily chosen Cartesian coordinates and their time derivatives, new variables of position, q_k, are introduced in such a way that each of the equations of motion depends only on one pair of variables $q_k, \dot{q}_k$. The following procedure is generally used for *conservative* systems: coordinates of position q_k are chosen in accordance with the symmetry properties of the force field—e.g. polar coordinates for a central force, parabolic coordinates for a superposition of a central and a homogeneous force field—and the kinetic energy T is expressed as function of the q_k and $\dot{q}_k$. Conjugate momentum variables p_k are then defined by the equations

$$p_k = \frac{\partial}{\partial \dot{q}_k} T(q_1, q_2, \ldots, \dot{q}_1, \dot{q}_2, \ldots). \tag{6}$$

If the total energy is then expressed as a function H of the "canonical" variables q_k, p_k, the equations of motion take the form

$$\dot{p}_k = -\frac{\partial}{\partial q_k} H(p_1, p_2, \ldots, q_1, q_2, \ldots) \tag{7a}$$

$$\dot{q}_k = \frac{\partial}{\partial p_k} H(p_1, p_2, \ldots, q_1, q_2, \ldots). \tag{7b}$$

The function H is called the *Hamiltonian.* For the conservative systems considered at present it has a constant value E, the total energy of the system. For non-conservative systems, the Hamiltonian can be defined in a more general way and is no longer equal to the energy. We also exclude in this chapter the inertia effects of an electric charge moving in a magnetic field. They will be mentioned in those cases where their influence is important.

A great variety of coordinate transformations can be found which retain the form of the canonic equations (7) in the new coordinates and are therefore called *canonic transformations.* For their systematic study the reader has to be referred to textbooks of Mechanics, and the relevant results will be quoted here without proof.

In some systems, introduction of suitably chosen variables q_k causes the function H to become a sum of functions H_k each of which depends on one pair of conjugate variables p_k, q_k only. Each of the equations (7) then splits up into f equations each of which contains one pair of variables only and can be solved independently as for a system of one degree of freedom. Examples of systems in which this simplest kind of *separation* is possible are coupled harmonic oscillators and generally any system of mass points acted upon by elastic forces. Vibrations of polyatomic molecules can be treated in this way.

For the motions of the electrons in an atom, however, a more general method of separation has to be used. A function S of the q_k, also containing f integration constants α_k is defined by the relations

$$p_k = \partial/\partial q_k\, S(q_1, q_2, \ldots, \alpha_1, \alpha_2, \ldots). \tag{8}$$

The Hamilton equation then becomes

$$H(\partial S/\partial q_k, q_k) = E, \tag{9}$$

the so-called *Hamilton–Jacobi equation.* If the coordinates q_k are appropriately chosen, the equation can often be solved by writing S as the sum of functions depending on the individual coordinates only:

$$S(q_1, q_2, \ldots) = S_1(q_1) + S_2(q_2) + \ldots \tag{10}$$

which splits (8) into f equations which can be solved separately like equations for one degree of freedom. If $\mathrm{d}S$ is the total differential indicating the change of S during the motion, i.e. with the values α_k kept constant, we find

$$\begin{aligned} \mathrm{d}S &= \partial S/\partial q_1\, \mathrm{d}q_1 + \partial S/\partial q_2\, \mathrm{d}q_2 + \ldots \\ &= p_1\, \mathrm{d}q_1 + p_2\, \mathrm{d}q_2 + \ldots, \end{aligned} \tag{11}$$

where the $p_1, p_2 \ldots$ depend only on the $q_1, q_2 \ldots$ respectively. S is known as the *action function* or, more strictly, the *contracted* action function. It is constantly increasing during the motion (see below).

3. Cyclic coordinates and angle variables

Among the most important features of the atom as a mechanical system are its *symmetry* properties. For a free atom, all directions in space are equivalent, and for an atom in a magnetic or electric field of axial symmetry, which includes homogeneous fields, all values of the aximuth φ about the field axis are equivalent. In this case the Hamiltonian function H does not contain φ explicitly. Quite generally, a coordinate having this property is called *cyclic,* and eq. (7a) shows that the conjugate momentum variable is constant. For a single particle in a central force field the angle of azimuth φ about any arbitrary axis is cyclic. If the kinetic energy T is expressed in terms of polar coordinates with the chosen axis as polar axis, the conjugate momentum variable p_φ is found from (6) to be the component of the angular momentum about this axis, $L_z = m\rho^2\dot{\varphi} = m(r \sin \vartheta)^2\dot{\varphi}$, and this is constant, according to (7a). Since the same applies to the choice of any other axis, the angular momentum vector **L** itself is constant. In external fields with only axial symmetry, the component L_z about the axis of symmetry is constant.

If a canonic transformation from the variables q_k, p_k into new variables can be found which causes H to be independent of all f new variables of position w_k, and if the conjugate, constant momenta are called J_k, eq. (7b) gives

$$\dot{w}_k = \partial H/\partial J_k = \text{const.} = \nu_k, \tag{12a}$$

$$w_k = \nu_k t + \text{const.} \tag{12b}$$

If any coordinate is expressed as function of these *angle variables* w_k, its periodic properties will appear as harmonic functions of the uniformly increasing w_k, and the values ν_k are, apart from constant factors, the frequencies of the Fourier expansion of the coordinate in question. The transformation achieving this is defined by

$$w_k = \frac{\partial}{\partial J_k} S(q_1, q_2, \ldots, J_1, J_2, \ldots) \tag{13}$$

together with (8) where $J_1, J_2 \ldots$ take the place of the $\alpha_1, \alpha_2 \ldots$ The constant momenta J_k have the dimension of an action and can be shown to be

$$J_k = \oint p_k \, \mathrm{d}q_k \tag{14}$$

where the circle indicates that the integral has to be taken over a complete period of the coordinate q_k. This definition adjusts the constants in such a way that the ν_k become numerically equal to the frequencies.

Equations (11) and (14) show that the action function S increases during the motion in such a way that one period in the coordinate q_k causes S to increase by J_k.

4. Central force fields*

The methods of the previous paragraph can best be illustrated by their application to what may be described as the fundamental problem of atomic systems. We assume a particle of mass m to be

* See e.g. G. Joos, *Theoretical Physics*, Blackie 1932, part II.

moving in a central force field derived from a potential $V(r)$ where r is the distance from a fixed centre. The symmetry of the field suggests the use of polar coordinates r, φ, ϑ (fig. II, 1), with an arbitrarily chosen axis, as separation variables. From the kinetic energy, expressed as function of these coordinates,

$$T = m/2(\dot{r}^2 + r^2\dot{\vartheta}^2 + r^2 \sin^2\vartheta\dot{\varphi}^2) \tag{15}$$

the conjugate momentum variables are found by means of (6),

$$p_r = m\dot{r}, \quad p_\vartheta = mr^2\dot{\vartheta}, \quad p_\varphi = mr^2 \sin^2\vartheta\dot{\varphi}. \tag{16}$$

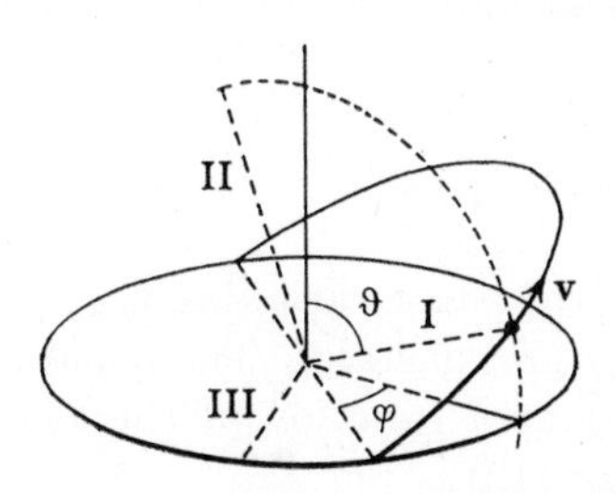

FIG. II, 1. Motion in a central force field, in polar coordinates.

The Hamiltonian, i.e. the energy as function of the p_k and q_k, is

$$H = \frac{1}{2m}(p_r^2 + p_\vartheta^2/r^2 + p_\varphi^2/r^2 \sin^2\vartheta) + V(r) \tag{17}$$

which, by substitution of (8), leads to the Hamilton–Jacobi equation

$$(\partial S/\partial r)^2 + \frac{1}{r^2}(\partial S/\partial\vartheta)^2 + \frac{1}{r^2 \sin^2\vartheta}(\partial S/\partial\varphi)^2 + 2m(V(r) - E) = 0 \tag{18}$$

where the energy, as a constant, is called E.

The substitution, corresponding to eq. (10),

$$S(r, \vartheta, \varphi) = S_r(r) + S_\vartheta(\vartheta) + S_\varphi(\varphi) \tag{19}$$

is now found to split the equation (18) into three equations; substitution in (18) and multiplication by $r^2 \sin^2\vartheta$ gives $(\partial S_\varphi/\partial\varphi)^2$ as equal to an expression which does not depend on φ and must therefore be constant ($= \alpha_\varphi^2$). This procedure, when continued, leads to the equations (20):

$$\begin{aligned} \partial S_\varphi/\partial\varphi &= \alpha_\varphi \\ (\partial S_\vartheta/\partial\vartheta)^2 + \alpha_\varphi^2/\sin^2\vartheta &= \alpha_\vartheta^2 \\ (\partial S_r/\partial r)^2 + \alpha_\vartheta^2/r^2 + 2m(V(r) - E) &= 0. \end{aligned} \tag{20}$$

The three integration constants are E, the energy; $\alpha_\varphi = p_\varphi$ which is the component of the angular momentum along the polar axis, and $\alpha_\vartheta{}^2$ which can be shown to be the square of the total angular momentum.

If the velocity **v** represented by its components in a Cartesian system whose axis I is in the direction of the radius, and II in the meridional plane (fig. II, 1), v_{I} does not contribute to the angular momentum about the origin, and the square of the total angular momentum is the sum of the squares of the angular momenta about II and III: $\mathbf{L}^2 = (mr^2\dot{\vartheta})^2 + (mr^2\dot{\varphi}\sin\vartheta)^2 = p_\vartheta{}^2 + p_\varphi{}^2/\sin^2\vartheta$.

The momenta $p_\varphi = \partial S_\varphi/\partial\varphi$, $p_\vartheta = \partial S_\vartheta/\partial\vartheta$ and $p_r = \partial S_r/\partial r$ are found from these equations as functions of α_φ, $\alpha_\vartheta{}^2$ and E, and the three integrals

$$\oint p_\varphi \mathrm{d}\varphi = J_\varphi, \quad \oint p_\vartheta \,\mathrm{d}\vartheta = J_\vartheta \quad \text{and} \quad \oint p_r \,\mathrm{d}r = J_r$$

can be calculated. Of the three variables, r, ϑ, φ, only φ is cyclic, and of the momenta only p_φ is constant. The equations can be solved for E which then appears as a function of the J_k, and the frequencies are found by means of (12).

5. Degeneracy

The actual results of these calculations will, of course, depend on the function $V(r)$. But quite independently of the nature of this function, it is found that E depends on J_ϑ and J_φ only in the form of the sum $(J_\vartheta + J_\varphi)$, so that according to (12a), $\nu_\vartheta = \nu_\varphi$.

This coincidence of two frequencies is called *degeneracy*. It causes the motion to take place in one plane, so that it can be described as a motion of only two degrees of freedom. But it is important to remember that there is an infinite number of possible orbital planes, a fact which has to be taken into account for any statistical calculation. This degeneracy is due to the isotropic nature of space, in the absence of external fields.

The definition of the w_k, ν_k and J_k is still arbitrary to a certain extent: instead of two frequencies ν_a and ν_b one can always introduce one of them, say ν_a and the difference $\nu_a - \nu_b = \nu'$. This is convenient whenever two frequencies are equal or nearly equal to one another. Thus, in the absence of external fields, ν_φ and ν_ϑ are equal and the substitution $\nu_3 = \nu_\varphi - \nu_\vartheta (= 0)$ or $w_3 = w_\varphi - w_\vartheta (= \text{const.})$ can be made. According to (12a) H is then independent of J_3.

Similarly, the substitution $w_2 = w_\vartheta - w_r$ is convenient if the potential of the force field is nearly (or exactly) proportional to $1/r$, since w_2 is then nearly (or exactly) constant. The orbit can then be described as a closed, elliptical orbit whose major axis rotates uniformly with the frequency ν_2 whose value is small or zero. The new J values can be shown to be the following:

$$\begin{aligned} w_1 &= w_r, \quad J_1 = J_r+J_\vartheta+J_\varphi \\ w_2 &= w_\vartheta-w_r, \quad J_2 = J_\vartheta+J_\varphi \\ w_3 &= w_\varphi-w_\vartheta, \quad J_3 = J_\varphi. \end{aligned} \tag{21}$$

The equations for the ν are exactly analogous to those for the w. These relations, together with eq. (12), show that

(a) If H depends on J_1 and J_2 only, but not on J_3, $\nu_\varphi = \nu_\vartheta$, and $\nu_3 = 0$, $w_3 =$ const. (motion in a fixed plane);

(b) if H depends on J_1 only, but not on J_2 or J_3, $\nu_r = \nu_\vartheta = \nu_\varphi$ and $\nu_2 = 0$, $w_2 =$ const. (motion in a closed orbit).

6. The orbital motion

The substitution of w_3 is equivalent to introduction of polar coordinates in the plane of the orbit. H now depends only on J_1 and J_2, the latter of which is equal to 2π times the total angular momentum $|\,\mathbf{L}\,|$. The frequency $\nu_1 = \mathrm{d}w_1/\mathrm{d}t$ measures the frequency at which r oscillates between its extreme values, and $\nu_2 = \mathrm{d}w_2/\mathrm{d}t$ is the difference of the frequencies of angular and radial motion, it is the frequency at which the perihelion precesses (fig. II, 2).

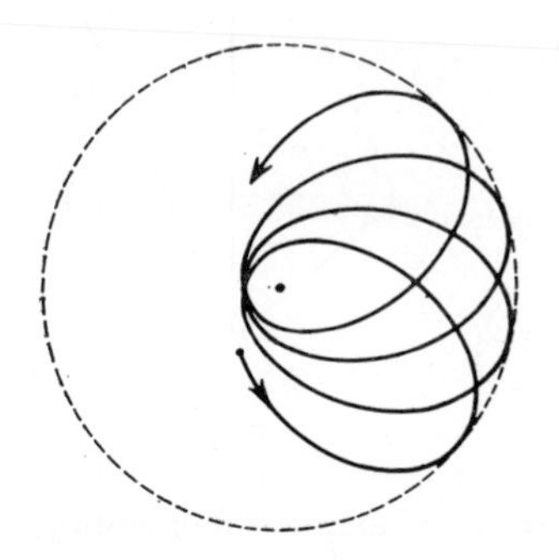

FIG. II, 2. Motion in the orbital plane.

Of special importance is the Coulomb potential function $V(r) = -C/r$; the energy H is in this case found to depend only on the sum $J_r+J_\vartheta+J_\varphi = J_1$ so that, according to (12), $\nu_\vartheta = \nu_r$ and, according to (21), w_2 is constant, so that $\nu_2 = 0$. Since radius and azimuth now change at the same frequency, the orbit is a closed curve, namely a Kepler ellipse. The system has become doubly degenerate: $\nu_1 = 0$ and $\nu_2 = 0$.

For any more general function $V(r)$, every coordinate can be calculated as a function of the angle variables $w = \nu t$. For a Cartesian coordinate in the orbital plane, one finds

$$x = \sum_n \exp(\pm 2\pi i \nu_2 t) D_n \exp[2\pi i(n\nu_1 t + \epsilon_n)] \qquad (n = 0, 1, 2, ..) \tag{22}$$

This is the Fourier expansion of x, in which ϵ_n are phase constants and therefore D_n real amplitudes. While ν_1 occurs with higher harmonics, ν_2 appears only as a fundamental frequency, since it is related to the uniform precession. Higher harmonics are always absent for a frequency associated with a cyclic variable, in the present case the angle of azimuth in the orbital plane.

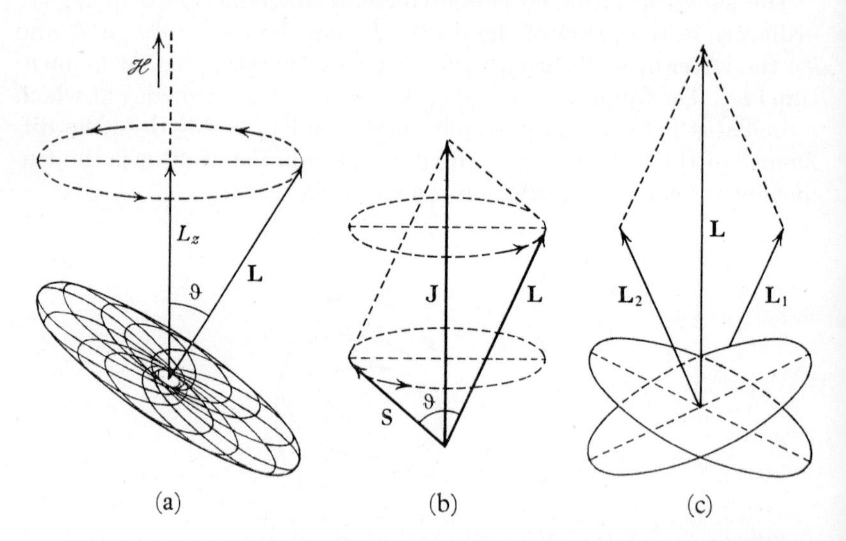

FIG. II, 3. Precession of angular momentum vectors. (a) orbital angular momentum in weak, external field, (b) spin–orbit–precession about resultant vector **J**, (c) two orbital angular momenta precessing about the resultant vector **L**.

An electron moving in a central force field would, according to classical concepts, emit a spectrum described by (22), and the intensities of the lines would be proportional to the squares of the D_n.

In the presence of an external magnetic or electric field which is either homogeneous or of axial symmetry, the angle of azimuth φ remains a cyclic variable if the polar axis is chosen to lie in the direction of the field. The component L_z of the angular momentum is constant. ν_ϑ is now no longer equal to ν_φ, and the orbit is not confined to a plane.

If, however, the field is weak enough, $|\nu_\vartheta - \nu_\varphi|$ is small and the motion can be regarded as taking place in a plane whose direction in space is changing slowly. The action of a magnetic field is particularly simple as it causes a steady torque on the orbit like that acting on a magnetic dipole which has the mechanical properties of a gyroscope. The result is a slow precession of the orbital plane (fig. II, 3a); the instantaneous vector of the angular momentum which is normal to the orbital plane precesses uniformly about the z-axis forming a constant angle ϑ with it. L_z thus remains constant while L_x and L_y are changing periodically in such a way that $|\mathbf{L}|^2$ remains constant.

7. Several particles with intrinsic spins

If several particles are subjected to forces acting between them, in addition to a central force, the total angular momentum $\mathbf{L}$ is constant in direction and magnitude, i.e. each of the three components

$$L_x = \sum_i m_i(y_i\dot{z}_i - z_i\dot{y}_i), \quad L_y = \ldots, \quad L_z = \ldots,$$

is constant.

The angular momentum of any one individual particle will not generally be constant, owing to the mutual interactions, and the motion may be very complicated. Often, however, these interactions are weak and consist of torques between the orbital motions of any two particles which then respond in the same way as spinning tops. The result is a gyroscopic motion which may be complicated in detail but consists in periodic changes of the directions of the axes of motion, leaving the absolute magnitudes $|\mathbf{L}^i|$ approximately constant.

To each constant angular momentum belongs a cyclic variable which is an angle of azimuth or some kind of average azimuth of several particles, but for many aspects of the theory these variables do not appear explicitly. This applies especially to the intrinsic angular momenta which must be attributed to electrons and to many nuclei. We can picture them as due to motions inside the particle, but its mechanical behaviour is fully described by the angular momentum vector $\mathbf{S}$ whose direction changes on account of interactions with other particles or with its own orbital motion while the relation $|\mathbf{S}| = \text{const.}$ holds rigorously. The letter $\mathbf{J}$ may be used to denote an angular momentum generally, either due to a spin or to orbital motion.

Let us first consider the magnetic interaction between two such vectors which may be, e.g. orbital momentum $\mathbf{L}$ and spin $\mathbf{S}$ of one electron. The interaction can be described as that between two magnetic dipoles, each of them being subject to a torque due to the magnetic field produced by the other. The effect is somewhat similar to that due to an external magnetic field and takes the form of a steady, *secular* perturbation. The magnetic torque is trying to increase or decrease the angle between $\mathbf{L}$ and $\mathbf{S}$ which react like gyroscopes by precessing at a uniform rate and at constant angle with one another about the direction of the vector resultant $\mathbf{J}$ which remains constant in magnitude and direction (fig. II, 3b). The precession is the faster the stronger the magnetic interaction. These facts are expressed by the vector relation

$$\mathbf{J} = \mathbf{L} + \mathbf{S} \tag{23}$$

For the orbital motions of two electrons, the electrostatic repulsion is usually the predominant interaction. Let us imagine the two electron orbits to be replaced by two negatively charged disks spinning in their own planes (fig. II, 3c). The electrostatic repulsion will then cause a torque trying to increase or decrease the angle between the vectors $\mathbf{L}_1$ and $\mathbf{L}_2$ of the angular momenta and will lead to a precession of these two vectors about the resultant $\mathbf{L}$. In contrast to the magnetic interaction, the torque will be independent of the sense of the rotations and thus of the signs of the vectors $\mathbf{L}_1$ and $\mathbf{L}_2$. If we try to make this picture more realistic by replacing each disk by a revolving, charged particle, the mathematical treatment, and in fact the appearance of the motion, becomes extremely complicated. In Quantum Mechanics the corresponding step is easier to perform, so that these complications of the classical motions fortunately need not be further considered.

If we consider three angular momenta each of them due to either intrinsic or orbital motion, any not too strong interaction causes the three vectors $\mathbf{J}_1$, $\mathbf{J}_2$, $\mathbf{J}_3$ to perform a complicated but slow movement about one another, and the total angular momentum is still the vector sum of the individual momenta:

$$\mathbf{J} = \mathbf{J}_1 + \mathbf{J}_2 + \mathbf{J}_3. \tag{24}$$

The movement becomes simple if the torque due to the force of interaction between two particles, e.g. 1 and 2, is much greater than the torques due to the interactions with 3. Then $\mathbf{J}_1$ and $\mathbf{J}_2$ precess about their common resultant $\mathbf{J}_{1,2}$, and the latter together with $\mathbf{J}_3$ perform a much slower precession about the total angular momentum $\mathbf{J}$. The more specific equations

$$\mathbf{J}_1+\mathbf{J}_2 = \mathbf{J}_{1,2}; \qquad \mathbf{J}_{1,2}+\mathbf{J}_3 = \mathbf{J} \tag{25}$$

then hold in place of (24), defining a new vector $\mathbf{J}_{1,2}$ as an approximately constant angular momentum.

B. THE OLD QUANTUM THEORY

1. Quantum conditions and correspondence principle

In classical Physics, the spectrum emitted by an atom would be derived in two steps: (1) application of the laws of mechanics leads to the orbital motions of the electrons which are multiply periodic as long as energy loss by radiation is slow enough to be neglected; (2) by the laws of electrodynamics, this model of an atom emits and absorbs light whose frequencies are equal to the Fourier frequencies of the classical motion, as in the imaginary experiment described in I.1.

As was pointed out before, these classical methods can explain neither the stability of atoms nor the type of atomic spectra actually observed. New laws of Physics had to be formulated which fitted atomic phenomena, but in such a way that they passed into the classical laws for macroscopic systems for which the latter laws were known to hold. Niels Bohr gave a precise form to these ideas in the *correspondence principle* which not only played an important part in the development of quantum laws but is also invaluable for the understanding of modern quantum mechanics.

In comparing the behaviour of an atom with that of a "model" such as Rutherford's model of the atom, to which we apply the laws of classical mechanics and electrodynamics, we find that frequencies and other quantities do not agree, but we can establish a clear one-to-one relationship of such values, so that, e.g. each actual frequency *corresponds* to a frequency of the classical model. In any imaginary process of scaling up the model to macroscopic dimensions while retaining its qualitative characteristics, the actual, quantum theoretical frequencies or other numerical values must pass into the classical values. We can perform such an imaginary transition either by increasing velocities or momenta and thereby increasing quantum numbers, or by reducing Planck's constant h to zero. Various examples in this and the following chapters will illustrate the working of the principle.

The older form of quantum theory due to Bohr, Sommerfeld, Wilson and others is based on two postulates, the first relating to the mechanics, the second to the electrodynamics of the atom.

(1) *Law of stationary states.* Of the states of motion which are possible in classical mechanics and which belong to a continuous range of values of the constants of motion, only certain *discrete* states are permitted in the atom as *stationary states,* so that only certain discrete values of the energy are possible as *quantised energy levels.*

(2) *Frequency relation.* The frequencies of emission and absorption of light by the atom are related to the energy values E_n of the stationary states by the equation

$$\nu_{n,n'} = (E_n - E_{n'})/h, \tag{26}$$

where h is Planck's constant.

These two postulates account at once for the existence of sharp spectral lines, the combination principle and the fact that spectral lines have definite excitation potentials, as established by Franck and Hertz.[I,6] It has, moreover, been found possible to formulate general rules of quantisation and of calculating the energy levels, in such a way that the calculated frequencies agree with those of the observed spectral lines in simple cases. A guide to finding these rules is provided by the comparison of (26) with the classical equation (12a). If we quantise the "phase integrals" J_k, i.e. if we postulate f quantum conditions for a system of f degrees of freedom by

$$J_k = \oint p_k \, dq_k = n_k \, h \,, \qquad (n_k = 0, 1, 2, \ldots) \tag{27}$$

the frequency relation

$$\begin{aligned} \nu_k &= (E_{n_k+1} - E_{n_k})/h \\ &= [H(J_1, J_2, \ldots, J_k + h, \ldots) - H(J_1, \ldots, J_k, \ldots)]/h \end{aligned} \tag{28}$$

corresponds to (12a) and becomes identical with it in the limit $h \to 0$. This limit can be defined more realistically by letting J_k, i.e. the quantum number n_k become very large.

This relation ensures that the quantisation rule (27) and the frequency relation (26) lead to frequencies which pass into the classical frequencies in the limit of large quantum numbers. Each quantum frequency arising from the change of one quantum number by 1 thus *corresponds*—with a specific meaning of this word—to a certain fundamental frequency of the classical system.

Frequencies arising from changes of one quantum number by 2, 3, .. *correspond* to the second, third, etc., harmonics of the classical frequency. Simultaneous changes of several quantum numbers *correspond* to combination frequencies (sums and differences) of the classical frequencies.

There are as many quantum numbers as degrees of freedom and as classical fundamental frequencies.

The quantities J_k have a property in classical mechanics which is closely connected with their quantisation: if any of the parameters of the system, such as force constants, are caused to change at a rate which is slow compared with the frequencies of the motion, the values J_k remain constant. If, e.g. the string of a conical point pendulum is shortened slowly, the energy of the system increases, on account of the work done against the centrifugal force, but the angular momentum remains constant.

This property of *adiabatic invariance* is consistent with the quantum conditions equating the J_k values to integral multiples of a universal constant. It led Ehrenfest to the *adiabatic principle* which applies to the response of quantised systems to gradual, non-periodic changes of parameters such as external fields. If the changes are slow in terms of the periods of the motion (*adiabatic*) the quantised value J_k remains the same, the system remains in the same quantum state as described by the set of quantum numbers. If the changes are rapid, transitions to another quantum state can occur. Quantum mechanics is able to describe such transitions quantitatively.

2. Angular momenta and the vector model

If the quantum conditions (27) are applied to any cyclic variable and its conjugate, constant angular momentum, the integration gives the factor 2π and leads to the result that any constant angular momentum, or component of a momentum, is a multiple of the fundamental unit $\hbar = h/2\pi$:

$$|\mathbf{J}_k| = j_k\hbar \tag{29}$$

where j_k is a quantum number.

If we thus quantise the momenta $|\mathbf{L}|$, L_z, $|\mathbf{S}|$, $|\mathbf{J}|$ in fig. II, 3 by assuming them to be integral multiples of $\hbar$, we obtain vector diagrams for the quantum numbers l, m, s and j (fig. II, 4a, c, d) and the following relations for the possible values of the quantum numbers:

$$m = l,\, l-1,\, l-2\, \ldots -l \tag{30}$$

$$j = l+s,\, l+s-1,\, \ldots |l-s|. \tag{31}$$

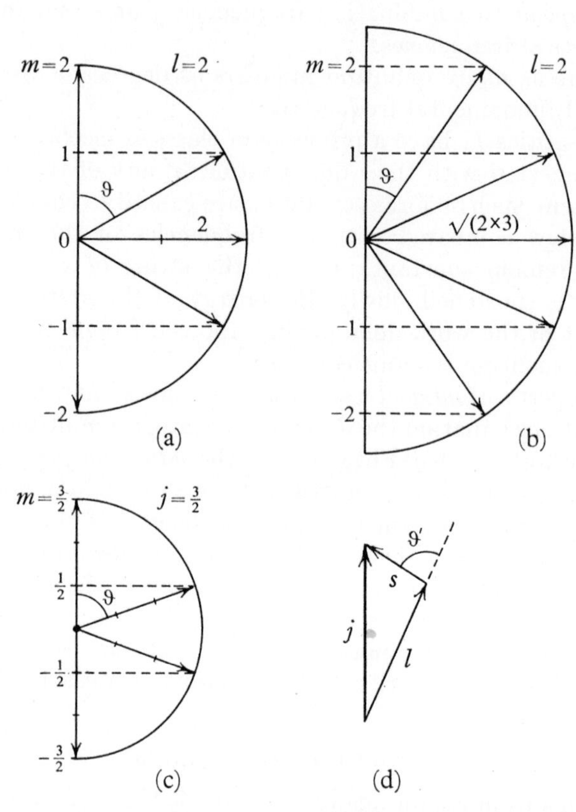

FIG. II, 4. Vector diagram of angular momenta in units of $\hbar$. The arrows in (a), (c) and (d) represent quantum numbers, those in (b) a fictitious absolute value $\sqrt{2(2+1)}$.

It should, however, be noted that the quantum conditions and the correspondence relations between classical and quantum frequencies remain unaffected if the more general assumption is made that only the differences between possible values of quantum numbers are integral. The relation (30) between m and l then permits half-integral as well as integral, but no other values. This is illustrated for a value 3/2 of the quantum number now called j (fig. II, 4c). In fact, half-integral numbers have to be used when intrinsic momenta such as the electron spin, are to be included.

Figure II, 4a, c, d shows that also the angles ϑ and ϑ' are quantised, and one derives directly from the diagram:

$$\cos\vartheta = m/l \tag{32}$$

$$\cos\vartheta' = \frac{j^2 - l^2 - s^2}{2ls}. \tag{33}$$

It will be noted that quantum numbers like m which describe components in a fixed direction can have positive or negative values, according to the sense of the precession. Quantum numbers describing absolute values of momenta, like l, j or s, have only positive values.

3. Degeneracy

The classical motion was described as degenerate (A.5 and 6) when two or more frequencies coincided or, with a suitably chosen definition of coordinates, one or more frequencies vanished. This implies, according to (12a) that the Hamiltonian becomes independent of one or more of the phase integrals J_k. Accordingly, degeneracy is characterised in quantum theory by the fact that the energy becomes independent of one or more quantum numbers, so that two or more quantum *states* have the same energy. One then speaks of singly or multiply degenerate levels.

Degenerate systems have the important property of being separable in several possible sets of coordinates. The harmonic, isotropic oscillator in a plane offers a simple example: the equations of motion can be solved either in Cartesian coordinates, with unlimited choice in the direction of the axes, or in polar coordinates. If, however, the potential of the force is slightly altered, by addition of either a non-isotropic term or of another central force term, the only possible separation coordinates are Cartesian coordinates in the first case, where the axes must be chosen according to the direction of the asymmetry, and polar coordinates in the second. The system is then no longer degenerate, and the orbit is not closed.

Another example is the motion under the influence of a Coulomb force (Kepler ellipse). The separation can be performed not only in polar coordinates (see p. 19) but in an infinite number of possible parabolic coordinate systems. If the force is altered by an additional central force, polar coordinates are the only possible separation coordinates. If, however, a homogeneous electric field is applied, only one parabolic coordinate system allows separation.

4. Electron spin and Pauli's principle

The multiplet structure of spectral terms and the facts of the anomalous Zeeman effect could not be explained by the quantum theory of the point electron. These difficulties were resolved by Goudsmit and Uhlenbeck who made the following *ad hoc* assumptions: (a) the electron possesses a mechanical angular momentum **S** of fixed absolute value whose component in any fixed direction can assume one of the two values $+\frac{1}{2}\hbar$ or $-\frac{1}{2}\hbar$. This behaviour can be described by a spin quantum number $s = \frac{1}{2}$ and a further quantum number m_s which can have one of the two values $\pm\frac{1}{2}$ and can be associated with the component of **S** in a fixed direction. In the older form of quantum theory, **S** would have the absolute value $\frac{1}{2}\hbar$.

This assumption gives the electron a fourth degree of freedom, though one of very peculiar nature.

(b) The electron possesses a magnetic moment, in the direction of the mechanical angular momentum, of one negative *Bohr magneton* (see p. 101 and 199),

$$\boldsymbol{\mu} = -\mathbf{S}e/mc. \tag{34}$$

In the old quantum theory, the interaction of the spin with the orbital motion of the same electron and with the spins of the other electrons are treated by the methods of vector addition outlined above. A comparison of the terms thus derived with those actually found in atoms led to a further, very fundamental postulate known as *Pauli's principle*: *No state exists in any quantised system in which two electrons are in the same quantum state*, i.e. two electrons can never have all four quantum numbers in common (see III, B).

C. THE METHODS OF QUANTUM MECHANICS

1. The Schrödinger method

a. Operators and commutation rules

The quantum postulates in their general form (p. 26) still apply in Quantum Mechanics as far as spectroscopic applications are concerned. The first interest of the spectroscopist lies in the energy states or *term-values*, and Quantum Mechanics has proved to be a more powerful method for calculating them than the older form of quantum theory. We can mainly confine ourselves to conservative systems, since loss of energy by radiation is so slow.

The Hamilton equation (9) can then be written in the form

$$H(p_k, q_k) = H(\partial S/\partial q_k, q_k) = E$$

where E is the constant value of the energy. If the q_k are the Cartetesian coordinates of the mass points, this equation becomes

$$\sum_k \frac{p_k^2}{2m_k} + V(q_k) = E. \tag{35}$$

The transition to Quantum Mechanics can be made by replacing the momentum coordinates p_k by *differential operators* $(\hbar/i)\partial/\partial q_k$ and letting them operate on a *wave-function* $\psi(q_k)$. This results in the time-independent Schrödinger equation

$$\left[\sum -\frac{\hbar^2}{2m_k}\nabla_k^2 + V(x_k, y_k, z_k)\right]\psi = E\psi \tag{36}$$

where ∇_k^2 stands for the *Laplace operator* $\partial^2/\partial x_k{}^2 + \partial^2/\partial y_k{}^2 + \partial^2/\partial z_k{}^2$. The substitution made is analogous to the classical relation $p_k = \partial S/\partial q_k$. Initially, we shall only consider the motion of a single particle so that q_k stands for one of the three coordinates of the same particle.

The solutions of the wave equation will depend entirely on the function V characterising the particular problem. We shall largely be concerned with motions of an electron in potential fields due to central forces of attraction, so that V vanishes at infinity and decreases with decreasing r. If certain restrictions of finiteness are imposed on ψ, it is found that for $E < 0$ such "well-behaved" solutions are only possible for definite (*discrete*) values of the energy E, the so-called *eigenvalues*. They are the stationary energy states of the atom. The solution belonging to an eigenvalue E is called the *eigenfunction*; it describes a stationary wave motion as function of the space coordinates. For $E > 0$, the energy can have all possible values; this is described as a continuous spectrum of eigenvalues. In this range, the ψ-functions are of a different type, corresponding to travelling waves and to non-periodic particle motions, such as hyperbolic orbits. We ignore, at present, the time dependence of the eigenfunctions.

If H is used as a symbol for the operator $(-\hbar^2/2m)\nabla^2 + V$, the Schrödinger equation

$$H\psi = E\psi \tag{37}$$

can be considered as an *eigenvalue equation*: if the operator H acting on the function ψ produces the function ψ itself, multiplied by a constant E, ψ is called an *eigenfunction* of the operator H belonging to the *eigenvalue* E.

TABLE 1

Classical variable	Schrödinger Operator
x, y, z	x, y, z
$p_x = m\dot{x}$	$p_x = -i\hbar\, \partial/\partial x$
$\mathbf{L}_x = yp_z - zp_y$	$\mathbf{L}_x = i\hbar(z\, \partial/\partial y - y\, \partial/\partial z)$
$\mathbf{L}_y = zp_x - xp_z$	$\mathbf{L}_y = i\hbar(x\, \partial/\partial z - z\, \partial/\partial x)$
$\mathbf{L}_z = xp_y - yp_x$	$\mathbf{L}_z = i\hbar(y\, \partial/\partial x - x\, \partial/\partial y)$
$= p\varphi = mr^2\dot{\varphi}$	$= -i\hbar\, \partial/\partial\varphi$
$\mathbf{L}^2 = \mathbf{L}_x{}^2 + \mathbf{L}_y{}^2 + \mathbf{L}_z{}^2$	$\mathbf{L}^2 = \mathbf{L}_x{}^2 + \mathbf{L}_y{}^2 + \mathbf{L}_z{}^2$
	$= -\hbar^2\left[\frac{1}{\sin\vartheta}\partial/\partial\vartheta(\sin\,\vartheta\partial/\partial\vartheta) + \frac{1}{\sin^2\vartheta}\partial^2/\partial\varphi^2\right]$
$p_r = m\dot{r}$	$p_r = -i\hbar(\partial/\partial r + 1/r)$
$p_r{}^2 = m^2\dot{r}^2$	$p_r{}^2 = -\hbar^2\frac{1}{r^2}\frac{\partial}{\partial r}\left(r^2\frac{\partial}{\partial r}\right)$
$H = \frac{1}{2m}(p_x{}^2 + p_y{}^2 + p_z{}^2) + V$	$H = -\frac{\hbar^2}{2m}\left(\frac{\partial^2}{\partial x^2} + \frac{\partial^2}{\partial y^2} + \frac{\partial^2}{\partial z^2}\right) + V$
$= \frac{1}{2m}(p_r{}^2 + \mathbf{L}^2/r^2) + V$	$= \frac{1}{2m}(p_r{}^2 + \mathbf{L}^2/r^2) + V$

Quantum mechanics provides general rules for associating a *differential operator* (this is meant to include algebraic operators) with each *dynamic variable*. A glance at the list of operators given in table II, 1, all of which refer to a single particle, will show the close relation with the corresponding expressions in the Hamiltonian form of classical mechanics. Different and more general forms of operators will be mentioned later.

The eigenvalues of an operator are interpreted as the possible values found in any measurement of the variable. The operand ψ which is, in the Schrödinger form of the theory, a function of the coordinates and the time, represents the state of motion of the system. If the function describing the state of a system is an eigenfunction ψ_n of the operator F, with the eigenvalue λ_n, the corresponding variable is known to have the value λ_n. In the general case, if the state is not an eigenstate of F, a measurement will give different values with certain probabilities.

If one of the eigenfunctions ψ_n of the Schrödinger equation, i.e. of the energy operator H, is also an eigenfunction of the operator F, so that

$$F\psi_n = \lambda_n \psi_n \tag{38}$$

where λ_n is a constant, the function ψ_n describes a state in which both the energy and the variable F are known to have the values E_n and λ_n respectively. It can be proved quite generally that two variables have *simultaneous* eigenfunctions if their operators *commute*, i.e. if $FH = HF$. This equation means that the successive application of the operators F and H to any function gives the same result if the operators are applied in the reverse order. The expression $FH-HF$ is often written as $[F, H]$ and described as *commutator*.

Commutation properties are closely connected with the *indeterminacy* relation, and the fact that two variables commute implies that their values can be measured independently, without restriction by indeterminacy. Trivial examples are coordinates of different particles which are entirely uncoupled; but also the different coordinates of one particle commute:

$$xy-yx = 0, \quad \text{or} \quad [x, y] = 0,$$

and also the linear momenta

$$[p_x, p_y] = 0. \tag{39}$$

On the other hand, conjugate variables have the commutator $i\hbar$:

$$[x, p_x] = i\hbar, \tag{40}$$

an equation which introduces the constant $\hbar$ and can be regarded as a quantisation rule. Quite generally, commutation rules can be derived from the *Poisson brackets* of Hamilton's theory. For one degree of freedom, the eigenvalues can be arranged in order of their values and numbered by a *quantum number* which can assume a succession of values differing by 1. For f degrees of freedom there are f quantum numbers, and every constant of the motion, such as the energy, can be expressed as a function of these f quantum numbers.

If m eigenfunctions $\psi_{n1}, \psi_{n2} \ldots \psi_{nm}$ have the same eigenvalue E_n of the energy, the system is called $(m-1)$ -fold degenerate. The Schrödinger equation, as a linear differential equation, is then solved by any linear function

$$\Phi_{nl} = c_1\psi_{n1} + c_2\psi_{n2} + \ldots c_m\psi_{nm} \tag{41}$$

with arbitrary constants c. The original set of m functions ψ_{nl} can thus be replaced by any new set of m linearly independent functions Φ_{nl}.

The non-degenerate solutions ψ_n of the wave equation have the important property of *orthogonality*, i.e.

$$\int \psi_n{}^*\psi_{n'}\,\mathrm{d}\tau \begin{array}{l} = 0 \quad \text{if} \quad n' \neq n \\ = \text{const}(= 1) \quad \text{if} \quad n' = n \end{array} \tag{42}$$

where $\mathrm{d}\tau$ is the volume element and the integration is to be extended over the whole of space. The same symbol $\mathrm{d}\tau$ is often used to indicate an element of a space of coordinates of several particles. Just as in the case of simple harmonic functions (see p. 15) this orthogonality allows any arbitrary function of the coordinates to be expanded in terms of the complete orthogonal set of functions. Also the formulae for the values of the expansion coefficients are the same as those applying to Fourier coefficients. If the value of the constant in (42) is made to be 1 by suitable choice of constants in the wave functions, these are said to be *normalised*. The set of functions is then called *orthonormal*.

In degenerate systems, the eigenfunctions belonging to the same eigenvalue are not always orthogonal to one another. Linear combinations Φ can then always be formed, by suitable choice of the constants in (41), which are orthogonal.

b. Systems with several particles

If one wishes to describe the motion of several particles whose interaction can be neglected, the potential energy V can be written as a sum $V_1 + V_2 + \ldots$ where each term depends only on one particle.

With this potential function, the wave equation (36) can be "separated" by the assumption of a solution

$$\Psi(1, 2, 3\ldots) = \psi_a(1)\psi_b(2)\ldots, \tag{43}$$

where each factor depends on the coordinates of one particle only. With this substitution, the wave equation is converted into a sum of expressions each of which depends only on one particle and must therefore, owing to the independence of the particles, be separately constant. This results in a set of single-particle wave equations. The assumption (43) is useful as a starting point for the treatment of several particles, e.g. of several electrons in one atom.

When interactions between electrons are included, the identity of the electrons leads to important *exchange* effects. Owing to this identity the Hamiltonian H is always symmetrical with regard to the different particles; this means that exchange of any two indices distinguishing two electrons must leave H unchanged. If this were not so, the Hamiltonian would ascribe different properties to individual electrons.

If Ψ (1, 2, 3, . .) is a wave function in which the figures 1, 2, . . symbolise the coordinates of the first, second, . . electron, an interchange of two numbers will generally change the function, though the eigenvalues will remain the same (*exchange degeneracy*). In the approximation discussed before, the function $\psi_a(1)\psi_b(2)$ is different from $\psi_a(2)\psi_b(1)$; in each case one of the electrons has arbitrarily been singled out by being attributed a certain quantum state. In fact, only functions free from this arbitrariness correspond to reality, namely functions which are either symmetrical or anti-symmetrical, i.e. functions which remain either unchanged or only change their sign when two particles are interchanged. In the present example of two electrons the linear combinations

$$\begin{aligned}\Psi_s &= (\psi_a(1)\psi_b(2)+\psi_a(2)\psi_b(1))/\sqrt{2}\\ \Psi_a &= (\psi_a(1)\psi_b(2)-\psi_a(2)\psi_b(1))/\sqrt{2}\end{aligned} \tag{44}$$

clearly fulfil these conditions. Instead of the two degenerate functions $\psi_a(1)\psi_b(2)$ and $\psi_a(2)\psi_b(1)$, these two independent linear combinations have to be used for describing the behaviour of the two identical particles.

c. *Spin functions*

If the particles are electrons, the existence of the spin cannot be disregarded even in many qualitative applications of the theory.

Dirac has shown that a relativistic formulation of the quantum-mechanical operators leads to the correct values of spin and magnetic moment of the electron, so that the spin has to be regarded as a relativistic effect, not as an additional property of the electron which has to be assumed *ad hoc*. Dirac's theory is mathematically complicated, and its use in this book will be restricted to chapter VI.

It is often sufficient to introduce the spin in a formal manner into the ordinary method of quantum mechanics in a way first proposed by Pauli and Darwin. In this theory, Schrödinger functions are used, and the effect of the spin on the orientation of the electron orbits is expressed by the use of certain linear combinations of the degenerate Schrödinger functions. The orientation of the spin is described by two linearly independent eigenfunctions like a motion of one degree of freedom whose quantum number is restricted to two values. The two states in which the component of the spin in a given direction is $\hbar/2$ or $-\hbar/2$ are described by two spin functions $v_{+\frac{1}{2}}$ and $v_{-\frac{1}{2}}$, often simply written $(+)$ and $(-)$. They are not otherwise specified except for general conditions of orthogonality and normality:

$$(+)^*(+) = 1, \qquad (-)^*(-) = 1, \qquad (+)^*(-) = 0. \tag{45}$$

The definition of the conjugate complex for these symbols will be given in II, 2. Spin functions do not depend on the space coordinates, and differential operators do not act on them. This makes them somewhat abstract, and one can just regard them as mathematical symbols with rules so devised that they produce the correct linear combinations of Schrödinger functions. This will best be seen when they are used in the form of column symbols.

For several electrons, symmetric and antisymmetric spin functions can be constructed in the same way as Schrödinger functions. For two electrons, of the four possible products

$$(+)(1)(+)(2), \qquad (-)(1)(-)(2), \qquad (+)(1)(-)(2), \qquad (+)(2)(-)(1)$$

only the first two have definite symmetry property. The following linearly independent antisymmetric and symmetric combinations of the last two can be constructed:

$$\begin{aligned} \chi_a &= \frac{1}{\sqrt{2}}[(+)(1)(-)(2)-(+)(2)(-)(1)] \\ \chi_s &= \frac{1}{\sqrt{2}}[(+)(1)(-)(2)+(+)(2)(-)(1)]. \end{aligned} \tag{46}$$

These, together with

$$\chi_s' = (+)(1)(+)(2)$$
$$\chi_s'' = (-)(1)(-)(2)$$

are the spin functions to be multiplied by Schrödinger functions in order to form the complete wave functions describing the state of the system.

A nomenclature is sometimes used in which the number of the particle is indicated by the order of the brackets, so that, e.g.

$$(+)(-) \equiv (+)(1)(-)(2).$$

The Pauli principle now takes the form of the postulate:

Only antisymmetric wave functions describe possible states of an atomic system.

A symmetric spin function thus always has to be combined with an antisymmetric space function and vice versa.

The identity of the postulate with the Pauli principle in the form stated before (p. 30) can be seen thus: if two electrons agree in all quantum numbers, an interchange of these electrons will leave the state of the system unaltered. This is just the state described by a symmetrical wave function and thus excluded.

d. Wave functions in central force fields

The mechanics of a single particle moving in a central force field forms the basis of almost every theoretical treatment of atomic states. We assume that the force is derived from a potential function $V(r)$ depending only on the distance r from a fixed origin. To take account of the spherical symmetry of the field, we introduce polar coordinates; the Laplace operator then becomes

$$\nabla^2 = \frac{1}{r^2}\left[\frac{\partial}{\partial r}\left(r^2\frac{\partial}{\partial r}\right) + \frac{1}{\sin\vartheta}\frac{\partial}{\partial\vartheta}\left(\sin\vartheta\frac{\partial}{\partial\vartheta}\right) + \frac{1}{\sin^2\vartheta}\frac{\partial^2}{\partial\varphi^2}\right]. \tag{47}$$

The wave equation can be solved by separation of the variables. The close similarity of the method with that used in classical mechanics (p. 19) becomes clear if ψ is regarded as an exponential function of S. We assume that the wave function is a product of a function depending solely on r and two functions depending on the angles ϑ and φ. Substitution of

$$\psi(r, \vartheta, \varphi) = R(r)\Theta(\vartheta)\Phi(\varphi) \tag{48}$$

in the wave equation (36) and division by $\psi/r^2 \sin^{-2}\vartheta$ splits the equation into the sum of a term $(1/\Phi)(\mathrm{d}^2\Phi/\mathrm{d}\varphi^2)$ and another term which is independent of φ. By the same arguments as used on p. 19, each of these terms must be equal to a constant. Continuation of the process, again as before, leads to the equations

$$\frac{\mathrm{d}^2\Phi}{\mathrm{d}\varphi^2} = -m^2\Phi \tag{49}$$

$$\frac{1}{\sin\vartheta}\frac{\mathrm{d}}{\mathrm{d}\vartheta}\left(\sin\vartheta\frac{\mathrm{d}\Theta}{\mathrm{d}\vartheta}\right) - \frac{m^2}{\sin^2\vartheta}\Theta = -\lambda\Theta \tag{50}$$

$$\frac{1}{r^2}\frac{\mathrm{d}}{\mathrm{d}r}\left(r^2\frac{\mathrm{d}R}{\mathrm{d}r}\right) - \frac{2m_0}{\hbar^2 r^2}\lambda R - \frac{2m_0}{\hbar^2}(V(r)-E)R = 0 \tag{51}$$

in close analogy with the classical equations (20).

The electron mass is now written as m_0 in order to distinguish it from the quantum number m.

Equation (49) is a simple eigenvalue equation and has the eigenfunctions

$$\Phi(\varphi) = \text{const.} \times \exp(im\varphi) \tag{52}$$

with the eigenvalues $m = 0, \pm 1, \pm 2, \ldots$, since only for integral values of m is the function Φ unaltered by an increase of φ by multiples of 2π. This is a condition for the function to be *well-behaved*, i.e. to allow of any physical interpretation. m is called the *magnetic* quantum number.

In a similar way (50) has the form of an eigenvalue equation. A mathematical discussion shows that the eigenvalues λ are given by

$$\lambda = (|m|+\mu)(|m|+\mu+1) \tag{53}$$

where $\mu = 0, +1, +2, \ldots$ Instead of this quantum number μ, another quantum number l is generally more convenient to use. It is the *asimuthal* quantum number, defined as the combination

$$\begin{aligned} l &= |m|+\mu \\ \lambda &= l(l+1). \end{aligned} \tag{54}$$

so that l is then an integral, positive number and subject to the condition

$$l = \geqq |m| \tag{55}$$

The eigenfunctions Θ which are solutions of (50) are the so-called *associated Legendre functions.* With the abbreviation $x = \cos\vartheta$

they can be defined by

$$P_l^m(x) = (1-x^2)^{|m|/2}\,\mathrm{d}^{|m|}/\mathrm{d}x^{|m|}P_l(x) \tag{56}$$

where the functions $P_l(x)$ are the *Legendre polynomials* defined by

$$P_l(x) = \frac{1}{2^l l!}\mathrm{d}^l/\mathrm{d}x^l(x^2-1)^l \tag{57}$$

The functions with even values of $\mu = l-|m|$ remain unchanged when x changes its sign, while those with odd values of μ change their sign. Each function has μ nodes in the whole range from $\vartheta = 0 \to \pi$. The functions

$$Y_{l,m}(\vartheta, \varphi) = NP_l^m(\cos\vartheta)\times\Phi(\varphi) \tag{58}$$

are called *spherical harmonics*; N is a normalising factor which is sometimes omitted as irrelevant.

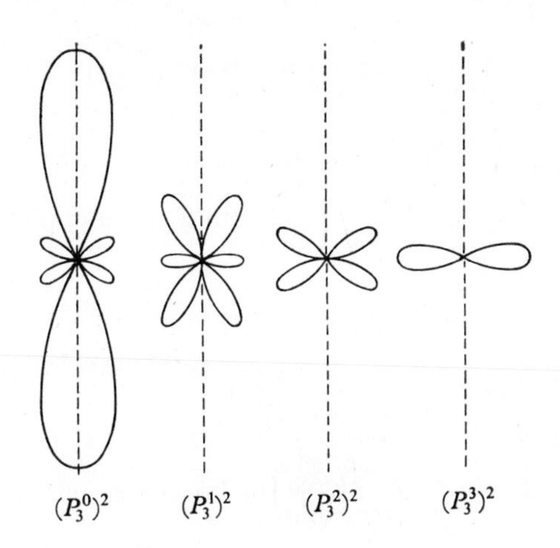

FIG. II, 5. Polar diagram of $(P_l^m)^2$ for $l = 3$, $m = 0, 1, 2, 3$. (L. Pauling & E. B. Wilson, *Introduction to Quantum Mechanics*, McGraw Hill, 1935).

Table II, 2 gives a list of the functions $Y_{l,m}$ up to $l = 3$, and fig. II, 5 shows the values $(P_3^m)^2$ in a qualitative polar diagram in a plane through the polar axis.

Equations 50 and 49 lead to a very important connection between the wave functions in central fields and the angular momentum vector $\mathbf{L} = \mathbf{r}\times\mathrm{d}\mathbf{r}/\mathrm{d}t$. Comparison with table II, 2 shows that the functions $Y_{l,m}$ are also eigenfunctions of the operator $\mathbf{L}^2$, with the eigenvalues $\hbar^2 l(l+1)$. Owing to the factor $\Phi(\varphi)$ they are also eigenfunctions of the operator $\mathbf{L}_z$, with the eigenvalues $m\hbar$:

$$\begin{aligned} \mathbf{L}^2 &= l(l+1)\hbar^2, \\ \mathbf{L}_z &= m\hbar. \end{aligned} \tag{59}$$

TABLE 2

Spherical harmonics

$$Y_{l,m} = \left[\frac{2l+1}{4\pi}\frac{(l-|m|)!}{(l+|m|)!}\right]^{1/2} P_l^m(\cos\vartheta)\exp(im\varphi)$$

$$Y_{0,0} = \sqrt{\left(\frac{1}{4\pi}\right)}$$

$$Y_{1,0} = \sqrt{\left(\frac{3}{4\pi}\right)}\cos\vartheta$$

$$Y_{1,\pm1} \quad \sqrt{\left(\frac{3}{8\pi}\right)}\sin\vartheta\exp(\pm i\varphi)$$

$$Y_{2,0} = \sqrt{\left(\frac{5}{16\pi}\right)}(3\cos^2\vartheta-1)$$

$$Y_{2,\pm1} = \mp\sqrt{\left(\frac{15}{8\pi}\right)}\sin\vartheta\cos\vartheta\exp(\pm i\varphi)$$

$$Y_{2,\pm2} \quad \sqrt{\left(\frac{15}{32\pi}\right)}\sin^2\vartheta\exp(\pm 2i\varphi)$$

$$Y_{3,0} = \sqrt{\left(\frac{7}{16\pi}\right)}(5\cos^3\vartheta-3\cos\vartheta)$$

$$Y_{3,\pm1} = \mp\sqrt{\left(\frac{21}{64\pi}\right)}(5\sin\cos^2\vartheta-\sin\vartheta)\exp(\pm i\varphi)$$

$$Y_{3,\pm2} = \sqrt{\left(\frac{105}{32\pi}\right)}\sin^2\vartheta\cos\vartheta\exp(\pm 2i\varphi)$$

$$Y_{3,\pm3} = \mp\sqrt{\left(\frac{35}{64\pi}\right)}\sin^3\vartheta\exp(\pm 3i\varphi)$$

If the operators $\mathbf{L}^2$ and $p_r{}^2$ from table II, 2 are introduced into the Schrödinger equation, this becomes

$$\left[\frac{1}{2m_0}p_r{}^2+\frac{\mathbf{L}^2}{2m_0 r^2}+V(r)\right]\psi = E\psi. \tag{60a}$$

Owing to the fact that the solutions (58) are eigenfunctions of $\mathbf{L}^2$, as well as of H, the operator $\mathbf{L}^2$ can be replaced in (60) by its eigenvalues $l(l+1)\hbar^2$ in turn, eliminating the variables ϑ and φ. The solutions of the resulting individual equations can only depend on r: expressing this fact by calling them $R(r)$, we get the radial equations, one for each value of l:

$$\left[\frac{1}{2m_0}p_r^2+\frac{l(l+1)\hbar^2}{2m_0r^2}+V(r)\right]R(r) = ER(r) \tag{60b}$$

which are identical with (51).

The solutions of the radial equations depend on the function $V(r)$ which is at present left unspecified. If $E < 0$, i.e. if the particle has not sufficient energy to escape, well-behaved solutions of $R(r)$ exist only for integral values of a new quantum number n_r ($= 0, 1, 2 \ldots$) which appears as a constant in the expression for E. It is equal to the number of nodes of $R(r)$ in the whole range of r from 0 to ∞.

The energy only appears in the radial equation and therefore only depends on l, apart from n_r; but to each value E_l belong $2l+1$ eigenfunctions ψ_{lm} since there are $2l+1$ possible values of m. The energy level is then $2l$-fold degenerate.

Each state described by a function ψ_{lm}, as an eigenfunction of $\mathbf{L}^2$ and $\mathbf{L}_z$, has a definite value $\hbar^2 l(l+1)$ of the square of the angular momentum $\mathbf{L}$ and a definite value $m\hbar$ of its component $\mathbf{L}_z$. This situation cannot be described rigorously by a vector diagram, since never more than one component of $\mathbf{L}$ can have a definite value. A diagram with a vector $\mathbf{L}$ of magnitude $\sqrt{\mathbf{L}^2} = \hbar\sqrt{l(l+1)}$ is sometimes drawn (fig. II, 4b) and often leads to correct results; for $m = l$, e.g. the vector $\mathbf{L}$ can be imagined as precessing at a small but finite angle about the z-axis, but such pictures cannot be taken too literally. One also uses vector diagrams of the quantum numbers l and m, as in the old quantum theory (fig. II, 4a), merely to illustrate the connection (55) between l and m.

A statement that $\mathbf{L}^2$ has a definite value, say $\hbar^2 3(3+1)$ merely means that values of $\mathbf{L}_z$ up to $3\hbar$ are observed. From the fact that all directions in space are equivalent, and that all states of different m have the same statistical weight 1, it follows that the average.

$$\overline{\mathbf{L}^2} = 3\overline{\mathbf{L}_z{}^2} = 3\hbar^2(2\times 3^2+2\times 2^2+2\times 1+0)/7 = 12\hbar^2$$

which is, in fact, equal to $\hbar^2 l(l+1)$ with $l = 3$.

In the absence of an external field, the choice of the z-axis is arbitrary and the physical significance of m is not immediately obvious.

But by the very act of measuring the component in any direction, say L_z, we define the z-axis by the conditions of the experiment, such as the magnetic field in a Stern–Gerlach experiment. According to quantum mechanics, the result of such an experiment will always be one of the eigenvalues $m\hbar$. If the differences between the energies of these states are small compared with the Bolzmann energy kT, all values of m will be observed with equal probability.

Whenever the Coulomb potential $V = \text{const.}/r$ serves as a first approximation, it is convenient to use, instead of n_r, the quantum number n defined by

$$n = n_r + l + 1. \tag{61}$$

The ranges and connections between the quantum numbers are summarised in the following list:

$$\begin{aligned} n &= 1, 2, 3, \ldots & n &= n_r + l + 1 = n_r + |m| + \mu + 1 \\ l &= 0, 1, \ldots n-1 & l &= |m| + \mu \\ m &= 0, \pm 1, \pm 2 \ldots \pm l & & \end{aligned} \tag{62}$$

e. Parity

If in the functions $Y_{l,m}$ the coordinates are changed in such a way that each point is replaced by a point opposite the origin and at equal distance from it (reflection at the origin $\mathbf{r} \to -\mathbf{r}$, or r, ϑ, $\varphi \to r$, $\pi - \vartheta$, $\varphi + \pi$), the functions are found to be either unchanged or changed in sign only. They are accordingly called *even* or *odd* functions. It is easy to verify that

$$\begin{aligned} Y_{lm}(\vartheta, \varphi) &= Y_{lm}(\pi - \vartheta, \varphi + \pi) \text{ for even } l \\ &= -Y_{lm}(\pi - \vartheta, \varphi + \pi) \text{ for odd } l. \end{aligned}$$

The function $R(r)$ is even, by its definition, so that the complete wave function ψ is of even or odd parity according to the "parity" of l. The reflection of the coordinates at the origin is often regarded as an operator having the two eigenvalues $+1$ and -1.

Eigenfunctions of several identical particles moving in a field of central symmetry and having arbitrary interactions between one another can also be shown to have definite parity; reflection of the coordinates of all particles at the origin leaves the wave function either unaltered or only changes its sign. Since the parity of a state is a definite property with distinct eigenvalues, any change in the interactions, which may be thought to be brought about by a slow, *adiabatic* process, cannot alter the parity which is therefore the same as that of the eigenfunction for independent particles. This is easily seen to be equal to the parity of the sum Σl for all particles.

f. Time dependent eigenfunctions

The atomic system has so far been assumed to be conservative, i.e. of constant energy. The wave functions could therefore be regarded as functions of the space coordinates only. The complete wave functions of any eigenstate of H with the eigenvalue E_k are obtained by multiplication of the space functions by the factor $\exp[i(E_k/\hbar)t]$. In all relations between ψ-functions belonging to the same energy state, the time factor could be omitted as a common factor.

For the description of phenomena in which the energy is not conserved, such as the radiation by an atom, the more general time-dependent Schrödinger equation

$$i\hbar\frac{\partial U}{\partial t} = -\frac{\hbar^2}{2m}\nabla^2 U \tag{63}$$

has to be used. If the potential function depends explicitly on the time, a further term $V(x, y, z, t)$ has to be added to the right side of the equation.

2. Matrix methods

a. Fundamental concepts

In wave mechanics the state of a conservative system is defined by the eigenfunction ψ_n where n stands for all quantum numbers required. An observable has generally no definite value, only an average or *expectation-value* can be stated, such as the distance of the electron from the origin

$$\bar{r} = \int r\psi_n{}^*\psi_n\, d\tau.$$

For the x-component of the dipole moment, e.g. we find

$$e\bar{x} = e\int x\psi_n{}^*\psi_n\, d\tau$$

which is not only constant but also vanishes. One cannot hope to come to a description of any process in which the energy is not conserved, such as radiation, unless the restriction to one single quantum state is abandoned. This leads to a generalised definition of the expectation value of an observable:

$$f_{nn'} = \int \psi_n{}^* f \psi_{n'} \, \mathrm{d}\tau. \tag{64}$$

Since f is an operator, the order of the factors under the integral is generally important. When applied to the dipole moment, $f = ex$, the order becomes irrelevant. The time factors of the ψ functions, $\exp(itE/\hbar)$, yield the factors $\exp[it(E_{n'} - E_n)/\hbar]$ in (64) and all these terms together give a correct description of the spectrum of the atom. We can then regard the observable ex as described by the array of all the two-index quantities

$$ex_{nn'} = e \int x \psi_n{}^* \psi_{n'} \, \mathrm{d}\tau \exp[it(E_{n'} - E_n)/\hbar] \tag{65}$$

which can be considered as the *elements* of a *matrix* in which each of the indices n and n can assume all possible values of the quantum numbers.

The state of a non-conservative system such as a radiating atom can be described by a linear combination of eigenfunctions of different energies:

$$\varphi = c_1 \psi_1 + c_2 \psi_2 + \,.\,. \tag{66}$$

and the usual definition

$$\bar{x} = \int x \psi^* \psi \, \mathrm{d}\tau$$

then leads to mixed terms $x_{nn'}$ as in (65) which are time-dependent unless $n = n'$. The possibility of the expansion (66) depends on the orthogonality of the eigenfunctions.

In conservative systems, such expansions are possible in cases of degeneracy. The matrix elements $f_{nn'}$ are then not time-dependent as long as the degeneracy exists. One often uses expansions in terms of eigenfunctions which belong to a lower degree of approximation, especially in perturbation theory (see section 3).

An eigenfunction of a *degenerate* system can be expanded in different ways, by the use of different sets of orthogonal functions. The ψ can then be regarded as a vector in the "space" of the set of eigenfunctions ψ_1, ψ_2, . ., with components c_1, c_2, . . . The same vector, i.e. the same state, has then a different set of components for every set of eigenfunctions, or every *representation*.

The matrix elements depend on the scheme of eigenfunctions chosen just as the components of a vector or tensor depend on the coordinate system. Each element of a matrix is associated with two definite eigenfunctions of a given representation by the relation (65), and thus with two sets of quantum numbers. One often uses the abbreviated notation

$$\int \psi^*_{njm} f \psi_{n'j'm'} \, d\tau \equiv \langle njm|f|n'j'm'\rangle \tag{67}$$

which can also be written

$$\Psi^* f \Psi$$

where Ψ is a "symbolic ψ" and Ψ^* is related to Ψ^* in a similar way as a conjugate to a complex number. The orthogonality is then expressed by

$$\Psi_n^* \Psi_{n'} = 0 \qquad \text{if} \quad n' \neq n.$$

In keeping with (67), the symbol of an eigenstate defined by the quantum numbers n, j, m is written:

$$|njm\rangle.$$

The symbolic notation avoids the cumbersome integrals, such as in the expression for the scalar product:

$$\int f(x, y, z) g(x, y, z) \, d\tau = (f, g).$$

As the result of a linear transformation, an eigenfunction in a scheme defined by quantum numbers n, l, m becomes a linear function of the eigenfunctions in the scheme defined by quantum numbers p, q, r. This is written

$$|nlm\rangle = \sum_{pqr} \langle pqr|nlm\rangle \times |pqr\rangle, \tag{68}$$

where the coefficients $\langle pqr|nlm\rangle$ form the elements of a *transformation matrix*.

Mathematical operations such as addition and multiplication can be defined in terms of the matrix elements:

$$\begin{aligned} (F+G)_{ik} &= F_{ik} + G_{ik}, \\ (FG)_{ik} &= \sum_j F_{ij} G_{jk} \end{aligned} \tag{69}$$

or

$$\langle i|FG|k\rangle = \sum_j \langle i|F|j\rangle \langle j|G|k\rangle,$$

but it is found that the fundamental rules of calculating with matrices hold independently of the system of representation used and can therefore be written in symbolic matrix equations just as vector equations can be written without specification of the coordinate system. As an example, the commutation rule for matrix addition can be written

$$F+G = G+F,$$

implying that in any representation used

$$(F+G)_{ik} = (G+F)_{ik}.$$

Multiplication, on the other hand, is not commutative.

For the details of matrix algebra, the reader must be referred to textbooks on quantum mechanics, and only a few definitions and results may be quoted in this context.

A *diagonal* matrix F is a matrix for which $F_{ik} = 0$ for $i \neq k$. The *Hermitian adjoint* matrix A^* of a matrix A results from interchanging rows and columns and substituting the conjugate complex for each element:

$$B = A^* \quad \text{if} \quad B_{ik} = A_{ki}^*.$$

In quantum mechanics, the relation of a matrix to its Hermitian adjoint matrix is the same as that of a function to its conjugate complex.

Observables are always represented by *Hermitian* matrices defined by the relation

$$f_{ik} = f_{ki}^* \tag{70}$$

for all elements. This causes the diagonal elements to be real, and also the matrix product $f \,.\, f^*$ to be real.

The unit matrix $\mathbf{1}$ is defined as a diagonal matrix whose elements are 1.

A matrix for which $AA^* = \mathbf{1}$ is called a *unitary* matrix. The transformation matrices mentioned above are unitary matrices.

Algebraic relations are not only independent of the representation, but they are also the same for matrices and differential operators referring to the same dynamic variables. It need therefore often not be stated if the symbols in an equation stand for matrices or differential operators.

b. Eigenvalues of matrices

If the functions ψ chosen for representing a differential operator F are eigenfunctions of F, all matrix elements except the diagonal elements vanish. If, e.g., F is chosen as the Hamiltonian operator H and the expansion functions are the eigenfunctions of H for the same system, we find

$$H_{ik} = \int \psi_i^* H \psi_k \, \mathrm{d}\tau = \int \psi_i^* E_k \psi_k \, \mathrm{d}\tau = E_k \int \psi_i^* \psi_k \, \mathrm{d}\tau. \tag{71}$$

$$H_{ik} = E_k \quad \text{for} \quad i = k, \quad \text{and} \quad H_{ik} = 0 \quad \text{for} \quad i \neq k.$$

The energy is then said to be diagonal in the ψ representation.

Though matrices can be considered as operators just as differential operators, the operand of a matrix is not a single wave function but a set of coefficients c_n in (66). They can be regarded as the components of a vector in the "space" of the functions ψ_n in (66). This symbolic operand Ψ is often written in the form of a column symbol, and Ψ^* as a row symbol:

$$\Psi = \begin{pmatrix} c_1 \\ c_2 \\ c_3 \\ \cdot \\ \cdot \end{pmatrix}, \qquad \Psi^* = (c_1^*, c_2^*, c_3^* \ldots)$$

and the rules by which a matrix acts on a column symbol, producing another column symbol is

$$\begin{pmatrix} F_{11} & F_{12} & \ldots \\ F_{21} & F_{22} & \ldots \\ \ldots & \ldots & \ldots \end{pmatrix} \begin{pmatrix} c_1 \\ c_2 \\ \ldots \end{pmatrix} = \begin{pmatrix} F_{11}c_1 & +F_{12}c_2 & \ldots \\ F_{21}c_1 & +F_{22}c_2 & \ldots \\ \ldots & \ldots & \ldots \end{pmatrix} \tag{72}$$

a rule which can be regarded as a special case of matrix multiplication (69). Similarly, the product $\Psi^*\Psi$ is $c_1c_1^* + c_2c_2^* + \ldots$ The fact, e.g. that Ψ_2 is an eigenfunction of H with the eigenvalue E_2 is expressed by the equation

$$\begin{pmatrix} E_1 & & & 0 \\ & E_2 & & \\ & & E_3 & \\ 0 & & & \ldots \end{pmatrix} \begin{pmatrix} 0 \\ 1 \\ 0 \\ 0 \end{pmatrix} = \begin{pmatrix} 0 \\ E_2 \\ 0 \\ 0 \end{pmatrix} = E_2 \cdot \begin{pmatrix} 0 \\ 1 \\ 0 \\ 0 \end{pmatrix}$$

which has the form of an eigenvalue equation.

If the energy matrix H is not given in the diagonal form, i.e. not in the representation of the eigenfunctions of H, the problem of determining the eigenvalues can be solved by considering the matrix equation

$$\begin{pmatrix} H_{11} & H_{12} & H_{13} & .. \\ H_{21} & H_{22} & H_{23} & .. \\ .. & .. & .. & .. \end{pmatrix} \begin{pmatrix} c_1 \\ c_2 \\ .. \end{pmatrix} = \lambda \begin{pmatrix} c_1 \\ c_2 \\ .. \end{pmatrix}$$

expressing the condition for the coefficients c_n of the expansion (66) to be so chosen that the linear combination is an eigenfunction of H. This condition is equivalent to the set of equations

$$\begin{aligned} H_{11}c_1 + H_{12}c_2 + \; .. &= \lambda c_1 \\ H_{21}c_1 + H_{22}c_2 + \; .. &= \lambda c_2 \\ .. \quad .. \quad .. \quad .. &= .. \end{aligned}$$

The condition for these to have non-trivial solutions is expressed by the determinant equation

$$\begin{vmatrix} (H_{11}-\lambda) & H_{12} & H_{13} & .. \\ H_{21} & (H_{22}-\lambda) & H_{23} & .. \\ H_{31} & H_{32} & .. & .. \end{vmatrix} = 0 \qquad (73)$$

known as the *secular equation.* It is of the order of the number of rows of the determinant and has as many solutions for the eigenvalue λ.

A relation which is of considerable importance in spectroscopic applications and which will often be used in the following chapters is the *diagonal sum rule.* It states that the sum of the diagonal elements of a matrix is equal to the sum of all its eigenvalues. This rule often simplifies the task of finding the eigenvalues of a matrix.

In Schrödinger's method, the operands are functions of the coordinates and the time, while the operators are symbols which have always the same form, independently of the nature of the particular system. In matrix mechanics, on the other hand, the operands are merely numbers, while the operators generally depend on the time and take on different forms for different representations and for different dynamical systems.

The time dependance of a variable F is described by the relation $\mathrm{d}F/\mathrm{d}t = (1/i\hbar)[F, H]$ provided that neither F nor H depend explicitly on t, so that in conservative systems which will mainly concern us, all constants of the motion commute with H. If two variables commute, and only then, it is possible to find a representation in which both are diagonal. This follows from the fact that diagonal matrices commute identically.

c. Angular momenta

It had been shown in II.A that spherical symmetry of the force field leads to the angle of azimuth φ about any arbitrary axis being a cyclic variable. Therefore, apart from the energy, also the angular momentum vector **J** is constant. It is to be noted, however, that any arbitrarily weak external field, e.g. a homogeneous magnetic field in the direction of the z-axis, leaves only one component J_z constant while J_x and J_y carry out simple harmonic motions with phase difference $\pi/2$ (uniform precession of **J**).

Quantum mechanics provides closely analogous results. If **J** is defined for a mass point as $\mathbf{r} \times \mathbf{p}$, the commutation relations (39) and (40) lead to the following commutation rules for **J**:

$$\begin{aligned} [J_x, J_y] &= i\hbar J_z; \quad [J_y, J_z] = i\hbar J_x; \quad [J_z, J_x] = i\hbar J_y, \\ [J_z, J^2] &= 0 \qquad [J_x, J^2] = 0 \qquad [J_y, J^2] = 0. \end{aligned} \tag{74}$$

This is seen as follows:

$$\begin{aligned} [J_x, J_y] &= (yp_z - zp_y)(zp_x - xp_z) - (zp_x - xp_z)(yp_z - zp_y) \\ &= yp_x(p_z z - zp_z) + xp_y(zp_z - p_z z) = i\hbar(xp_y - yp_x). \end{aligned}$$

The rules in the second line of (74) follow similarly with

$$J^2 = J_x{}^2 + J_y{}^2 + J_z{}^2.$$

The relations (74) show that it is possible to find a representation in which $\mathbf{J}^2$ and one of the components of **J** are diagonal, as well as H, but not one in which two components are simultaneously diagonal.

While **J** is identified with the orbital angular momentum $\mathbf{r} \times \mathbf{p}$ the eigenvalues of J_z (or J_x or J_y) are confined to the values $m\hbar = 0$, $\pm\hbar$, $\pm 2\hbar$ . . $\pm j\hbar$ with *integral* m, and those of $\mathbf{J}^2$ to $\hbar^2 j(j+1)$ with the *integral* values $j = 0, 1, 2 \ldots$ But a more general definition of **J** is possible if the permutation relations (74) alone are considered as fundamental postulates. It is then found that the permissible eigenvalues of $\mathbf{J}^2$ include *half-integral* values of j: $0, \frac{1}{2}, 1, \frac{3}{2} \ldots$ The values of m are always restricted to $+j$, $+j-1$, $+j-2$, . . $-j$. As an illustration of the methods used in this more general theory the eigenvalues of J_z may be chosen as an example.

We assume that $\psi_{\alpha,\lambda,m}$ is an eigenfunction of J_z with the eigenvalue $m\hbar$:

$$J_z\psi_{\alpha,\,\lambda,\,m} = m\hbar\psi_{\alpha,\,\lambda,\,m}.$$

The suffixes α and λ indicate eigenvalues of H and $\mathbf{J}^2$. We introduce the non-Hermitian "displacement operators"

$$J_+ = J_x + iJ_y, \qquad J_- = J_x - iJ_y. \tag{75}$$

With the use of (74) we find the commutation rule

$$J_zJ_+ - J_+J_z = \hbar J_+ \text{ or } J_+J_z = (J_z - \hbar)J_+.$$

When applied to the function $\psi_{\alpha,\lambda,m}$, now written simply ψ_m, this gives

$$(J_z - \hbar)J_+\psi_m = J_+J_z\psi_m = m\hbar J_+\psi_m,$$
$$J_zJ_+\psi_m = (m+1)\hbar J_+\psi_m.$$

The application of the operator J_+ to the eigenfunction ψ_m has thus produced a new function $(J_+\psi_m)$ which is an eigenfunction of the operator J_z with the eigenvalue $(m+1)\hbar$. By continuation of this process a "ladder" of eigenfunctions and eigenvalues $m\hbar$, $(m+1)\hbar$, $(m+2)\hbar$ . . can thus be built up. Similarly, application of the operator J_- leads to eigenfunctions with the eigenvalues $(m-1)\hbar$, $(m-2)\hbar$, . .. The action of these operators can be written

$$\begin{aligned} J_+\psi_m &= \text{const.} \times \psi_{m+1}, \text{ or } J_+|j, m\rangle = \text{const.} \times |j, m+1\rangle \\ J_-\psi_m &= \text{const.} \times \psi_{m-1}, \text{ or } J_-|j, m\rangle = \text{const.} \times |j, m-1\rangle. \end{aligned} \tag{76}$$

By somewhat similar arguments it can be shown that the upper and lower limits of m are $+j$ and $-j$ respectively, where the positive number j is connected with the eigenvalue of $\mathbf{J}^2$ by $\mathbf{J}^2 = j(j+1)\hbar^2$. The conditions that the values of m are equally spaced and that $|m_{\text{max}}| = |m_{\text{min}}| = j$ can only be fulfilled by integral or half-integral values of m and j.

This generalisation of the quantum theory of angular momenta makes it possible to include the intrinsic spin of the electron.

Table II, **3** shows the matrices of the angular momenta for $j = \frac{1}{2}$, 1 $\frac{3}{2}$ in the representation in which J_z and J^2 are diagonal.

The fact that a variable such as $\mathbf{J}^2$ commutes with the energy H or with another variable such as J_z or with both is often used for simplifying matrix equations. In the appropriate representation both, or all three, matrices are diagonal and their rows and columns can be numbered in order of the eigenvalues of $\mathbf{J}^2$ and J_z. In algebraic relations between diagonal matrices the diagonal elements do not get "mixed" so that the eigenvalues, i.e. the diagonal elements, can in turn be substituted for the matrix symbol, thus leading to a number of separate equations. In this way useful relations can often be obtained by means of quite elementary calculations.

TABLE 3

Angular momentum matrices

$j = \frac{1}{2} \quad (m = \frac{1}{2}, -\frac{1}{2})$

$$\mathbf{J}^2 = \tfrac{3}{4}\hbar^2\begin{pmatrix} 1 & 0 \\ 0 & 1 \end{pmatrix} \qquad J_z = \tfrac{1}{2}\hbar\begin{pmatrix} 1 & 0 \\ 0 & -1 \end{pmatrix}$$

$$J_x = \tfrac{1}{2}\hbar\begin{pmatrix} 0 & 1 \\ 1 & 0 \end{pmatrix} \qquad J_y = \tfrac{1}{2}\hbar\begin{pmatrix} 0 & -i \\ i & 0 \end{pmatrix}$$

$j = 1 \quad (m = 1, 0, -1)$

$$\mathbf{J}^2 = 2\hbar^2\begin{pmatrix} 1 & 0 & 0 \\ 0 & 1 & 0 \\ 0 & 0 & 1 \end{pmatrix} \qquad J_z = \hbar\begin{pmatrix} 1 & 0 & 0 \\ 0 & 0 & 0 \\ 0 & 0 & -1 \end{pmatrix}$$

$$J_x = \frac{\hbar}{\sqrt{2}}\begin{pmatrix} 0 & 1 & 0 \\ 1 & 0 & 1 \\ 0 & 1 & 0 \end{pmatrix} \qquad J_y = \frac{\hbar}{\sqrt{2}}\begin{pmatrix} 0 & -i & 0 \\ i & 0 & -i \\ 0 & i & 0 \end{pmatrix}$$

$j = \frac{3}{2} \quad (m = \frac{3}{2}, \frac{1}{2}, -\frac{1}{2}, -\frac{3}{2})$

$$\mathbf{J}^2 = \tfrac{15}{4}\hbar^2\begin{pmatrix} 1 & 0 & 0 & 0 \\ 0 & 1 & 0 & 0 \\ 0 & 0 & 1 & 0 \\ 0 & 0 & 0 & 1 \end{pmatrix} \qquad J_z = \tfrac{1}{2}\hbar\begin{pmatrix} 3 & 0 & 0 & 0 \\ 0 & 1 & 0 & 0 \\ 0 & 0 & -1 & 0 \\ 0 & 0 & 0 & -3 \end{pmatrix}$$

$$J_x = \tfrac{1}{2}\hbar\begin{pmatrix} 0 & \sqrt{3} & 0 & 0 \\ \sqrt{3} & 0 & 2 & 0 \\ 0 & 2 & 0 & \sqrt{3} \\ 0 & 0 & \sqrt{3} & 0 \end{pmatrix} \qquad J_y = \tfrac{1}{2}\hbar\begin{pmatrix} 0 & -i\sqrt{3} & 0 & 0 \\ i\sqrt{3} & 0 & -2i & 0 \\ 0 & 2i & 0 & -i\sqrt{3} \\ 0 & 0 & i\sqrt{3} & 0 \end{pmatrix}$$

The commutation relations (74) lead to a further important result: in the scheme in which $\mathbf{J}^2$ and J_z are diagonal, the matrices J_x and

J_y have only elements connecting states in which m differs by ± 1, e.g.

$$\begin{aligned} \langle jm+1|J_x|jm\rangle &= \tfrac{1}{2}\hbar\sqrt{[(j-m)(j+m+1)]}, \\ \langle jm+1|J_y|jm\rangle &= -\tfrac{1}{2}i\hbar\sqrt{[(j-m)(j+m+1)]}, \\ \langle jm|J_z|jm\rangle &= \hbar m. \end{aligned} \tag{77}$$

These formulae contain the quantum-mechanical description of the uniform precession. The off-diagonal elements are time-dependent if the energy depends on m (e.g. finite field strength in Zeeman effects). The factor i in the second equation contains the phase difference $\pi/2$ between the x- and y-components in the precessional motion.

For a single particle, the commutation rules allow only one of the components of $\mathbf{J}$ to be represented in diagonal form, simultaneously with $\mathbf{J}^2$. One describes $\mathbf{J}^2$ and J_z as a *complete set* of commuting variables and the two quantum numbers j and m as a complete set of quantum numbers, as far as the description of the angular momentum is concerned.

A system of two particles, without interaction between them, is more highly degenerate and allows a wider choice of representations. One can give a complete description of the angular momenta either by specifying $\mathbf{J}_1^2$, J_{1z}, $\mathbf{J}_2^2$, J_{2z} or $\mathbf{J}_1^2$, $\mathbf{J}_2^2$, $\mathbf{J}^2$, J_z, where $\mathbf{J} = \mathbf{J}_1+\mathbf{J}_2$. Each scheme is characterised by a quadruple of quantum numbers: j_1, m_1, j_2, m_2 and j_1, j_2, j, m. According to the interactions considered either one or the other scheme will be appropriate, i.e. will cause the energy matrix to be diagonal, and it is necessary to study the connection between the two, the quantum-mechanical theory of *vector addition*.

In the absence of interaction, the two schemes are alternative descriptions of a degenerate system, and each eigenfunction of a state in one can be written as a linear combination of eigenfunctions in the other, and vice versa.

The theory of *transformation* providing such relations can fortunately be simplified by a general consideration which allows some conclusions to be reached without calculations. In the operator equation

$$J_z = J_{1z}+J_{2z} \tag{78}$$

the right-hand side is diagonal in the first representation and therefore also the left side must be diagonal in the first, as well as in the second representation. Any linear relation between eigenfunctions in the two schemes can therefore only connect states for which $m = m_1+m_2$. We also make use of the fact that the total number of states must be the same in both schemes.

We consider a definite pair of j_1 and j_2 since these quantum numbers are the same in both schemes. The largest eigenvalue of J_z must arise from the largest eigenvalues of J_{1z} and J_{2z}, so that there is only one combination of quantum numbers m_1 and m_2 adding up to $m = j_{\max}$, namely $m_1 = j_1$ and $m_2 = j_2$. This is a single state and must therefore be represented by the same eigenfunction in both schemes, and the same applies to the state $m = -j_{\max}$. For any other value of m there will be several pairs m_1, m_2 adding up to m.

These relations can best be illustrated by a concrete example; we choose arbitrarily $j_1 = 1, j_2 = \frac{1}{2}$; this gives the scheme of table II, 4.

TABLE 4

Addition of angular momenta

$j_1 = 1, \quad j_2 = \frac{1}{2}$

	m_1	m_2		m	j	
a	1	$\frac{1}{2}$	...	$\frac{3}{2}$	$\frac{3}{2}$	α
b	1	$-\frac{1}{2}$	...	$\frac{1}{2}$	$\frac{3}{2}$	β
c	0	$\frac{1}{2}$		$\frac{1}{2}$	$\frac{1}{2}$	γ
d	0	$-\frac{1}{2}$	...	$-\frac{1}{2}$	$\frac{3}{2}$	δ
e	-1	$\frac{1}{2}$		$-\frac{1}{2}$	$\frac{1}{2}$	ϵ
f	-1	$-\frac{1}{2}$	...	$-\frac{3}{2}$	$\frac{3}{2}$	φ

With the use of Roman and Greek letters, as in the table, for the eigenfunctions in the two schemes, the following linear relations exist:

$$\begin{aligned} \alpha &\equiv a \\ \varphi &\equiv f \\ \beta &\equiv C_1 b + C_2 c \\ \gamma &\equiv C_3 b + C_4 c. \end{aligned} \tag{79}$$

The values j in table II, 4 are found as follows: $m = 3/2$ as the largest value of m must be due to $j = 3/2$, and the same applies to

$m = -3/2$; but $j = 3/2$ gives rise to two other values of m: $\frac{1}{2}$ and $-\frac{1}{2}$. This leaves two further values $m = \frac{1}{2}$ and $-\frac{1}{2}$ to be accounted for, and they must be attributed to $j = \frac{1}{2}$. This accounts for all permissible values of j from j_1+j_2 to j_1-j_2.

Quantum mechanics thus leads to the same states which one can derive from a vector diagram in which the lengths of the vectors are equal to the quantum numbers, as in Bohr–Sommerfeld's theory. It leads much further, however, by providing quantitative relations between the states in the two schemes. The coefficients C_1, C_2, C_3, C_4 in (79) are numerical constants which can be calculated from the values of the quantum numbers involved. In conventional symbols these relations can be written

$$|j_1 j_2 j m\rangle = \sum_{m_1 m_2} \langle m_1 m_2 | j m\rangle | j_1 j_2 m_1 m_2\rangle$$

The third equation (79), e.g. becomes

$$\begin{aligned} |1\,\tfrac{1}{2}\,\tfrac{3}{2}\,\tfrac{1}{2}\rangle &= \langle 1\,-\tfrac{1}{2}|\tfrac{3}{2}\,\tfrac{1}{2}\rangle|1\,\tfrac{1}{2}\,1\,-\tfrac{1}{2}\rangle + \langle 0\,\tfrac{1}{2}|\tfrac{3}{2}\,\tfrac{1}{2}\rangle|1\,\tfrac{1}{2}\,0\,\tfrac{1}{2}\rangle \\ &= \sqrt{(\tfrac{1}{3})}|1\,\tfrac{1}{2}\,1\,-\tfrac{1}{2}\rangle + \sqrt{(\tfrac{2}{3})}|1\,\tfrac{1}{2}\,0\,\tfrac{1}{2}\rangle, \\ C_1 &= \sqrt{\tfrac{1}{3}}, \qquad C_2 = \sqrt{\tfrac{2}{3}} \end{aligned}$$

where the transformation coefficients used in the second line have been taken from tables in Condon and Shortley, p. 76.

The example just discussed shows how the electron spin can be fitted into matrix mechanics if the operators, or matrices, $\mathbf{J}_2$ J_{2z} are identified with $\mathbf{S}$, S_z and the quantum numbers j_2, m_2 with s and m_s, while $\mathbf{J}_1$ is the orbital angular momentum of a p-electron.

If j_1 and j_2 are both chosen as $\frac{1}{2}$, the results can be applied to the spins of two electrons. In the nomenclature of p. 36, the spin functions of the first representation are the simple products on the left of the following table II, 5, and the eigenfunctions of the second representation, on the right, are identical with the functions defined on p. 36 and 37.

The spin eigenfunctions can be written as column symbols:

$$(+) = \begin{pmatrix}1\\0\end{pmatrix} \qquad (-) = \begin{pmatrix}0\\1\end{pmatrix} \qquad (+)^* = (1\;0)$$

The orthogonality relations (45) follow from the rules of matrix multiplication.

A complete wave function (Pauli function) consists of products of spin functions and Schrödinger functions (f, g) and can be written in various forms. For a single particle, e.g.

$$(+)f + (-)g \equiv \begin{pmatrix}1\\0\end{pmatrix}f + \begin{pmatrix}0\\1\end{pmatrix}g \equiv \begin{pmatrix}f\\0\end{pmatrix} + \begin{pmatrix}0\\g\end{pmatrix} \equiv \begin{pmatrix}f\\g\end{pmatrix}.$$

TABLE 5

Addition of two spins

	m		S	
(+) (1) (+) (2)	1	. . .	1	χ'_s
(−) (1) (+) (2)	0	. . .	1	χ_s
(+) (1) (−) (2)	0		0	χ_a
(−) (1) (−) (2)	−1	. . .	1	χ_s''

The gyroscopic motion resulting from the coupling of any two angular momenta has again an analogue in the quantum-mechanical formalism. Each of the operators representing, e.g. the angular momentum **L** and the spin **S**, obey commutation rules with regard to the vector resultant **J**, of the following kind:

$$[J_x, S_x] = 0 \qquad [J_x, S_y] = i\hbar S_z \qquad [J_x, S_z] = -i\hbar S_y. \tag{80}$$

Of the matrix elements $\langle jm|\mathbf{S}|j'm'\rangle$ only those are finite for which

$$j'-j = \Delta j = \pm 1 \text{ or } 0.$$

The difference 1 of the quantum number j again corresponds to simple harmonic motion due to uniform precession about **J**.

3. Approximation methods

a. Perturbation theory

Rigorous mathematical solutions of the quantum-mechanical equations in atomic systems are restricted to hydrogen-like atoms. Their importance is two-fold; on the one hand they allow us to test the fundamental assumptions of the theory; on the other hand they form the starting points for methods of approximation on which most of our interpretation of atomic spectra are based.

The mathematical details of the various methods are outside the scope of this book. A brief outline of the Schrödinger perturbation theory will be included in this chapter, on account of its usefulness even for a merely qualitative discussion of spectra, Zeeman effects and Stark effects. Some of the great number of other approximation methods will be mentioned together with their applications.

Perturbation theories are based on the fact that certain interactions can often be regarded as small compared with others: in most atoms, magnetic interactions between spin and orbit of an electron can be regarded as small compared with electrostatic interactions between electrons and between electrons and the nucleus; magnetic interactions with the nucleus can generally be considered to be small even compared with magnetic spin–orbit interactions of the electron. Finally, still smaller interactions with weak external fields can be added as a further step.

We first consider a *non-degenerate state.*

The Hamiltonian H of the perturbed system is expressed as the sum

$$H = H_0 + H' \tag{81}$$

where H_0 is the "unperturbed" Hamiltonian for which the wave equation can be solved and H' is the "perturbing" potential. H' can be expressed in terms of a parameter λ so that it vanishes for $\lambda \to 0$. For a non-degenerate state the eigenfunction ψ_m will agree with the eigenfunction u_m of the unperturbed Hamiltonian for $\lambda = 0$ and can be written

$$\psi_m = u_m + \lambda\psi' + \lambda^2\psi'' \,.., \tag{82}$$

while

$$W_m = E_m + \lambda W_1 + \lambda^2 W_2 \,.., \tag{83}$$

where E_m is the energy of the corresponding unperturbed state and W_m the total energy. These expressions (81), (82) and (83) are substituted in the wave equation and, as a most essential step, ψ' is expanded in terms of the set of unperturbed eigenfunctions:

$$\psi' = \sum a_n u_n. \tag{84}$$

Comparison of terms with equal powers of λ and the use of the orthogonality of the functions u_n leads to the following results:

$$a_m = 0, \qquad a_n = \frac{H'_{nm}}{E_m - E_n} \quad \text{for } n \neq m \tag{85}$$

$$W_m = E_m + H'_{mm} + \sum_n{}' \frac{|H'_{mn}|^2}{E_m - E_n} \tag{86}$$

zero- first- second- order term,

$$\psi_m = u_m + \sum_k{}' \frac{H'_{km} u_k}{E_m - E_k} + \ldots \tag{87}$$

where the $'$ in the summation sign indicates that the terms with $m = n$ and $m = k$ respectively are to be excluded, and where

$$H'_{km} \equiv \int u^*_k H' u_m \, d\tau \tag{88}$$

are the elements of the *perturbation matrix*. The parameter λ has now been put equal to 1, i.e. it has been included in the functions ψ', ψ'' . . and W_1, W_2 . . .

The first-order perturbation energy is given by the integral

$$H'_{mm} = \int u^*_m H' u_m \, d\tau$$

involving only the one unperturbed term. It is the quantum-mechanical average of the *perturbed energy operator applied to the unperturbed eigenfunction*. To this approximation we can speak of the perturbation of the one single term. Only in the second-order approximation is the calculation of the energy affected by other terms of the unperturbed system, through the non-diagonal elements H'_{nm}. The *eigenfunction* ψ_m is, however, in the first term of the perturbation "mixed" with the eigenfunctions u_k of those terms which are connected with u_m by the perturbation matrix element H'.

In most applications to atomic spectra the first-order perturbation energy for non-degenerate rates vanishes for reasons of symmetry.

The sign of the second-order term in (86) is determined by the denominator; the effect is a *mutual repulsion* of the perturbing terms which is the greater the closer they are together, owing to the denominator $E_m - E_n$. Though the sum in (86) and (87) extends formally over all states, it is usually sufficient to consider one or a few closely adjacent energy levels.

More important is the perturbation theory of *degenerate states*. If we assume that one of the unperturbed energy levels is twofold degenerate, it can be described by two different eigenfunctions u_k and u_m or by any two linear combinations of these. We further assume that the perturbation removes the degeneracy, resulting in two perturbed states of different energy, having the eigenfunctions ψ_k and ψ_m. According to the basic assumptions of perturbation theory, the two unperturbed functions have to be so chosen that they become identical with the perturbed functions in the limit $\lambda \to 0$. The first step in the perturbation treatment therefore consists in choosing linear combinations of the unperturbed, degenerate functions in such a way that the terms H'_{km} in (88) vanish for $k \neq m$: we have to diagonalise the sub-matrix H'_{mk} for all the degenerate states, in this case only two. If only the perturbed energy values are required, it will suffice to solve the *secular equation* (see p. 48) of this sub-matrix, in the present case the equation

$$\begin{vmatrix} H'_{mm} - W & H'_{mk} \\ H'_{km} & H'_{kk} - W \end{vmatrix} = 0. \qquad (89)$$

The two roots of this equation are the two values W of the first-order perturbation energy (λW_1 in eq. (83)).

The generalisation for more than two-fold degeneracy is obvious. The diagonalisation need not include states which are degenerate with the state considered if H' does not connect them with the latter. On account of this, the symmetry properties of the perturbation often leads to great simplifications.

If the degeneracy is only removed in second order, a similar, more complicated diagonalisation has to be performed.

b. The variation principle

Perturbation methods lead to accurate results only when the zero-order function is a fairly close approximation, so that the perturbation terms are comparatively small. Among the numerous other methods of finding approximate solutions of the wave equation, only one may be mentioned here, partly on account of its importance in atomic theory and partly because its basic ideas are easy to understand, namely the variation method.

A function $\psi(x,y,z)$ which is a solution of the Schrödinger equation, with an eigenvalue E, has an important extremum property. Let $\Delta\psi$ indicate a *variation* of the function, i.e. a change of its value not due to a change in the coordinates but due to a change in one of the parameters, such as the effective charge, or to an addition of another function, but subject to the condition that the normalisation relation

$$\int \psi^* \psi \, \mathrm{d}\tau = 1$$

remains valid. It can then be proved that the relation exists:

$$\Delta \int \psi^* H \psi \, d\tau = 0. \tag{90}$$

It states that for a continuous range of functions including ψ, the above energy integral has an extremum value for that function ψ which is an eigenfunction of H.

For the eigenfunction of the ground state, the statement of the variation principle goes further: for any arbitrarily chosen function $\varphi(x, y, z)$ for which

$$\int \varphi^* \varphi \, d\tau = 1,$$

the value of the integral

$$E' = \int \varphi^* H \varphi \, d\tau \tag{91}$$

is always larger than the energy E of the ground state. If E' is the energy calculated with an approximate wave function the true energy is an absolute minimum of E'. If two different functions are used for approximating the ground level, the lower of the two values of E' must be closer to the true energy. This is a valuable means of comparing two different trial functions, but there is, unfortunately, no direct way of judging the closeness of approximation to the true energy.

Since the energy is a stationary value, an appreciable change of the eigenfunction, if properly chosen, will cause only a negligible change in the energy. A ψ function which is a very good approximation for the energy of a state may be a very poor approximation for other purposes; it may be inaccurate at very small distances and thus be unsuitable for calculations of hyperfine structures or other effects involving the vicinity of the nucleus, or it may be inaccurate at very large distances and be unsuitable for calculations of Van der Waals forces.

Applications of the variation method will be discussed in III.B and IV.B. Especially important is its use in connection with the method of the *self-consistent field* which will be described in IV.B.

D. THE RADIATION OF ATOMS

1. Electric dipole radiation

a. Classical theory

If the energy levels in an atom are known, the Bohr frequency relation supplies immediately the values of the possible frequencies of the spectral lines. This information is, however, incomplete in several respects. Do all possible term differences lead to spectral lines? What are their relative strengths, both in emission and in absorption and what is the state of polarisation of the light? The answers to these questions can only be given by a more detailed theory of the interaction of electromagnetic radiation with atoms. This theory also leads to an understanding of the widths and shapes of spectral lines, a subject which will be treated in chapter VII.

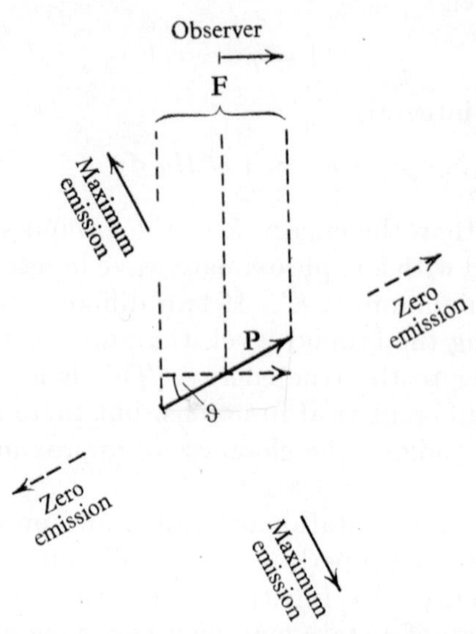

FIG. II, 6. Radiation from oscillating point-dipole.

We consider an electric dipole whose direction is fixed and whose dipole moment oscillates harmonically with frequency ν_0 and amplitude P_0. The radiation field at a point distant R from the dipole, where R is much greater than the wavelength of the light emitted, can be calculated and the result is easy to visualise: the observer stationed at that point will "see" an oscillating dipole of amplitude $P_0 \cos \vartheta$ (see fig. II, 6). The electric vector of the radiation field will be in the plane of the paper and of magnitude

$$F = \frac{4\pi^2 \nu_0{}^2}{c^2 R} P \cos \vartheta. \tag{92}$$

It vanishes for an observer in the direction of P and is strongest in the meridional plane. We complete this kinematic model by assuming the oscillator to consist of a point having the mass m of an electron and the electronic charge $-e$, so that $\mathbf{P} = -e\mathbf{r} = -e\mathbf{r}_0 \cos 2\pi\nu_0 t$, where the value r_0 of the amplitude is of atomic dimensions (fig. II, 7). The energy of this classical *electron oscillator* is

$$E = 2\pi^2 m\nu_0^2 r_0^2. \tag{93}$$

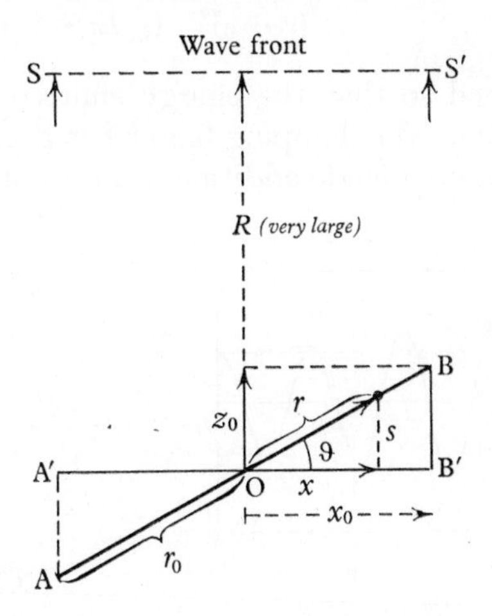

FIG. II, 7. Radiation from classical electron oscillator.

We assume at first that, apart from the restoring force which is specified by the values of ν_0 and m, the oscillator is only subject to the electromagnetic forces due to its own motion. These cause a loss of energy by radiation at a rate which is found by integration of Poynting's vector over the whole sphere:

$$-\mathrm{d}E/\mathrm{d}t = \frac{16\pi^4 e^2}{3c^3}\nu_0^4 r_0^2 = E\gamma_0, \tag{94}$$

with the definition

$$\gamma_0 = \frac{8\pi^2 e^2 \nu_0^2}{3mc^3}. \tag{95}$$

The integration of (94) gives an exponential decrease (fig. II, 8b)

$$E = E_0 \exp(-\gamma_0 t) \tag{96}$$

with the damping factor $-(1/E)\,\mathrm{d}E/\mathrm{d}t = \gamma_0$.

These relations contain the classical description of the process of *spontaneous emission.* On account of the damping, the light emitted is not strictly monochromatic; the intensity distribution is found by a Fourier analysis of the damped wave motion (fig. II, 8c), with the result

$$I(\nu) = I_0 \frac{(\gamma/4\pi)^2}{(\nu-\nu_0)^2+(\gamma/4\pi)^2} \tag{97}$$

where $I(\nu)$ is defined so that the energy emitted in the frequency interval $d\nu$ is $I(\nu)\,d\nu$. The damping factor has been written γ which may differ from γ_0 and include additional causes of damping.

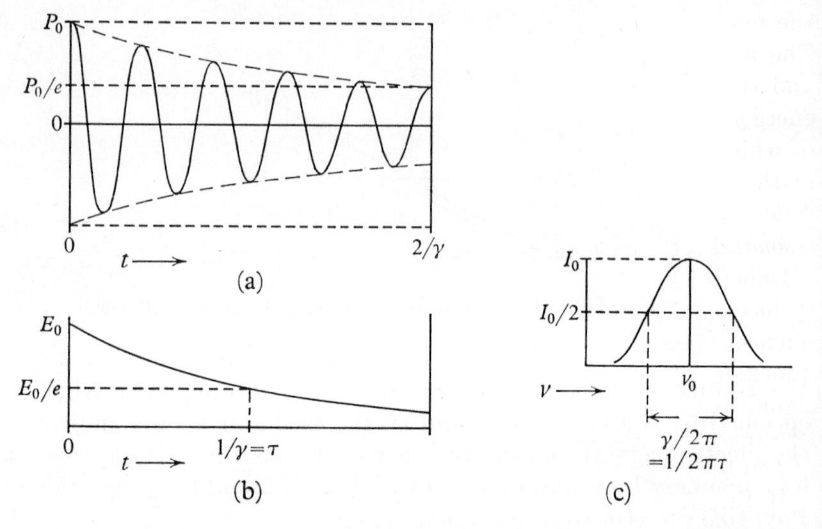

FIG. II, 8. Damped oscillation; exponential decrease of (a) amplitude and (b) energy; (c) spectral distribution of emitted radiation.

The half-value width of the distribution (97), defined as the difference of the frequencies of half of the maximum intensity, is

$$\Delta\nu_{\frac{1}{2}} = \gamma/2\pi. \tag{98}$$

In the optical range of the spectrum γ_0 is very small compared with ν so that, generally speaking, the damping of atomic radiation is very small and spectral lines are very narrow. For $\lambda = 5900$ A, or $\nu = 5{\cdot}1\times10^{14}\ \text{sec}^{-1}$, one finds $\Delta\nu = 1{\cdot}02\times10^{7}\ \text{sec}^{-1}$ or $\Delta\tilde{\nu} = 0{\cdot}34\times10^{-3}\ \text{cm}^{-1}$. According to (95) and owing to the relation

$\Delta\lambda = -c\Delta\nu/\nu^2$, the wavelength difference corresponding to γ_0 is independent of the frequency and has the value $\Delta\lambda = 1{\cdot}18 \times 10^{-4}$ A.

If atoms are exposed to the influence of an external, alternating electric field such as that of an incident light wave, the phenomena of absorption, scattering and dispersion are observed. For frequencies approaching a characteristic frequency of the atom, the phenomena are enhanced and the two latter are described as fluorescence and anomalous dispersion. These phenomena are regarded classically as due to the *forced oscillations* of the electron oscillator. Initially the external field causes an increase or decrease of the amplitude, according to its previous value, but finally a stationary state of oscillation at constant amplitude is established in which the radiated energy is being steadily replaced from the external field. In this final state the emitted light has exactly the frequency of the incident wave. The amplitude of the forced oscillation depends on the frequency and the value of γ. It determines both the rate of absorption of energy from the incident wave and the rate of scattering, the latter of which is most easily measured by its effect on the primary wave in the form of refraction, i.e. as the refractive index. The amplitude reaches its maximum value for $\nu = \nu_0$, a condition described as *resonance*.

Generalising our previous assumptions we now consider each atom to have f harmonic, classical oscillators for any one resonance frequency and, purely formally, do not restrict f to integral values. The dimensionless quantity f is known as *oscillator strength*; its physical significance will be discussed below (p. 67). Measurements of the two related phenomena, absorption and dispersion, allow the value of f to be determined.

If the intensity of a collimated beam of monochromatic light decreases from I_0 to I over the length of path s, the absorption coefficient α is defined by

$$I = I_0\, e^{-\alpha s}. \tag{99}$$

The value of α is then found to be

$$\alpha = \frac{Ne^2 f}{4\pi mc} \times \frac{\gamma}{(\nu-\nu_0)^2 + (\gamma/4\pi)^2}, \tag{100}$$

where N is the number of atoms per cc., and the total absorption α_0 is found by integration:

$$\alpha_0 = \int_0^\infty \alpha\, d\nu = \pi f N e^2/mc. \tag{101}$$

This relation which allows the determination of f from α_0 is independent of γ and is therefore still valid if any other causes of damping or broadening are present, such as Doppler broadening and pressure broadening (see chapter VII). In general, the measurement of α_0 requires spectroscopic resolution high enough to follow the change of α with ν. For weak absorption, however, i.e. if $\alpha s \ll 1$ for all values of ν, the total loss of energy becomes directly proportional to α_0. The relation (99) then takes the form

$$I = I_0(1-\alpha s),$$

giving

$$\int (I_0 - I)\,\mathrm{d}\nu = I_0 s \int \alpha(\nu)\,\mathrm{d}\nu = I_0 s \alpha_0. \tag{102}$$

The value of α_0 can then be determined from the loss of intensity without any spectroscopic resolution other than that required for separating one spectral line from the others.

The refractive index n can be written, provided $n-1$ to be $\ll 1$,

$$n-1 = \frac{Ne^2 f}{2\pi m}\frac{\nu_0{}^2-\nu^2}{(\nu_0{}^2-\nu^2)^2+\nu^2\gamma^2/4\pi^2} \approx \frac{Ne^2 f}{4\pi m\nu_0}\frac{\nu_0-\nu}{(\nu_0-\nu)^2+(\gamma/4\pi)^2}. \tag{103}$$

The second part of the equation is based on the approximation $|\nu-\nu_0| \ll \nu_0$. For frequencies not too close to ν_0 one can neglect the damping term and write

$$n-1 = \frac{Ne^2 f}{2\pi m(\nu_0{}^2-\nu^2)}. \tag{104}$$

If several oscillators of different frequencies are present, the right-hand side of (103) and (104) has to be replaced by a sum. In the immediate vicinity of one characteristic frequency, however, the term containing this frequency dominates, so that its f-value can be determined from the anomalous dispersion caused by this spectral line (see VII.D).

b. Emission and absorption of quanta

The attempt to combine the classical view of radiation with the quantum theory of the atom meets with fundamental difficulties. If the atom can only exist in stationary states of well-defined energy E_j, the process of emission must be instantaneous and can be described as quantum jump. This, however, clashes with the fundamental mathematical fact that emission of a spectral line of width

$\Delta\nu$ sec^{-1} requires a finite duration of the wave train of the order of magnitude $\tau = 1/\gamma = 1/(2\pi\nu)$ sec. In which state would the atom then be at the moment when the wave train has carried away half of its original energy of excitation? The answer, based on the statistical interpretation of quantum mechanics due to Born, Bohr, Heisenberg and others, would be this: A measurement of the energy of the atom at this moment would give the result 0 or E_j with equal probability. It is only the statistically defined average energy $\bar{E}$ which decreases exponentially according to classical laws.

Let N_j be the number of atoms in the excited state E_j. When considering the properties of the wave train, we have to ascribe the decrease of the energy N_jE_j to the factor E_j as in the classical theory. If, on the other hand, we take the extreme quantum view and regard E_j as a quantised, fixed value, we must ascribe the decrease to the factor N_j:

$$\begin{aligned} \mathrm{d}E/\mathrm{d}t &= -\mathrm{d}(N_jE_j)/\mathrm{d}t \\ \text{classically:} \quad &= -N_j\,\mathrm{d}E_j/\mathrm{d}t = -N_jE_j\gamma \\ \text{quantum theory:} \quad &= -E_j\,\mathrm{d}N_j/\mathrm{d}t = -N_jE_jA_{ji}. \end{aligned}$$

The duality of the description is connected with the fact that time and energy are conjugate variables and do not commute (principle of *indeterminacy*).

In the extreme quantum view the emission of radiation is similar to a radioactive decay:

$$N_j = N_j^0\, e^{-A_{ji}t}. \tag{105}$$

In this sense A_{ji} is called the *spontaneous transition probability* from the state j to the lower state i which may be the ground state. It measures the relative rate of decrease of energy in the same way as γ measures it in classical theory, and the two quantities can be related to one another by means of the correspondence principle.

To the processes of interaction of atoms with an external radiation field must correspond processes in quantum theory; there must be transition probabilities proportional to the density ρ_ν of radiation, defined as energy density per dν, as well as spontaneous transition probabilities. The relation between these can best be seen from Einstein's derivation of Planck's law of black body radiation.

If N_j and N_i are the numbers of atoms per cc. in the states j and i, the latter of which may be the ground state, the number of transitions per second is

(I) from j to i: $\quad N_jA_{ji}$ (spont. emission) $\quad + \quad N_jB_{ji}\rho_\nu$ (induced emission)

(II) from i to j: $N_i B_{ij} \rho_\nu$
(absorption)

If we equate (I) and (II) for the state of equilibrium, the use of the Bolzmann equation

$$\frac{N_i}{N_j} = \frac{g_i}{g_j}\exp[(E_j - E_i)/kT] = \frac{g_i}{g_j}\exp(h\nu/kT) \tag{106}$$

leads to the relation

$$\rho_\nu = \frac{A_{ji}}{(g_i/g_j)\exp(h\nu/kT)B_{ij} - B_{ji}} \tag{107}$$

where g_j and g_i are the *statistical weights* of the two states. The condition that for small values of $h\nu/kT$ the equation should pass into the classical formula of Rayleigh and Jeans

$$\rho_\nu = \frac{8\pi}{c^3}\nu^2 kT \tag{108}$$

can easily be derived from (107) by substituting $1 + h\nu/kT$ for $\exp(h\nu/kT)$:

$$B_{ij}g_i/g_j = B_{ji} \quad \text{and} \quad A_{ji} = B_{ij}(g_i/g_j)8\pi h\nu^3/c^3. \tag{109}$$

Einstein's B values are closely related to the oscillator strength f. According to the definition of α_0 (p. 63) the energy absorbed per cm^3 is $\alpha_0 \rho_\nu c$. Equating this quantity with the corresponding quantum-theoretical rate of absorption of energy gives

$$\alpha_0 \rho_\nu c = \frac{\pi e^2}{m} f N_i \rho_\nu = N_i B_{ij} h\nu \rho_\nu, \tag{110}$$

$$f_{ij} = B_{ij}hm\nu/(\pi e^2) = A_{ji}(g_j/g_i)mc^3/(8\pi^2\nu^2 e^2) = A_{ji}(g_j/g_i)/(3\gamma_0). \tag{111}$$

The more specific nomenclature f_{ij} has been introduced here instead of f.

In a gas at high temperature or in an electric discharge the number of atoms in excited states can become appreciable, and negative terms corresponding to induced emission have to be included in the dispersion formula. The influence of all transitions from the excited terms j to higher terms k and lower terms i is expressed in the following approximate dispersion formula in which the damping term has been omitted in the denominator:

$$n-1 = \frac{e^2}{2\pi m}\sum_j N_j\left(\sum_k \frac{f_{jk}}{\nu_{jk}^2-\nu^2}+\sum_i \frac{f_{ji}}{\nu_{ji}^2-\nu^2}\right) \tag{112}$$

where the f_{ji} are negative strength factors and are related to the positive f_{ij} by $f_{ij} = -f_{ji}(g_j/g_i)$. Therefore

$$A_{ji} = -3\gamma_0 f_{ji} = 3\gamma_0 f_{ij}(g_i/g_j). \tag{113}$$

This dispersion formula connecting transition probabilities with the refractive index is due to Kramers and Heisenberg and formed one of the first steps in the evolution of quantum mechanics.

The f values as here defined result from averaging over all directions or quantum-theoretically over all values of the quantum number m. f values relating to transitions between individual subterms with fixed m are sometimes used and the relations between them become somewhat different.*

c. The f sum rules

The classical treatment as based on the concept of the harmonic oscillator can be extended to periodic motions of a general kind. For a single degree of freedom we can expand the expression for the coordinate by a Fourier series. The square of the coefficient of the nth harmonic then appears, in the expressions for absorption, emission and dispersion, as the strength f of a *virtual oscillator* of frequency $n\nu_0$. For periodic motions of several degrees of freedom, the harmonics of all fundamental frequencies and their combinations can similarly be regarded as due to virtual oscillators.

The correspondence principle relates these oscillators to definite quantum-mechanical frequencies and allows qualitative, and sometimes quantitative conclusions to be drawn, from the classical model, on transition probabilities and f-values.

A useful relation which was found in this way and later proved rigorously by means of quantum mechanics is the *f-sum rule.*(1,2) For a spectrum which can be attributed to a single electron, it states that the sum of the f values of all transitions from one term is equal to 1. For an excited term the negative f values of the downward transitions have to be included:

$$\sum_{k,i}(f_{jk}+f_{ji}) = 1. \tag{114}$$

* See e.g. H. Bethe, *Handb. d. Phys.*, Springer 1930, **24,** I.

If the spectrum can be ascribed, to a sufficient degree of approximation, to z electrons, the sum is equal to z. These will generally be z equivalent electrons, i.e. of the same quantum numbers n and l, such as the two valency electrons of Ca.

A partial sum rule including only transitions with the same change of the quantum number l is often useful in single electron spectra:[3,4]

$$\sum_{n'} f_{n\,l \to n',l-1} = -\frac{1}{3}\frac{l(2l-1)}{2l+1} \tag{115a}$$

$$\sum_{n'} f_{n,l \to n',l+1} = \frac{1}{3}\frac{(l+1)(2l+3)}{2l+1} \tag{115b}$$

These relations show that for $l \to l-1$ the downward transitions prevail, resulting in a negative value of the sum, while for $l \to l+1$ the upward transitions prevail.

d. Selection rules from correspondence principle

Most of the selection rules for dipole radiation can be derived by means of the correspondence principle (see p. 26). If a certain quantum number corresponds to a classical frequency which appears only as the fundamental in the Fourier expansion of the coordinates as functions of time, only changes of ± 1 are "allowed" in the sense that they give rise to electric dipole radiations and therefore to lines of appreciable strength. This restriction applies to all cyclic variables arising from axial symmetry. If we disregard the spin, the motion of a single electron in a central field is confined to a plane; the Hamiltonian is independent of the angle of azimuth which therefore gives rise to a uniform precessional motion of frequency ν_2 (see p. 21) which appears only as fundamental in the Fourier analysis. This leads to the selection rule

$$\Delta l = \pm 1 \tag{116}$$

for the azimuthal quantum number l (k in Sommerfeld's notation). As there is no coordinate of the motion not containing ν_2, the transition $\Delta l = 0$ is not allowed.

For the quantum number m associated with uniform precession of the orbital axis about the fixed direction of an external field, analagous arguments lead to the selection rules

$$\Delta m = \pm 1 \text{ and } \Delta m = 0. \tag{117}$$

The former is derived from the x, y coordinates of the motion, the latter from the z coordinate which clearly contains the frequencies of the orbital motion but not that of the precession about the field- (z-) axis.

This theory can also be expected to describe the polarisation of the radiation correctly; the first two transitions will emit light of the same polarisation as a point charge uniformly rotating to the right or left respectively in the x, y plane, the third transition will act like a point charge oscillating in the z direction only. This is completely verified in Zeeman spectra (see p. 97). If we include the effect of the spin on the motion of the electron, the only rigorously constant angular momentum is **J**. Very often, however, the spin–orbit coupling is so weak that the orbital motion is practically the same as it would be in the absence of the spin, except for the slow precession of the orbital plane; the quantum number l then retains its significance and its selection rules are unaffected.

The quantum number j is associated with the uniform precession about **J**, and therefore subject to the selection rule

$$\Delta j = 0, \pm 1.$$

In contrast to the rule for l, the transition $\Delta j = 0$ is allowed. It arises from the z component of the orbital motion which does not vanish because the orbital plane is no longer at right angles to the direction of **J** defining the z axis. The value of z is not affected by the precession and therefore is associated with $\Delta j = 0$.

For several electrons, the total angular momentum **J** is still constant, and the associated quantum number J obeys the same rule as j, with an additional rule, forbidding $J = 0 \rightarrow J' = 0$, which will be derived below (p. 77).

Other selection rules in spectra with several electrons depend on the coupling conditions and will be discussed in the appropriate chapters (III.B.6 and V.A).

e. Selection rules and line strength in quantum mechanics

In quantum mechanics the coordinate x (or y, z) is replaced by a *matrix* (see II.C.2). Each of its elements

$$x_{ji} = \int x\psi_j{}^*\psi_i \, \mathrm{d}\tau \times \exp[it(E_j - E_i)/\hbar] \tag{118}$$

takes the place of a Fourier component of the classical coordinate x.

The classical mean square of the coordinate of the harmonic oscillator $\overline{x^2} = \frac{1}{2}x_0^2$ corresponds to

$$2|x_{ji}|^2 = 2\left|\int x\psi_j{}^*\psi_i\,\mathrm{d}\tau\right|^2.\text{(*)}$$

For $i = j$, the value x_{ji} becomes equal to

$$\int x\psi_j{}^*\psi_j\,\mathrm{d}\tau.$$

Since $\psi_j{}^*\psi_j$ is the probability density, the integral, multiplied by e, is the dipole moment due to a charge density ρ, i.e.

$$\int x\rho\,\mathrm{d}\tau.$$

Since $\psi_j{}^*\psi_j$ is an *even* function of x and contributions from positive and negative values of x mutually cancel out in the integration, the value x_{jj} vanishes. This result expresses the fact that atoms in stationary states have no dipole moment.

If we apply (94) to the radiation in one spectral line which corresponds to one Fourier component of x, y, z for each of the classical coordinates

$$x_0,\ y_0,\ z_0\ (r_0{}^2 = x_0{}^2+y_0{}^2+z_0{}^2),$$

we find the quantum-theoretical value of the rate of energy emission in the transition $j \to i$:

$$-\left(\frac{\mathrm{d}E}{\mathrm{d}t}\right)_{ji} = \gamma_{ji}E_{ji} = A_{ji}h\nu = \frac{16\pi^4e^2\nu_{ji}{}^4}{3c^3}\times 4(|x_{ji}|^2+|y_{ji}|^2+|z_{ji}|^2),$$

and finally

$$A_{ji} = \frac{64\pi^4e^2\nu^3}{3hc^3}\sum_{m_i}(|x_{ji}|^2+|y_{ji}|^2+|z_{ji}|^2). \tag{119}$$

In applying this formula, one has to remember that one energy level, in the absence of external fields, generally consists of several (g) states. The symbol Σ_{m_i} indicates that the sum of all matrix elements connecting one state of the upper level j with all states of the lower level i is to be taken. Condon and Shortley define, for reasons of symmetry between the two levels, the *strength* of the line as the sum

* The factor 2 has to be introduced because one classical Fourier component corresponds to two matrix elements; see W. Heitler, *Quantum theory of radiation*, Oxford University Press 1936, p. 106.

over all states of both levels,

$$S_{ij} = S_{ji} = e^2 \sum_{m_i m_j} (x_{ji}^2 + y_{ji}^2 + z_{ji}^2), \tag{120}$$

so that

$$A_{ji} = \frac{1}{g_j} \frac{64\pi^4\nu^3}{3hc^3} S_{ji} = \frac{2{\cdot}02 \times 10^{18}}{g_j \lambda'^3} S_{ji}', \tag{121}$$

where λ' is measured in Angstrom units and S' in atomic units of $a_0^2 e^2$ (see p. 413). As before, e is given in e.s.u.

From (111) and (121) follows

$$f_{ij} = \frac{8\pi^2 \nu m}{3he^2 g_i} S_{ji} = \frac{304}{g_i \lambda'} S_{ji}', \tag{122}$$

in the same units.

The relation (119) has also been derived by the more rigorous theory of quantum electrodynamics in which the radiation field and the atom are treated as a single quantised system. Further results of this theory, especially certain shifts of lines and relations for the finite widths of levels will be mentioned in III.A.5, VI.B.3 and VII.B.

From eq. (118), the *selection rules* for dipole radiation can be easily derived by means of symmetry considerations. We first disregard the spin and consider a single electron in a central field of force, and a transition in which the quantum number m changes from m to m'. The eigenfunction consists of the product $R(r)\Theta(\vartheta)\Phi(\varphi)$. Since $z = r\cos\vartheta$ does not depend on φ, its matrix element can be written as the product

$$\begin{aligned} z_{mln\ m'l'n'} = & \int_0^\infty R_{nl} R_{n'l'} r^3\, \mathrm{d}r \times \\ & \times \int_0^\pi P_l^m P_{l'}^{m'} \cos\vartheta \sin\vartheta\, \mathrm{d}\vartheta \int_0^{2\pi} \exp[i(m-m')\varphi]\, \mathrm{d}\varphi \end{aligned} \tag{123}$$

the last factor of which vanishes except for $m = m'$. We have therefore the selection rule

$$\Delta m = 0$$

for the emission of light whose electric vector oscillates in the direction of the external field, i.e. in the direction of the z-axis.

Let $\rho = r \sin \vartheta$ be the projection of the vector r on the x, y plane; we can then introduce, instead of x and y, the coordinates $q^+ = x+iy$ and $q^- = x-iy$:

$$q^+ = \rho \cos \varphi + i\rho \sin \varphi = \rho\, e^{i\varphi}$$

$$q^- = \rho \cos \varphi - i\rho \sin \varphi = \rho\, e^{-i\varphi}.$$

Since ρ is independent of φ, the matrix elements of q^+ and q^- contain the factors

$$\int_0^{2\pi} \exp[i(m-m'+1)\varphi]\,\mathrm{d}\varphi \quad \text{and} \quad \int_0^{2\pi} \exp[i(m-m'-1)]\varphi\,\mathrm{d}\varphi$$

respectively. The first of these vanishes except for $m = m'-1$, the second except for $m = m'+1$, giving the selection rules

$$\Delta m = +1 \text{ and } \Delta m = -1$$

for a dipole radiation emitting right- and left-circularly polarised light in the z direction, or plane polarised light, with the electric vector at right angles to the field direction, in the x, y plane.

The quantum number m and the polarisation rules assume physical significance only if a fixed direction in space is defined by the conditions of the experiment. This occurs in Zeeman- and Stark-effects where the direction of the external field defines the polar axis, and in phenomena of scattering and fluorescence where the electric vector of the incident radiation can be used for defining an axis.

The selection rules for l can be derived from the properties of the functions $P_l{}^m$; the second integral in (123) can easily be shown to vanish unless $l' = l \pm 1$, leading to the rule

$$\Delta l = \pm 1.$$

It can also be derived from the *parity* properties of the functions $P_l{}^m$ which depend only on l. Any linear coordinate has obviously odd parity. When the product of two wave functions and a coordinate q (x, y or z) is integrated over the whole of space, any two elements $d\tau$ diametrically opposite to one another will cancel out if the product has odd parity, i.e. if the product of the two wave functions has even parity. Thus only states of opposite parity can lead to dipole radiation. This would still allow transitions $\Delta l = 3, 5 \ldots$, but any such transitions could not be reconciled with the selection rules for m since the states of largest m could then not radiate at all.

The *parity property* remains well defined for a free atom with any number of electrons with any kind of coupling. This leads to the rigorous selection rule due to *Laporte*:[5]

Only transitions between even and odd terms are allowed for dipole radiation.

In every spectrum, however complicated, all terms can be divided in two classes according to their parities, and only combinations between *even* and *odd* terms are observed as lines of appreciable intensity. Terms are *even* and *odd*, respectively, if the sum $\Sigma_i l_i$ of the absolute values of the l for all electrons is even or odd. Odd parity of a term is often indicated by a superscript ^{0} in the term symbol, such as $P_{\frac{1}{2}}{}^0$.

If we wish to calculate absolute values of transition probabilities or ratios of intensities of lines involving different pairs of values of n or l, the radial integrals have to be calculated. This has been done rigorously for hydrogen-like spectra. For other atoms the radial wave functions $R(r)$ can only be computed by laborious and usually not very accurate methods of approximation.

A much more restricted task is the calculation of intensity ratios for lines of a multiplet or for the components of a Zeeman pattern. This problem had been solved by correspondence methods, before the advent of quantum mechanics, in a remarkably complete form (see p. 189). The procedure was based on a Fourier analysis of the orbital motion, taking into account the precession of the orbital axis about the vector **J** and of the latter about the direction of the external field. Even the adjustment of the results for small quantum numbers could be made without recourse to quantum mechanics which did not, in fact, exist then.

In quantum mechanics the Fourier analysis of the precessional motions is replaced by commutation relations. It is found that the radius vector **r** which determines the electric dipole moment of the radiating electron obeys the commutation rules (80) with regard to **L** and, in the presence of a spin momentum **S**, also with regard to **J**. This is based on the assumption that **r** and **S** commute. It leads immediately to two important results:

(a) The sum

$$\sum_{m'} |\langle jm|\mathbf{r}|j'm'\rangle|^2 \tag{125}$$

is independent of m. This expresses the fact that the radiation emitted in one spectral line, in the absence of external fields, is independent of the orientation of the angular momentum. In the Zeeman effect it leads to the equality of certain sums of intensities.

(b) The sum (125) taken over j' as well as m' is also independent of j. This means that the total radiation of an atom in a state of given j, leading to all possible values j', is independent of the orientation of the orbital plane with regard to the spin.

In this and the preceding statement it is taken for granted that the other pairs of quantum numbers defining the two states, such as n and l for one or for several electrons, have always the same values, which means that one is concerned with a given multiplet.

The result (b) leads to *sum rules* for lines within a multiplet (see p. 170). Beyond these sum rules, the relative values of all the matrix elements of **r** can be calculated without any special assumptions apart from the commutation properties and lead to formulae for the relative intensities of all lines of multiplets and of all Zeeman components (see pp. 190 and 210).

The assumption that the operator **r** of one electron commutes with its spin momentum **S** is justified to the extent to which the influence of the spin on the Schrödinger functions, i.e. on the shapes of the orbit, can be neglected. This will be a good approximation for weak coupling between spin and orbit.

2. Radiation of higher order[6,7,8,9]

a. Classical theory

The electrostatic potential produced at a point distant R from the origin, by a distribution of fixed charges near the origin can be expanded in a series of terms describing the effect of (1) the net charge, (2) the resulting dipole, (3) the resulting quadrupole . ., and the contribution of each of these terms decreases with the inverse first, second . . power of R. The effect of periodically moving charges has to be described by retarded potentials taking account of the finite speed of propagation of the action, and only these retarded potentials give rise to the radiation field, a periodically varying field whose amplitude is inversely proportional to R.

An expansion of the retarded potentials leads to effects of successively decreasing magnitude, the first of which is identical with that caused by a mathematical point dipole and which was treated in the last section. The next term depends on the angles in the same way as a potential caused by an electrostatic quadrupole and the corresponding radiation is described as *electric quadrupole radiation*. The expansion parameters are not, as in the static case, the ratios x/R, y/R . ., where x, y . . are the coordinates of the charges, but x/λ, y/λ . . where λ is the wavelength of the emitted radiation.

The complete calculation is rather involved, but the relation between quadrupole- and dipole-radiation will become clear from the following calculation which is not rigorous but contains the essentials of the theory.

A point charge e may be oscillating along AOB so that the displacement from O is given by $r = r_0 \sin 2\pi\nu t$.

At a sufficiently distant point, the electric vector of the dipole radiation field is given, as shown in elementary textbooks, by

$$E = \frac{e}{c^2R}\ddot{r}\cos\vartheta = \frac{4\pi^2\nu^2 er}{c^2R}\cos\vartheta, \tag{126}$$

an approximation which replaces the real oscillator by its projection on the wavefront, A′OB′ (fig. II, 7). We now compare the radiation of the real oscillator AOB with that of the oscillator A′OB′ by taking into account the fact that the disturbance arrives in the plane SS′ earlier, with a decrease in retardation of $\delta = s/c = (r\sin\vartheta)/c$; we shall thus have taken the retardation into account to the second-order approximation considering that the dipole radiation contains the retardation in the first order. The field strength E' in the radiation field then becomes

$$E' = \frac{4\pi^2\nu^2 e}{c^2R} r_0 \cos\vartheta \sin 2\pi\nu(t-\delta)$$

$$= \frac{4\pi^2\nu^2 e}{c^2R} r_0 \cos\vartheta(\sin 2\pi\nu t\cos 2\pi\nu\delta - \cos 2\pi\nu t\sin 2\pi\nu\delta).$$

With the substitution of $\delta = 1/c(r_0 \sin\vartheta \sin 2\pi\nu t)$ and of the first two terms in the expansion of $\cos 2\pi\nu\delta$ and $\sin 2\pi\nu\delta$, this becomes

$$E' = \frac{4\pi^2\nu^2 e}{c^2R}[r_0\cos\vartheta\sin 2\pi\nu t - (r_0^2/c)\cos\vartheta\sin\vartheta \times 2\pi\nu\sin 2\pi\nu t\cos 2\pi\nu t]$$

$$= \frac{4\pi^2\nu^2 e}{c^2R}[r_0\cos\vartheta\sin 2\pi\nu t - (\pi r_0^2\cos\vartheta\sin\vartheta/\lambda)\sin 4\pi\nu t] \tag{127}$$

$$= \frac{4\pi^2\nu^2 e}{c^2R}[x_0\sin 2\pi\nu t - (\pi x_0 z_0/\lambda)\sin 4\pi\nu t].$$

While the first term is the dipole radiation which would be emitted by an oscillator A′OB′, the second term describes a radiation with frequency 2ν resulting from modulation of the lateral oscillation of frequency ν by the longitudinal component z of the vibration with the same frequency. The amplitude of the quadrupole term vanishes at $\vartheta = 0$, $\pi/2$, π and $3\pi/2$ and has maxima at $\vartheta = \pi/4$, $3\pi/4$, $5\pi/4$ and $7\pi/4$. It is smaller than the amplitude of the dipole radiation in the ratio of the dimension of the oscillator to the wavelength of the emitted light (factor $\pi z_0/\lambda$ in eq. 127). For atoms, the intensity of the quadrupole radiation can be expected to be of the order of $(\pi a/\lambda)^2$ weaker than that of dipole radiation if a is the atomic radius; this is about 10^{-7} for visible light.

If a second charge e is assumed to oscillate along AOB with a phase difference of π compared with the first, the dipole terms cancel while the quadrupole terms due to the two charges have the same sign and phase. This is a model of a light source emitting quadrupole radiation only. The most symmetrical charge distribution of finite quadrupole moment is a non-spherical distribution of axial symmetry. If the z axis is chosen as the axis of symmetry, the quadrupole moment is defined as

$$Q = \overline{z^2} - 1/3\overline{r^2}.$$

The general expression for the amplitude of the quadrupole radiation contains all the products xy, yz ...

b. Quantum theory and selection rules

In quantum mechanics the matrix elements of the transition probability of quadrupole radiation are of the form

$$Q_{ji} = \int \psi_j^* xy \psi_i \, \mathrm{d}\tau. \tag{128}$$

The selection rule for the quantum number l for electrical quadrupole radiation follows from the parity properties in a similar way as that for dipole radiation. The product of two coordinates such as xy, is clearly *even*, so that the matrix element does not vanish if the two wave functions have the same parity. This leads to the selection rule for a single electron

$$\Delta l = 0, \pm 2. \tag{129}$$

For several electrons, the same argument leads to a rule analogous to the Laporte rule for dipole radiation.

Only transitions from even to even or from odd to odd terms give rise to quadrupole radiation.

The complete expansion of the radiation field produced by a system of moving charges contains not only terms described as electric dipole, quadrupole-, octopole-radiation, but also terms described as magnetic dipole-, quadrupole . . radiations. *Magnetic dipole radiation* appears as a term of the same order of magnitude as electric quadrupole radiation, but the relation between its magnetic vector H and the magnetic dipole of the atom is the same as that between the corresponding electrical quantities in electric dipole radiation. The parity property of the angular momentum operator giving rise to the magnetic moment is even, the same as that of the products xy, — so that only even–even and odd–odd transitions are allowed as in electric quadrupole radiation. For a single electron or for several electrons with Russell–Saunders coupling (see p. 250) the selection rule in the absence of electric fields is for *magnetic dipole radiation*:

$$\Delta L = 0, \qquad \Delta J = \pm 1 \tag{130}$$

where $L \equiv l$ and $J \equiv j$ for a single electron. The magnetic field of this type of radiation is coupled with the oscillatory components of **L** and **S** at right angles to **J**, due to the precessional motion; this implies an oscillating magnetic dipole moment at right angles to **J**. Transitions $\Delta J = 0$ are allowed as magnetic dipole radiation only in the presence of an external magnetic field; they are then due to the precession about the direction of the field, with the selection rule $\Delta m = \pm 1$. If the nucleus has a magnetic moment, transitions $\Delta J = 0$ can be caused as magnetic dipole radiation by the precession of the nuclear spin and **J** about the over-all resultant of the angular momentum **F** (see chapter VI). Both these transitions are of radio- or microwave-frequency (see below).

For the quantum number J describing the total angular momentum of one or several electrons, including their spin momenta, the selection rules for all types of radiation follow directly from a general result of quantum electrodynamics: the quantum emitted in electric dipole radiation can be shown to have the angular momentum $\hbar$, that emitted in electric quadrupole radiation the angular momentum $2\hbar$. Conservation of the total angular momentum of atom and radiation field leads to the condition which, in terms of a Bohr–Sommerfeld vector diagram (fig. II, 9), can be written $|\mathbf{J}-\mathbf{J}'| = n\hbar$, where $n = 1$ and 2 for electric dipole- and quadrupole-radiation respectively. In terms of quantum numbers, it implies $J+J' \geqq n$, or the selection rules

electric dipole radiation: $\Delta J = 0, \pm 1$,
but $J = 0 \rightarrow J' = 0$ forbidden;
electric quadrupole radiation: $\Delta J = 0, \pm 1, \pm 2$,
but $J = 0 \rightarrow J' = 0, \quad J = \frac{1}{2} \rightarrow J' = \frac{1}{2}$
and $J = 0 \leftrightarrow J' = 1$ forbidden.

Spectral lines which are not allowed as electric dipole transitions are often referred to as *forbidden lines*, but this term is sometimes used in a wider sense including transitions which are forbidden as dipole radiation only to a first approximation for a certain kind of coupling between the electrons; examples are *intercombination* lines (see p. 148). But even the appearance of lines which are strictly forbidden as dipole radiation is not always due to higher-order radiations. Transition rules hold strictly only for free atoms, and external fields due to ions in a discharge or even to neighbouring neutral atoms can give rise to *enforced* transitions of observable intensity.

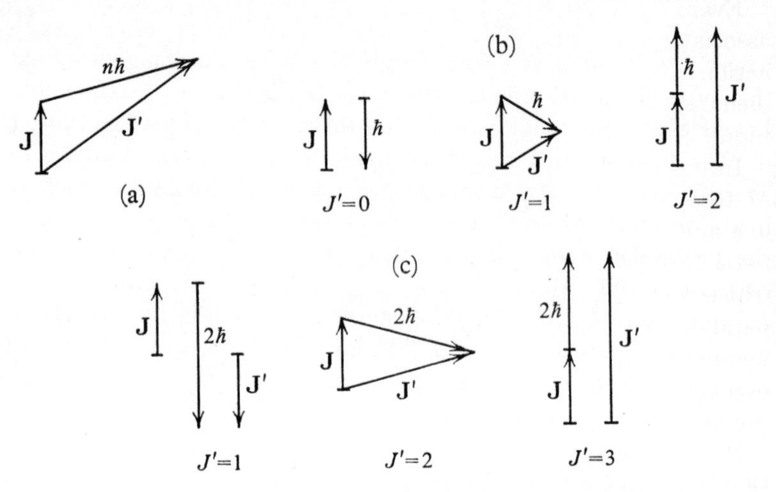

FIG. II, 9. Selection rules for J from conservation of angular momentum; (a) principle of method, (b) electric dipole radiation for $J = 1$, (c) electric quadrupole radiation for $J = 1$.

The nature of a forbidden line can often be established only by means of observations of the Zeeman effect which shows typically different patterns for the different types of radiation (see p. 215).

The total angular momentum of the atom is generally identified with the resultant of the momenta of all the electrons, including their spin momenta. This is not strictly permissible for atoms whose nuclei possess a spin. On account of this, lines which are forbidden by the J-selection rule are sometimes observable as weak lines emitted only by the odd isotopes of the element (see VI.B.7).

Magnetic dipole radiation has found extensive application in the field of radiofrequency resonance. Classically, the Larmor precession of an electron spin or orbit about the z axis can be regarded as a magnetic oscillator in the x, y plane; it acts as a harmonic oscillator in an important respect: the frequency is independent of the amplitude (angle with the z-axis). The spontaneous emission of radiation of the Larmor frequency is entirely negligible, owing to the smallness of the frequency. But absorption of energy from a radio-frequency field having the same frequency is quite strong and is used in the *atomic beam resonance* method and in *paramagnetic resonance*. Quantum-theoretically, it can be regarded as the absorption of a quantum $h\nu$ causing magnetic dipole transitions between two Zeeman levels.

Even for the precessional frequencies of *nuclear spins* which are associated with very weak magnetic moments this magnetic interaction with the radiation is detectable and is used in the measurement of nuclear magnetic moments by the molecular beam technique[(10)] and of the resonance techniques in bulk matter.[(11,12)]

In optical spectra, magnetic dipole lines are due to superposition of frequencies of precession of spins and orbital moments about one another and frequencies of other electronic motions. The magnetic nature of the radiation is revealed by the Zeeman effect in which transitions $\Delta m = \pm 1$ are polarised with the electric vector parallel to the external field, because it is always normal to the magnetic vector which is primarily causing the emission and is oscillating in the x, y direction. The transitions $\Delta m = 0$ appear polarised with the electric vector in the x, y plane.

3. Intensities and conditions of excitation

a. Absorption and dispersion

The observed intensity of a spectral line is the result of two factors; one of these is a characteristic property of the atom and can be described as transition probability, line strength or f-value. The other factor depends on the conditions of excitation; this will only be discussed briefly in order to indicate the problems arising when line strengths are to be compared with observed intensities.

The conditions are most clearly defined in measurements of absorption. But this method is mainly confined to lines whose lower term is the ground state or a state of comparatively low energy, unless the temperature is very high. Lines whose lower term is metastable can often be observed in absorption in electric discharges. According to (101), the total absorption α_0 gives directly the strength of the line or its f-value. This relation is valid even if the width is due mostly to Doppler effect or to any other cause of broadening which only changes the frequency distribution but not the value of the dipole matrix element. Pressure broadening generally conforms with this requirement provided the action of the colliding atom does not cause a forbidden transition to become allowed.

The measurement of α_0 requires a resolving power high enough to resolve the contours of the line, and it is for this reason that strongly pressure-broadened lines have been used for such measurements. The alternative method of using weak absorption has been mentioned in section 1.

The indirect method of determining the line strength by interferometric measurement of the anomalous dispersion is generally more accurate. The refractive index is measured as function of frequency for values of ν close to the resonance frequency ν_0 and the value of f found from (103). This method has also been applied to excited states in glow discharges, but the presence of the negative dispersion makes the interpretation of such results difficult.

It is sometimes possible to measure the radiation width of a spectral line directly, by eliminating the influence of the Doppler broadening, and to deduce the line strength from it. (See VII.D.)

Magnetic rotation has also been suggested as a means of measuring f-values of absorption lines,[13] but little use appears to have been made of it.

b. Excitation

In emission spectra, the simplest conditions of excitation are those in which the excited states of the atoms are approximately in thermal equilibrium and the number of atoms in any given state is proportional to the statistical weight and the Boltzmann factor according to (106); the comparison of intensities of lines gives then directly the ratio of their transition probabilities. In speaking of the ratio of the intensities of two lines without specifying the conditions, one tacitly assumes temperature equilibrium at infinite temperature so that the Boltzmann factor is equal to 1. The latter part of the assumption is sufficiently realised in practice whenever the transitions originate from levels whose energies differ very little from one another.

A condition which can be regarded as an extreme opposite to thermal equilibrium is that in which for two levels the ratio of the rates at which atoms are excited into these levels has a fixed value β. If, e.g. only one transition is possible from each level, the number of quanta in the two lines will be simply in the ratio β, regardless of the transition probabilities. The intensity ratio is then merely a function of the *excitation* probabilities of the two levels. The argument assumes that energy is only lost by radiation.

If atoms are excited by electron impacts at very low gas pressure, so that the number of collisions during radiation is negligible, intensity ratios very different from those of thermal equilibrium are sometimes observed. This effect appears to be confined to terms of different l, because the excitation probability depends strongly on the value of l; the transition rule $\Delta l = \pm 1$ retains some degree of validity even for the transition between two levels caused by electron impact.

Almost ideal conditions of thermal equilibrium were established by A. S. King and his co-workers in an extensive series of measurements of absorption- and emission-spectra of metallic vapours, with the use of an electrically heated graphite furnace.[14] A large number of relative f-values have been measured in this way and the variation of line intensities as functions of temperature have greatly contributed to the term analysis of complex spectra.

In flames and in certain parts of electric arcs the excitation corresponds approximately to thermal equilibrium.

In most sources of atomic spectra, such as glow discharges, the conditions of excitation are more complicated and it is not always possible to connect observed intensities with transition probabilities; only if two lines have a common upper level their intensities will always be in the ratio of their transition probabilities.

The higher the average speed of the electrons, the more strongly will states of high energy be excited and the more will spectra of higher states of ionisation predominate. In comparing spectra of condensed sparks with those of uncondensed sparks or arcs it is usually possible to distinguish between lines of the neutral atom and those of the first, second, etc., ionised states. In a similar, less pronounced way, the cathode dark space will favour ionic spectra.

In some forms of gas discharge the kinetic energies of the electrons are distributed approximately according to Maxwell's distribution law. One can then use the concept of an *electron temperature* which may have a very much higher value than that defined by the kinetic energy of the atoms and molecules. The population of the excited states will then be nearly in accordance with the electron temperature, because excitation by electron impacts is a process of high probability and causes the excitation energy to be in equilibrium with the kinetic energy of the electrons rather than that of the atoms.

If two terms have nearly the same energy and only differ in their values of J or F, their population in gas discharges corresponds very closely to statistical equilibrium and the ratio of the intensities of transitions from these terms is equal to the ratio of their statistical weights multiplied by the transition probability. This fact forms the basis of the intensity sum rules (see p. 170).

c. Metastable levels and self-absorption

This quasi-statistical equilibrium between states of similar energy does not usually exist if one of them is *metastable*. A term is called metastable if all transition probabilities to lower states are extremely small, owing to the operation of some selection rule. The concentration of atoms in a metastable state is abnormally high in a discharge; they lose their energy generally by collisions, not by radiation.

Metastability of a term is a matter of degree, depending on the values of the transition probabilities and the gas density. At extremely low densities as they exist in stellar nebulae and in the higher atmospheres of sun and earth, a state which can only emit quadrupole radiation with a lifetime of perhaps 1/10 second will lose all its energy by radiation; the intensity of the line will only depend on the rate at which the term is excited (possibly by emission from higher terms), and the line may appear quite strong while it is unobservably weak in laboratory sources.

Many prominent lines of nebulae and of the solar corona, once ascribed to hypothetical elements nebulium and coronium, have been interpreted in this way by Bowen[15] and by Grotrian[16] and Edlén[17] as lines of extremely low transition probability, of common elements and their higher stages of ionisation. The full understanding of the emission process as electric quadrupole and magnetic dipole radiation is due to Rubinowicz.[9]

While the strongest nebular lines are due to neutral atoms or to atoms which have lost one or two electrons, such as O, O II, O III, N, N II, N III, the corona lines are emitted by highly stripped atoms such as Ca XII, Fe XIII, Ni XVI. Also some auroral lines have been ascribed to forbidden transitions.

The intensity of spectral lines whose lower level is the ground term or a metastable term is strongly affected by re-absorption of the light. Since stronger emission lines are more strongly absorbed, *self-absorption*[18] tends to level out intensity differences. The stronger the absorption the more the *radiation energy* approaches equilibrium with the *excitation* energies of the atoms. In the extreme case, the gas radiates as a body which is "black" in the small wavelength ranges covered by absorption lines. Two closely adjacent lines then always appear of practically equal intensity, a condition which can easily be observed in any commercial sodium lamp, for the two yellow lines of sodium.

For the strongest lines connected with the ground state, often described as resonance lines, the self-absorption is generally so strong that it is very difficult to measure the intensity ratio without self-absorption. This difficulty specially arises in hyperfine structure research.

Since the excitation is usually weaker near the walls of the discharge tube, self-absorption by the "cooler" atoms near the wall can cause a decrease of intensity near the centre of the line, the so-called *self-reversal* which often makes a line appear as a well-resolved doublet.

The condition of thermal equilibrium between atoms and radiation field is quite distinct from the thermal equilibrium among the different states of excitation of the atoms. The former is fulfilled in the limiting case of an infinitely thick layer of gas in thermal equilibrium, the latter for a thin layer in which thermal equilibrium between the different states of excitation has been established by collision processes.

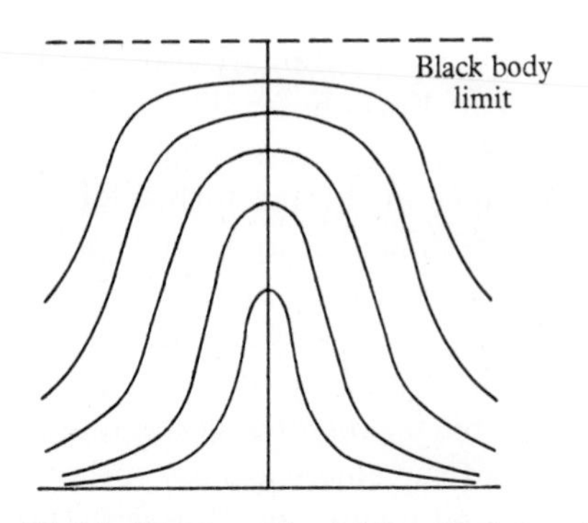

FIG. II, 10. Line profiles with self-absorption.

The approach to equilibrium with the radiation is illustrated in fig. II, 10, showing qualitatively the appearance of a resonance line for different thicknesses of the gas layer or different densities of atoms. The temperature or current density is hereby assumed to be constant throughout the body of the gas.

CHAPTER

III. The Simple Spectra

A. HYDROGEN AND THE HYDROGEN-LIKE IONS*

1. Series structure and the Bohr–Sommerfeld theory

a. The series in hydrogen

Electric discharges in hydrogen emit both the molecular spectrum, often called many-line spectrum, and the spectrum of the atom. The stronger the excitation the more pronounced is the atomic spectrum. As the recombination of the atoms occurs mainly at the walls of the discharge tube, surface conditions and diameter of the discharge tube have a great influence on the relative intensity of the two spectra. Plate 1(1) shows the spectrum of a hydrogen discharge in a tube whose walls were covered with ice in order to retard the recombination of atoms. The molecular spectrum is virtually absent and a large number of lines of the *Balmer series* can be seen. The original spectrograms also show a faint continuous spectrum beyond the series limit. The colour of this type of hydrogen discharge is purple, owing to the predominance of the α line.

In the spectra of the sun and of many stars the Balmer lines appear as absorption lines. The Fraunhofer lines C, F and f are the hydrogen lines H_α, H_β and H_γ.

A further series of lines in the far ultra-violet, the *Lyman series*, has its first and strongest member at 1215 A and converges to a limit at 912 A. Further series have been found in the infra-red range of the spectrum. Table III, 1 gives the wavelengths of some of these lines and of their limits. The Lyman-α line, λ1215, can be photographed in the solar spectrum by means of spectrographs carried in rockets.

The wave-numbers of the entire atomic spectrum of hydrogen can be accurately represented (neglecting fine structure, however) by the formula

$$\tilde{\nu}_{n'n} = R_{\mathrm{H}}(1/n^2 - 1/n'^2) \tag{1}$$

where n and n' are integral numbers and $n = 1, 2, 3, 4, 5$ for the *Lyman*-, *Balmer*-, *Paschen*-, *Brackett*- and *Pfund*-series and where

* See also G. W. Series, *The Spectrum of Atomic Hydrogen*, Oxford Univ. Press 1957.

$n < n'$. With the use of the combination principle, these facts can be expressed by the relation

$$T_n = R_H/n^2 \qquad (n = 1, 2, ..) \tag{2}$$

for the term values whose differences give the wave-numbers of the lines. Fig. III, 1 shows the term diagram.

TABLE 1

Wavelengths of hydrogen lines in A.U.

	$n =$ 1	2	3	4	5
$n' =$	Lyman	Balmer	Paschen	Brackett	Pfund
2	α1215·66				
3	β1025·72	α6562·8			
4	γ 972·53	β4861·3	18751		
5	δ 949·74	γ4340·5	12818	40500	
6	ϵ 937·80	δ4101·7	10938	26300	74000
∞	912	3648	8208	14600	22800

b. Circular motion

In his first theory of the hydrogen atom, Bohr(2) assumed arbitrarily that the electron moved in a circular orbit, according to the laws of classical mechanics; these are simply expressed by equating the centrifugal force mv^2/r to the electrostatic attraction Ze^2/r^2, for a nucleus of charge Ze. The total energy can then be written as a function of one constant such as radius, period or angular momentum p_φ. Since the latter is, in Bohr's quantum theory, equal to $nh/2\pi$, any of the constants can be expressed as function of the quantum number n. From $mv^2/r = e^2Z/r^2$, the kinetic energy is found to be

$$T = mv^2/2 = e^2Z/2r = -V/2,$$

so that the total energy is

$$E = T+V = T-2T = -T = -e^2Z/2r, \quad \text{and } r = e^2Z/mv^2.$$

Introducing the quantum condition

$$\oint p_\varphi \, d\varphi = 2\pi m r^2 \dot{\varphi} = 2\pi m r v = 2\pi e^2 Z/v = nh \tag{3}$$

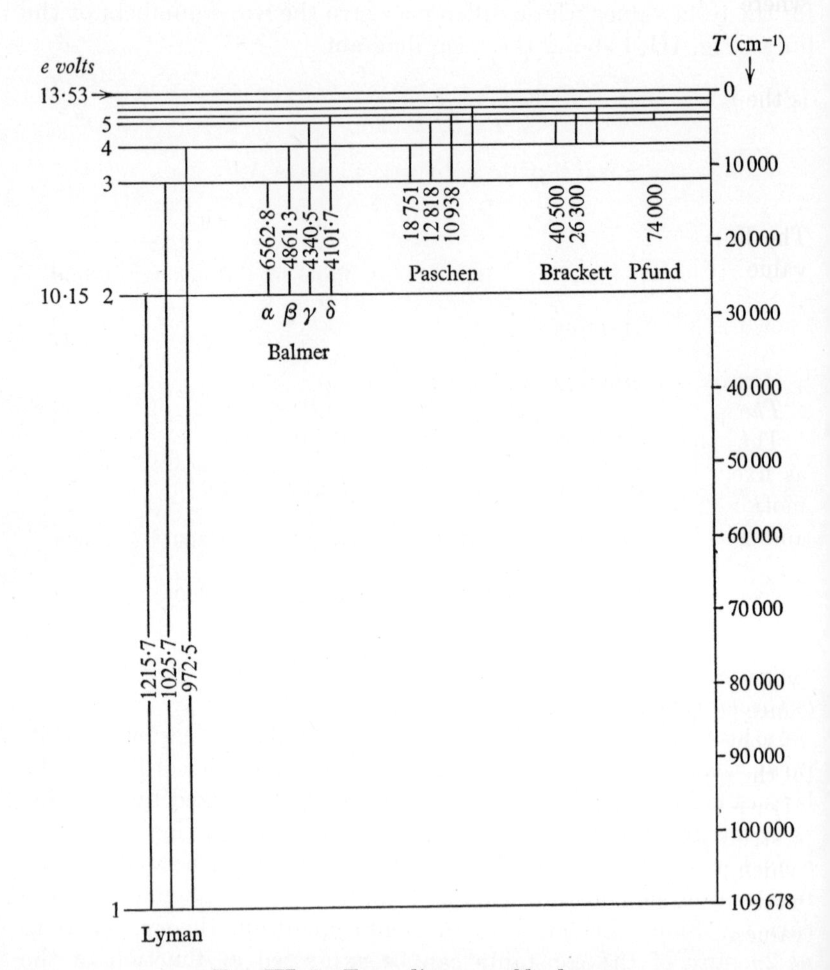

FIG. III, 1. Term diagram of hydrogen.

we find the energy of the stationary states as function of the quantum number n,

$$E_n = -mv^2/2 = -\frac{2\pi^2 m e^4 Z^2}{h^2 n^2}. \tag{4}$$

The radius of the nth orbit is

$$r_n = \frac{nh}{2\pi mv} = \frac{n^2h^2}{4\pi^2me^2Z} = \frac{n^2}{Z}a_0,$$

where

$$a_0 = h^2/4\pi^2me^2 = \hbar^2/me^2 = 0{\cdot}5292\times 10^{-8}\ \text{cm} \tag{5}$$

is the radius of the first Bohr orbit. The velocity in the nth orbit is

$$v_n = \frac{2\pi e^2Z}{nh}. \tag{6}$$

The Bohr frequency relation $h\nu_{nn'} = E_{n'}-E_n$ leads to (1), with the value

$$R_\infty = \frac{2\pi^2me^4}{ch^3}. \tag{7}$$

c. The nuclear mass effect

The suffix ∞ in (7) indicates that the calculation treats the nucleus as fixed. It is easy to refine the theory by taking account of the motion of the nucleus about the common centre of gravity of nucleus and electron. This leads to the formula

$$R_\text{H} = \frac{2\pi^2me^4}{ch^3(1+m/M)} = R_\infty/(1+m/M) \tag{8}$$

where the reduced mass $\mu = Mm/(M+m)$ has taken the place of m. Since the mass of the proton, M, is 1836 times as great as the mass m of the electron, the correction factor is very small. The agreement of the experimental value of R with that calculated from the formula (7) with the use of the known values of the universal constants was a striking success of Bohr's theory. The experimental value of R_H which is one of the most accurately known universal constants, was later used for determining other universal constants.[3] An accurate value, deduced by least squares adjustment of various constants[4] is

$$R_\infty = 109737{\cdot}31\ \text{cm}^{-1};$$

its accuracy is claimed to be $0{\cdot}01\ \text{cm}^{-1}$.

The theory can be applied to the hydrogen-like ions He^+, Li^{++}, etc., with the nuclear charge numbers $Z = 2, 3, \ldots$, and gives the term values

$$T_n = Z^2R/n^2. \tag{9}$$

This equation shows that for He^+ the terms of even n agree, apart from the difference in the value of R, with the hydrogen terms of half the value of n. The spectrum of He^+ thus shows lines very close to the Balmer lines of H, with other lines, due to odd values of n, between them. The latter were originally thought to be hydrogen lines with half integral quantum numbers (Pickering series). It was found later that the lines of He^+ do not exactly coincide with the corresponding hydrogen lines. The ratio of their wave-numbers was found to agree with the value $(1+m/M_{\mathrm{H}})/(1+m/M_{\mathrm{He}}) = 1{\cdot}000407$ predicted by the theory.

The factor $1/(1+m/M)$ in the Rydberg constant, which also appears in the more rigorous, wave-mechanical theory, causes the lines of deuterium to be displaced compared with those of light hydrogen and led to the discovery of deuterium. This is the simplest example of a type of *isotope shift* which is entirely due to the difference in mass (see VI.C).

The accurate measurement of either the displacement of the lines of He^+ compared with those of H or of the lines of D and H relative to one another allows an accurate determination of m/M_{H} or, with the use of Faraday's constant, of the value of e/m. One verifies easily the relation

$$m = \frac{M_{\mathrm{H}} M_D(\nu_D - \nu_{\mathrm{H}})}{M_D \nu_{\mathrm{H}} - M_{\mathrm{H}} \nu_D}.$$

If the atomic weights of the hydrogen and deuterium nuclei are indicated by H^+ and D^+, and Faraday's constant by $eN = F$ we find

$$e/m = \frac{F(\mathrm{D}^+ \nu_{\mathrm{H}} - \mathrm{H}^+ \nu_D)}{\mathrm{H}^+ \mathrm{D}^+ (\nu_D - \nu_{\mathrm{H}})}. \tag{10}$$

Very accurate measurements of this kind have been made especially by Houston,[5] using He^+, and by others[6,7] using the isotopes of hydrogen.

The stronger lines of the ions He^+, Li^{++} . . move progressively farther into the ultra-violet where absolute measurements are difficult, mainly owing to the lack of reliable wavelength standards. In fact, the calculated wavelengths of the hydrogen-like spectra of ions are often used as standards for the measurement of other lines. But for the spectra of He^+ and Li^{++} the formula (9) has been accurately tested.

Table III, 2[8,9] gives the calculated wavelengths of the observed resonance lines ($n = 1$, $n' = 2$), the values of R and the ground terms. The latter include a small correction due to relativity effects, according to the fine structure formula (III.A.5).

TABLE 2

	Resonance line in A (calc.)	R	Value of ground term in cm^{-1}
H I	1215·664	109677·7	109679·2
He II	303·779	109772·4	438913
Li III	134·994	109728·8	987677
Be IV	75·925	109730·8	1756065
B V	48·585	109731·9	2744207
C VI	33·734	109732·4	3952252
N VII	24·779	109733·1	5380416
O VIII	18·967	109733·7	7028916

d. Elliptical orbits

In spite of the striking numerical agreement, the simple Bohr theory must be regarded as unsatisfactory in the following respects:

(a) the restriction to circular orbits is entirely arbitrary,

(b) application of the correspondence principle would allow only the transitions $\Delta n = 1$ to be observed as strong lines.

(c) the theory does not account for the "fine structure" of the Balmer lines first observed by Michelson as a splitting of width $\approx 0{\cdot}3\ \mathrm{cm}^{-1}$.

These objections were met by an extension of the theory due to W. Wilson(10) and mainly Sommerfeld.(11) Only a very brief account of this theory, which has now been superseded by wave-mechanics, may be given here.

Sommerfeld treated the electron as free to move in three dimensions according to the methods outlined in section II. If the degeneracy due to the isotropic nature of the potential is taken into account at the outset, the motion is confined to a plane and the axis of the polar coordinate system can be chosen at right angles to the orbital plane. According to the remaining two degrees of freedom there are two J-values each of which is quantised individually by the relations

$$J_\varphi = \oint p_\varphi \mathrm{d}\varphi = 2\pi p_\varphi = kh \tag{11}$$

$$J_r = \oint p_r \mathrm{d}r = n_r h. \tag{12}$$

Only the first of the two quantum numbers, k, has a simple physical significance, as a measure of the angular momentum in units of $h/2\pi = \hbar$. The particular choice of the potential $V(r) = -Ze^2/r$ is found to have the result that the two fundamental frequencies generally found in a system of two degrees of freedom coincide, so that the orbit becomes a closed curve, the well-known Kepler ellipse (as φ runs through one period of its values, say $0 \to 2\pi$, r also performs one period, e.g. from $r_{\max}$ to $r_{\min}$ and back to $r_{\max}$). Since $\nu_r = \partial E/\partial J_r = \nu_\varphi = \partial E/\partial J_\varphi$, the energy must be a function of $J_r + J_\varphi$ or of $k + n_r$. The calculation gives the value

$$E = -\frac{2\pi^2 m Z^2 e^4}{h^2(k+n_r)^2} = -\frac{RZ^2hc}{(k+n_r)^2} \tag{13}$$

where the value R is defined as before. This result is identical with that of Bohr's simple theory if we put

$$n = k + n_r.$$

As the quantum numbers k and n_r are simply defined as integral numbers, this substitution only expresses the trivial fact that an integral number is the sum of two other integral numbers, and it restricts the range of possible values of the quantum numbers. In order to exclude pendulum orbits passing through the nucleus, Sommerfeld postulated that $k \geqq 1$, and therefore $n \geqq 1$. The radial quantum number n_r can assume the value 0 for which the orbit is circular.

Both theories ascribe to the ground state of hydrogen a circular orbit, $n = 1$, $k = 1$, of radius a_0, given in (5), and of angular momentum $\hbar$.

The length of the major axis of the ellipse is found to depend only on $n = k + n_r$ so that all orbits belonging to the same energy have the same major axis. The value of the major semi-axis is

$$a = a_0 n^2/Z. \tag{14}$$

The frequencies of the orbital motion are not directly observable. How far they can be regarded as "real" is best demonstrated by the application of Bohr's correspondence principle. The relation (1) for the quantum frequencies for $n' - n = 1$ must become identical with the classical fundamental frequencies given by (II.12a) in the limit $n \to \infty$. This can be verified for hydrogen, most easily for the circular orbits. The orbital frequency is found to be

$$\nu_{\text{cl}} = 2Rc/n^3. \tag{15}$$

The quantum frequency for $\Delta n = 1$ is

$$\nu_{\text{qu}} = Rc[1/n^2 - 1/(n+1)^2] \tag{16}$$

which passes into (15) in the limit of large n. One finds, e.g. that the classical frequencies for the 9th and 10th orbit are 9·02 and $6{\cdot}58 \times 10^{12}$ sec^{-1} while the quantum frequency of the transition is $7{\cdot}7 \times 10^{12}$ sec^{-1}. The physical reality which demonstrates itself as the frequency of the observable spectral lines is connected with *two* quantum states; and with increasing n, two adjacent states ($\Delta n = 1$) differ less and less from one another so that the classical concept of motion in one orbit becomes increasingly applicable.

Though the detailed features of the orbits are not directly observable and remain a somewhat abstract aspect of the theory, Sommerfeld succeeded in deriving an important result, showing that some degree of reality had to be ascribed even to details of the elliptical orbits. The speed of the electron in the lower quantum orbits is not quite negligible compared with the speed of light so that relativistic effects must be expected. In applying relativistic corrections to the theory of elliptical orbits, Sommerfeld was able to explain quantitatively those features of the fine structure which were known at the time (see III.A.5a).

The continuous spectrum which, under suitable conditions, can be observed beyond the limit of the Balmer series must be attributed to a transition from a state of positive energy of the electron to the quantum state $n = 2$. Similar continua beyond the series limit have been observed in many spectra. The emission of these continua results from recombination of an electron with a positive ion. The energy $h\nu$ of the light quantum is made up from the potential energy $h\nu_{\text{lim}}$ and the kinetic energy $h(\nu - \nu_{\text{lim}})$ of the electron at infinity. If the lower state of a series is the ground state of the atom, the absorption also shows a continuous spectrum beyond the series limit. It is due to photo-ionisation of the atom.

In the theory of Bohr and Sommerfeld, an energy state beyond the limit of the term series, i.e. with $E > 0$, has to be visualised as a hyperbolic orbit.

2. The wave mechanical theory

a. Energy levels, quantum numbers and angular momenta

The wave mechanical description of hydrogen and the hydrogen-like ions forms a special case of the central force problem, with the potential $V(r) = -Ze^2/r$. The solutions contain the factor

$P_l^m(\vartheta)\exp(im\varphi)$ which was discussed in II, and which describes how the wave function depends on the angles. The radial factor $R(r)$ is the solution of the radial equation (II.38) with the Coulomb potential,

$$\left(\frac{1}{2m}p_r^2+\frac{l(l+1)\hbar^2}{2mr^2}-\frac{Ze^2}{r}\right)R = ER. \tag{17}$$

For the condition $E < 0$ describing states of a bound electron, in which we are primarily interested, the equation is found to have *well-behaved* solutions only for definite values of E. If the radial distance is expressed by the dimensionless quantity

$$\rho = 2Zr/na_0$$

where a_0 is the Bohr radius $\hbar^2/me^2$ (see eq. 5) the solutions take the form of *associated Laguerre functions*

$$R_{nl} = \exp(-\rho/2)\rho^l L_{n+l}^{2l+1}. \tag{18}$$

L_{n+l}^{2l+1} are polynomials in ρ of degree $n-l-1$, where n and l are integral numbers, and $l \leqq n-1$. The eigenvalues are found to depend on n only and are identical with the values E_n given by (4) and (13). The nuclear mass correction leads again to (8).

Some of the functions R_{nl} are listed in table III, 3; the normalising factors, ensuring that

$$\int_0^\infty R_{nl}^2 r^2\,\mathrm{d}r = 1$$

have been omitted in (18). They are given by

$$N = -\left[\left(\frac{2Z}{na_0}\right)^3 \frac{(n-l-1)!}{2n\{(n+l)!\}^3}\right]^{\frac{1}{2}}. \tag{19}$$

In fig. III, 2 R_{nl} is plotted as function of ρ. A list of some of the complete wave functions for hydrogen and hydrogen-like ions is given in table III, 4.

The wave functions vanish for certain values of r which then define *nodal spheres.* Owing to the factor P_l^m the functions also have *nodal cones*, defined by the values of ϑ for which the function vanishes. If the complex functions $\exp(im\varphi)$ are replaced by real, linear combinations $\sin m\varphi$ and $\cos m\varphi$, the wave function vanishes also in certain meridional *nodal planes.*

The number of nodal planes, nodal cones and nodal spheres is equal to the value of m, $\mu(= l-|m|)$ and $n_r(= n-l-1)$ respectively.

TABLE 3

Radial functions R_{nl} for hydrogen, He^+ etc.

$n = 1,\qquad R_{10} = (Z/a_0)^{3/2} 2 \exp(-\rho/2)$

$n = 2,\qquad R_{20} = \dfrac{(Z/a_0)^{3/2}}{2\sqrt{2}}(2-\rho)\exp(-\rho/2)$

$$R_{21} = \frac{(Z/a_0)^{3/2}}{2\sqrt{6}}\rho\exp(-\rho/2)$$

$n = 3\qquad R_{30} = \dfrac{(Z/a_0)^{3/2}}{9\sqrt{3}}(6-6\rho+\rho^2)\exp(-\rho/2)$

$$R_{31} = \frac{(Z/a_0)^{3/2}}{9\sqrt{6}}(4-\rho)\rho\exp(-\rho/2)$$

$$R_{32} = \frac{(Z/a_0)^{3/2}}{9\sqrt{(30)}}\rho^2\exp(-\rho/2)$$

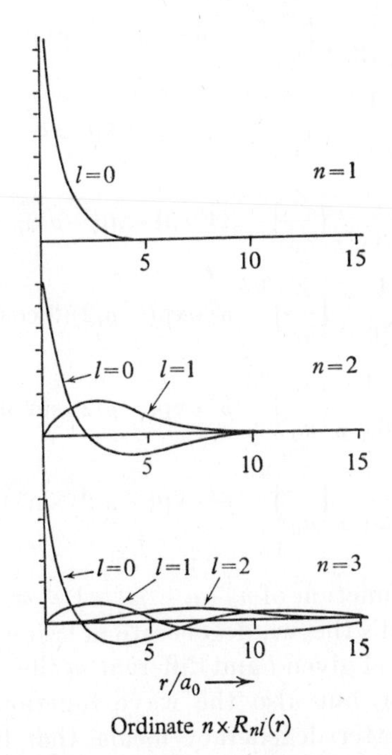

FIG. III, 2. Radial wave functions of hydrogen for $n = 1$, 2 and 3.

The quantum numbers m, μ and n_r appearing as primary quantum numbers associated with the coordinates φ, ϑ and r have thus a direct physical significance as the number of nodes in the range of the respective coordinates.

TABLE 4

Hydrogen-like wave functions (normalised)

$$\psi_{1s}(m=0) = \frac{1}{\sqrt{\pi}}\left(\frac{Z}{a_0}\right)^{3/2} \exp(-\rho/2)$$

$$\psi_{2s}(m=0) = \frac{1}{4\sqrt{(2\pi)}}\left(\frac{Z}{a_0}\right)^{3/2} (2-\rho)\exp(-\rho/2)$$

$$\psi_{2p}(m=0) = \frac{1}{4\sqrt{(2\pi)}}\left(\frac{Z}{a_0}\right)^{3/2} \rho\exp(-\rho/2)\cos\vartheta$$

$$\psi_{2p}(m=\pm 1) = \frac{1}{8\sqrt{\pi}}\left(\frac{Z}{a_0}\right)^{3/2} \rho\exp(-\rho/2)\sin\vartheta\exp(\pm i\varphi)$$

$$\psi_{3s}(m=0) = \frac{1}{18\sqrt{(3\pi)}}\left(\frac{Z}{a_0}\right)^{3/2} (6-6\rho+\rho^2)\exp(-\rho/2)$$

$$\psi_{3p}(m=0) = \frac{1}{18\sqrt{(2\pi)}}\left(\frac{Z}{a_0}\right)^{3/2} (4-\rho)\exp(-\rho/2)\cos\vartheta$$

$$\psi_{3p}(m=\pm 1) = \frac{1}{36\sqrt{\pi}}\left(\frac{Z}{a_0}\right)^{3/2} (4-\rho)\exp(-\rho/2)\sin\vartheta\exp(\pm i\varphi)$$

$$\psi_{3d}(m=0) = \frac{1}{4\sqrt{(6\pi)}}\left(\frac{Z}{a_0}\right)^{3/2} \rho^2\exp(-\rho/2)(3\cos^2-1)$$

$$\psi_{3d}(m=\pm 1) = \frac{1}{36\sqrt{\pi}}\left(\frac{Z}{a_0}\right)^{3/2} \rho^2\exp(-\rho/2)\sin\vartheta\cos\vartheta\exp(\pm i\varphi)$$

$$\psi_{3d}(m=\pm 2) = \frac{1}{72\sqrt{\pi}}\left(\frac{Z}{a_0}\right)^{3/2} \rho^2\exp(-\rho/2)\sin^2\vartheta\exp(\pm 2i\varphi)$$

The energy is a function of $n_r+\mu+|m|+1 = n$, so that the motion is, as in Sommerfeld's theory, degenerate in two ways: not only have the wave functions of given l and different m the same energy (as for every central field), but also the wave functions of given n and different l. This latter degeneracy means that the choice of these eigenfunctions is, to a large extent, arbitrary. One can, in fact, separate the wave equation of hydrogen in other coordinates, e.g. in parabolic coordinates and find solutions of the appropriate separate equations. Each of the resulting wave functions could be expressed as a linear combination of the degenerate polar wave functions. But the advantage of the Schrödinger polar wave functions lies in the fact that they are also eigenfunctions of the operator L^2, the square of the angular momentum, and of L_z (see p. 40) so that, for a state described by a Schrödinger polar function, the total angular momentum and its z component have well-defined values.

The most striking contrast to Bohr and Sommerfeld's theories is the result that the hydrogen atom in the ground state has the angular momentum zero ($l = 0$) and is completely spherically symmetrical. The old quantum theory ascribed an angular momentum of $\hbar$ to the ground state. Atomic beam deflection experiments of the Stern Gerlach type showed conclusively that the angular momentum of the normal hydrogen atom is $\hbar/2$, indicating the incompleteness of either theory (see section A.5).

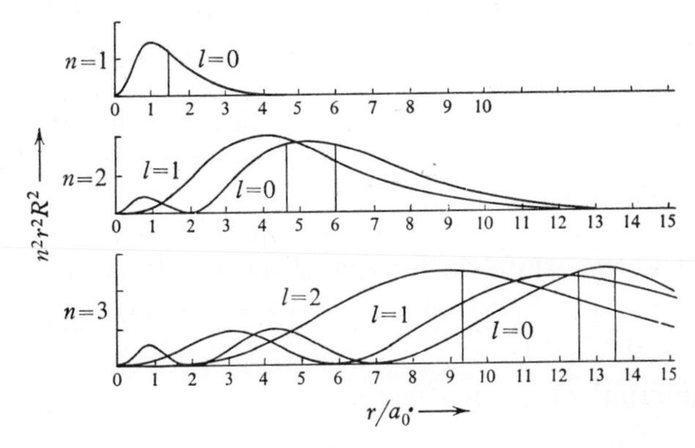

FIG. III, 3. Radial charge distribution in hydrogen for $n = 1$, 2 and 3. (L. Pauling, *Proc. Roy. Soc.* **114,** 181, 1927).

b. The charge distribution

The statistical particle density, or the probability of finding the electron in unit volume near a given point is proportional to R_{nl}^2. For comparison with the old quantum theory it is more appropriate to consider the probability of finding the electron within unit element of distance r. This value, const. $\times\, r^2 R_{nl}^2$, is plotted in fig. III, 3 as function of r/a_0. For convenience of scale, the ordinates have been multiplied by n^2. The most probable distance in the ground state agrees with the value a_0, the radius of the first Bohr orbit. The average value of the distance is indicated, in each curve, by a vertical line. The increase of the average distance with n and the dependence of the curves on l show marked similarities of the wave-mechanical charge distribution to that of an elliptical motion averaged over an imaginary precession of the ellipse about the nucleus.

For all wave functions with $l = 0$ (s-states) P_l^m is constant and the ψ function and the charge distribution are spherically symmetrical. For $l > 0$, ψ depends on the angles, but addition of the values ψ^2 for all states with the same l shows that

$$\sum_{m=-l}^{+l} \Theta_{lm}^2(\vartheta)\Phi_m^*\Phi_m = \text{const.} \tag{20}$$

If $2l+1$ electrons are imagined to occupy all these states of different m, but without interacting with one another, the resulting charge distribution would be spherically symmetrical. In this form, the statement refers to a very hypothetical case, but the fact that it applies to all central force fields, not only to the specific Coulomb potential, makes the fact very important for the description of other atoms.

It must be stressed that the functions plotted in fig. III, 3 contain the factor r^2 and must therefore become zero for $r = 0$, but this does not mean that the charge density itself vanishes at this point. The latter value, determined by $\psi_{(0)}{}^2$, is of importance for hyperfine structures and isotope shifts in heavier elements. It is finite at $r = 0$ only for s terms ($l = 0$) as can be seen from fig. III, 2.

3. The normal Zeeman effect

a. Lorentz's theory

As far as the fine structure of the lines can be disregarded, the model of the point electron is not only able to explain the structure of the spectra of hydrogen and the hydrogen-like ions, but also the effect of magnetic and electric fields on the lines. The fine structure can only be neglected if it is small compared with the effects caused by the external fields. We shall, in this paragraph, accept this assumption; the limits of its validity will become clear later when the connection of effects in strong fields with those in weak fields is discussed for other atoms.

In fields of more than about 10000 gauss, the hydrogen lines show the *normal Zeeman effect* which is, in other atoms, characteristic of

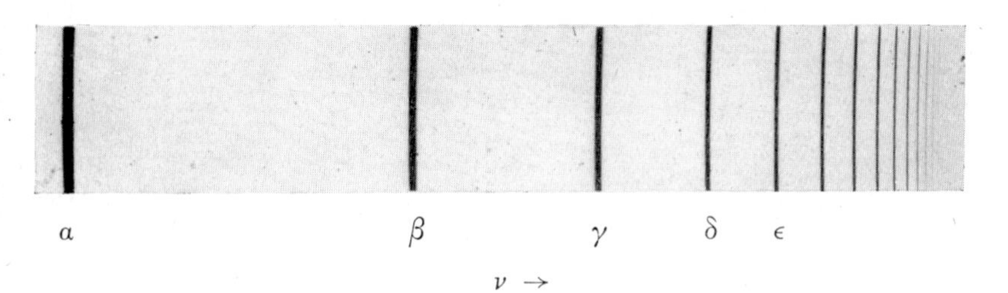

PLATE 1. Balmer series. (G. Herzberg, ref. III. 1).

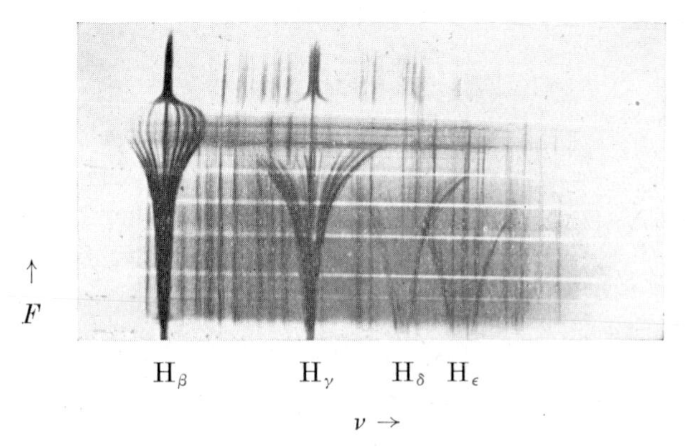

PLATE 2. Balmer lines in strong electric fields.
(H. R. v. Traubenberg, R. Gebauer & G. Lewin, ref. III. 23).

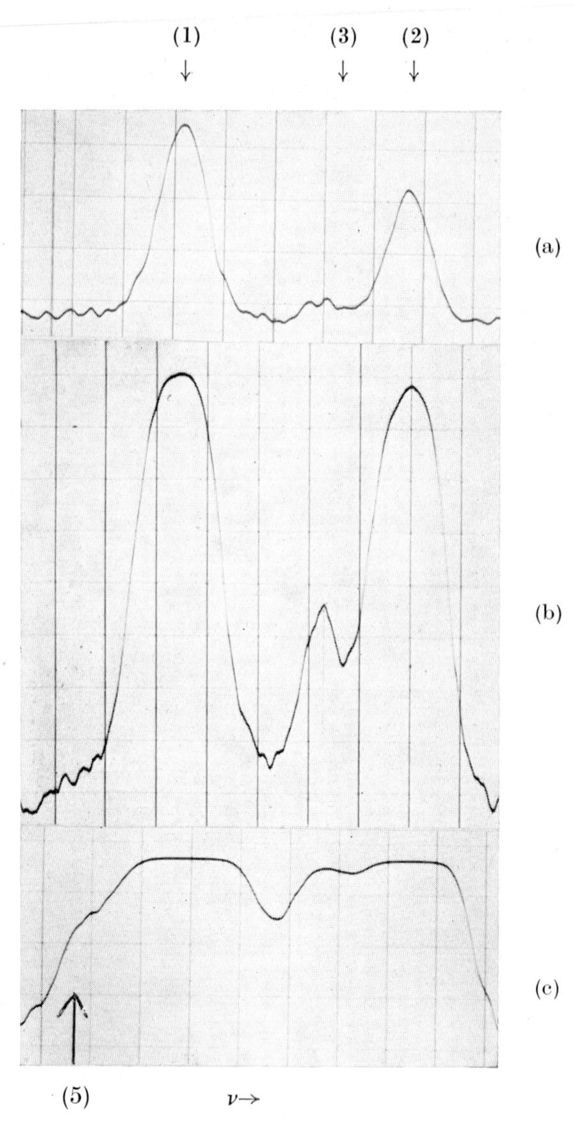

PLATE 3. Fine structure of H α; discharge tube cooled with liquid hydrogen; photometer tracing of Fabry-Perot fringes. (H. G. Kuhn & G.W. Series, ref. III. 32).

$\nu \rightarrow$

PLATE 4. Absorption spectrum of Na; quartz spectrograph.

singlet lines. The appearance is that of the so-called *Lorentz triplet* shown in fig. III, 4 (bottom part). The σ-*components* appear circularly polarised in longitudinal observation, i.e. in the z direction, and plane polarised, with the electric vector in the x, y plane, in transverse observation. The π-*component* is absent in longitudinal observation, and plane polarised, with the electric vector in the z direction, in transverse observation.

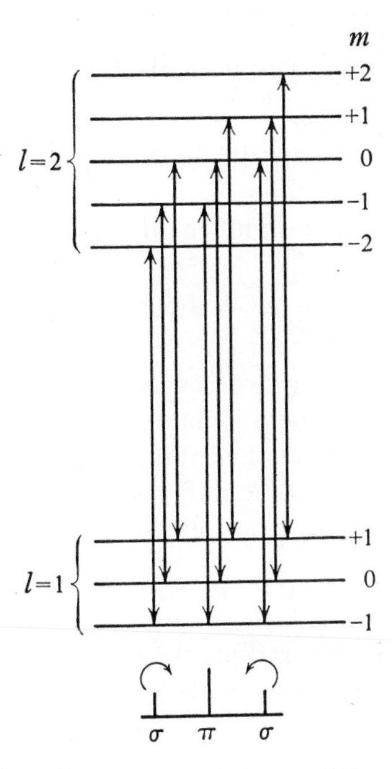

FIG. III, 4. Term diagram of normal Zeeman effect.

It is instructive to consider at first the action of a magnetic field on a specially simple classical motion. Following Lorentz, we assume a particle of mass m_0 which is attracted towards a fixed centre by a force $\mathbf{f} = -k\mathbf{r}_0$ where k is a constant and $\mathbf{r}_0$ the distance from the centre. This model is described as an isotropic, harmonic oscillator. The most general type of motion of the particle takes place in an ellipse whose plane may be defined as the x, y plane, and we choose the initial conditions in such a way that the orbit is a circle of radius r_0. The angular velocity ω is then given by

$$m_0\omega^2 r_0 = kr_0.$$

If we assume the particle to carry a charge of $-e$ e.s.u. and a homogeneous magnetic field of strength $\mathscr{H}$ gauss to be applied in the z direction, the field causes an additional radial force $\pm e\omega r_0\mathscr{H}/c$. We now imagine the angular velocity to be at the same time changed to a value ω' so as to balance the additional magnetic force by an additional centrifugal force while r_0 is kept constant. The condition for this is

$$m_0\omega'^2 r_0 - m_0\omega^2 r_0 = e\omega' r_0\mathscr{H}/c.$$

If $|\omega'-\omega| = |\Delta\omega| \ll \omega$, it can be written as

$$\Delta\omega = \pm\frac{e\mathscr{H}}{2m_0c} \quad \text{or} \quad \Delta\nu = \pm\frac{e\mathscr{H}}{4\pi m_0c}\,\text{sec}^{-1}, \tag{21}$$

corresponding to the wave-number difference

$$\Delta\tilde{\nu} = \pm\frac{e}{4\pi m_0c^2}\mathscr{H} = 4{\cdot}6688\times10^{-5}\mathscr{H}\ \text{cm}^{-1}. \tag{22}$$

Instead of considering these two possible stationary states of motion, we now imagine that the field is made to grow slowly from 0 to the value $\mathscr{H}$ and calculate the change of motion. According to Faraday's law of induction, the induced e.m.f. and the electric field of strength F (e.s.u.) corresponding to it are given by

$$2\pi r_0Fc = \pm\pi r_0^2\times\mathrm{d}\mathscr{H}/\mathrm{d}t, \quad \text{or} \quad F = \pm(r_0/2c)\,\mathrm{d}\mathscr{H}/\mathrm{d}t.$$

Neglecting changes in the radius, we can calculate the increase (or decrease) in speed caused by this field F which is tangential to the orbit:

$$m_0\,\mathrm{d}v/\mathrm{d}t = r_0m_0\,\mathrm{d}\omega/\mathrm{d}t = \pm e(r_0/2c)\,\mathrm{d}\mathscr{H}/\mathrm{d}t,$$

and integrating from $\mathscr{H} = 0$ to $\mathscr{H}$,

$$\Delta\omega = \pm e\mathscr{H}/2m_0c \text{ or } \Delta\nu = \pm e\mathscr{H}/4\pi m_0c.$$

This shows that the motion with the frequency increased by $\Delta\nu_L$ and approximately constant radius can be brought about by slow (adiabatic) increases of H in the otherwise free system. The approximations made are equivalent to omission of terms proportional to $\mathscr{H}^2$.

Any arbitrary state of motion of a three-dimensional, isotropic, harmonic oscillator can be expressed as a superposition of a linear oscillation in the z direction and two rotations, of opposite sense, in the x, y plane. A large number of such oscillators, with random distribution of their initial conditions, placed in a magnetic field, would therefore, according to classical physics, emit radiation whose frequencies and state of polarisation are those of the Lorentz triplet described. The z-component of the motion is unaffected by the field and gives rise to the π-component. The frequency difference measured in normal Zeeman effects such as that in hydrogen agrees exactly with (21).

Also the intensity ratio of the components can be derived from the classical theory, by a simple statistical consideration: in transverse observation, e.g. viewed in the x direction, the σ light is produced by the vibration in one degree of freedom only (one can imagine the rotating particle to be replaced by two oscillations in the x and y directions with phase difference $\pi/2$) and both σ components together must therefore have the same intensity as the π component. This agrees with the observed facts and with the obvious requirement that for vanishing field strength all three components must add up to unpolarised light.

For similar reasons the total intensity emitted in the z-direction must be the same as that in the x- or y-direction. It follows that the σ-components must appear twice as intense in longitudinal as in transverse observation. Vanishing field strength then leads to isotropic emission of unpolarised radiation.

The change in angular velocity which causes the Zeeman effect is also responsible for the phenomenon of *diamagnetism*. The additional angular momentum was seen to be independent of the direction or magnitude of the original motion; it is, by the assumptions made, small compared with the angular momentum of the field-free motion.

The increase of the angular momentum as the result of a slow increase of $\mathscr{H}$ appears to contradict Ehrenfest's law of the adiabatic invariance of angular momenta (p. 27). This contradiction is due to the fact that our definition of momenta and angular momenta as given in II holds strictly only in the absence of magnetic fields. In their presence, a term $A_x e/c$ is to be added to the momentum $m\dot{x}$ where A_x is a component of the vector potential. This causes an additional term $er^2\mathscr{H}/2c$ in the angular momentum just cancelling the increase in the purely mechanical angular momentum due to the change in ω.

As the result of the application of the field, the energy changes by an amount equal to the change in kinetic energy, since the radius and therefore the potential energy remains approximately constant, namely by $\Delta E = \frac{1}{2}m_0\Delta(v^2) = \frac{1}{2}m_0 r^2 2\omega\Delta\omega$, which becomes, after substitution of $\Delta\omega$ from (21) and with the use of the symbol L for the angular momentum $mr^2\omega$,

$$\Delta E = \frac{e}{2m_0c}\mathscr{H}L. \tag{23}$$

This expression allows an alternative interpretation of the energy change. A charged particle moving in an orbit of area A with period τ is equivalent to a magnetic dipole of moment $\mu = Ae/\tau c$. But according to the law of areas, $\mathrm{d}A/\mathrm{d}t = L/2m_0$ for any arbitrary motion under the influence of a central force. The magnetic moment of the orbit is therefore

$$\boldsymbol{\mu} = \mathbf{L}\frac{-e}{2m_0c}. \tag{24}$$

The increase of kinetic energy according to (23) can thus be regarded as the magnetic energy $-\mu\mathscr{H}$ of the magnetic dipole of moment μ in the field $\mathscr{H}$ parallel to its direction.

b. Larmor's theorem and magnetic levels

It is possible to generalise these calculations and to apply them to the orbits of Sommerfeld's theory, by the use of *Larmor's theorem.* In our simple model, the additional radial force due to a small increase $\Delta\omega$ in angular velocity can be written in the form $m_0r_0(\omega+\Delta\omega)^2 - m_0r_0\omega^2 \approx 2m_0r_0\omega\Delta\omega = 2m_0v\Delta\omega$, which is a special form of the general expression for the *Coriolis force*

$$\mathbf{f}_C = 2m_0\mathbf{v}\times\boldsymbol{\Delta\omega} = 2m_0\mathbf{v}\times\boldsymbol{\omega}_\mathrm{L} \tag{25}$$

which acts on a particle moving with velocity $\mathbf{v}$ relative to a coordinate system rotating with an angular velocity described by the vector $\boldsymbol{\omega}_\mathrm{L}$. The force exerted by a magnetic field $\mathscr{H}$ on a charge e moving with velocity $\mathbf{v}$ is described by an expression containing a similar vector product

$$\mathbf{f}_M = -\frac{e}{c}\mathbf{v}\times\mathscr{H}. \tag{26}$$

It follows from a comparison of these vector equations that for any orbital motion of a particle in a central field, the effect of an external magnetic field $\mathscr{H}$ can be exactly compensated by a uniform rotation about the direction of $\mathscr{H}$ superimposed on the field-free motion, of appropriate angular velocity $\boldsymbol{\omega}_L$. From the condition $\mathbf{f}_C = \mathbf{f}_M$ follows the value of the *Larmor frequency* $\nu_L = \omega_L/2\pi$,

$$\nu_L = \frac{e}{4\pi m_0c}\mathscr{H}. \tag{27}$$

This fact can also be expressed by the statement that for an observer in a frame of reference rotating with frequency ν_L, the magnetic field is non-existent.

The Larmor theorem only holds rigorously if ν_L is small compared with the frequencies of the motion. The rotation is described as *Larmor precession.* The theorem implies that the angle ϑ between $\mathscr{H}$ and the orbital axis, which gives also the instantaneous direction of the angular momentum, remains constant during the precession. For $\vartheta = 0$ and a circular orbit, we regain exactly our earlier results of the circular movement of the oscillator where the precession merely consists in a change of speed of rotation.

For the general case of a movement in any kind of orbit, it can again be shown that the increase in energy which is, to a first approximation, exclusively an increase in kinetic energy, is equal to the magnetic energy described by the scalar product

$$E_{\text{mgn.}} = -\boldsymbol{\mu}\cdot\mathscr{H} = \frac{e}{2m_0c}\mathbf{L}\cdot\mathscr{H} = \frac{e}{2m_0c}\mathscr{H}L_z. \tag{28}$$

This relation forms an example of perturbation theory in classical mechanics: the first-order perturbation energy is obtained by applying the perturbing field to the unperturbed motion. In the present case this is simply the magnetic energy given by (28). The increase of an apparently potential energy is again, as in (23) in fact an increase of kinetic energy of a frictionless motion.

Referring to eq. II (59) and to fig. II, 4, we can introduce the quantum condition

$$L_z = m\hbar \tag{29}$$

and find the quantised levels of the magnetic energy

$$E_m = m(\hbar e/2m_0c)\mathscr{H} = m\mu_0\mathscr{H} \tag{30}$$

where μ_0 is known as the *Bohr Magneton.*
Each level of given l splits into $2l+1$ equidistant levels with the *magnetic* quantum numbers $m = l, l-1, \ldots -l$. Moreover, the spacing of these levels is given by $\mathscr{H}$ and a universal constant alone:

$$\Delta E = \frac{e}{4\pi m_0c}h\mathscr{H} = h\nu_L \tag{31}$$

and is the same for all levels. This is due to the fact that the ratio of magnetic moment to angular momentum, the so-called *magneto-gyric ratio*, is the same for all orbital motions of an electron in any kind of force field, namely

$$\mu/|\mathbf{L}| = -\gamma_0 = -e/2m_0c. \tag{32}$$

As a consequence of these equal spacings, all transitions with the same Δm give the same frequency $\nu = \nu_0 + \gamma_0 \mathscr{H} \Delta m$. The selection rule for m leads to only three frequencies (see fig. III, 4):

$$\Delta m = 0 \qquad \nu = \nu_0 \qquad (\pi \text{ component})$$

$$\Delta m = \pm 1 \qquad \nu = \nu_0 \pm \frac{e}{4\pi m_0 c}\mathscr{H} \qquad (\sigma \text{ components})$$

which represent again the normal Lorentz triplet. It can be regarded as a modulation of the unperturbed frequency by the Larmor frequency $\nu_L = He/4\pi m_0 c$.

The intensity ratio of the components can, in this theory, only be derived from the correspondence principle associating $\Delta m = 0$ with the z component of the classical motion and $\Delta m = \pm 1$ with the motion in the x, y plane, so that the results of the classical theory are simply taken over.

c. Wave mechanical treatment

In wave mechanics, the Zeeman effect offers a specially simple example of a perturbation of a degenerate system. The perturbation term is found in analogy to (28) where operators have to be substituted for $\mathbf{L}$ and L_z, so that

$$H' = \frac{e}{2m_0c} L_z \mathscr{H} = \gamma_0 L_z \mathscr{H} = -\gamma_0 \mathscr{H} i\hbar\, \partial/\partial\varphi. \tag{33}$$

When the matrix elements $\langle nl'm'|H'|nlm\rangle$ are formed for all degenerate states of different l and m:

$$\gamma_0 \mathscr{H} \iiint R_{nl'} P_{l'}^{m'} \Phi i\hbar\, \partial/\partial\varphi R_{nl} P_l^m \Phi r^2 \,\mathrm{d}r \sin\vartheta \,\mathrm{d}\vartheta \mathrm{d}\varphi,$$

two simplifications arise immediately. The operator $\partial/\partial\varphi$ acts only on Φ and gives the factor.

$$m \int_0^{2\pi} \exp[i\varphi(m-m')]\,\mathrm{d}\varphi \begin{array}{ll} = m & \text{for } m = m' \\ = 0 & \text{for } m \neq m'. \end{array}$$

The terms dependent on r and ϑ form separate integrals, and the orthogonality of the functions $\Theta(\vartheta)$ causes H' to vanish unless $l' = l$. This means that the matrix H' is, in the given representation, diagonal in l and m, and we merely have to replace the operator L_z in (33) by any one of its eigenvalues $m\hbar$ in turn, in order to find the perturbation energy of the first order. The possible values of m are $l, l-1, \ldots -l$. The result,

$$E' = \gamma_0 \mathscr{H} m\hbar = m \frac{eh}{4\pi m_0 c} \mathscr{H} \tag{34}$$

agrees with that of the old quantum theory.

The degeneracy in l remains unchanged, the perturbation does not "mix" states of different l. It removes the degeneracy in m by adding to each term the energy (34) dependent on m. It must be remembered that this theory neglects the effects of the spin leading to fine structure and is therefore valid only for sufficiently strong fields.

Since l and m remain "good" quantum numbers in the perturbed system, the selection and polarisation rules derived in II.D for unperturbed wave functions remain valid. The same applies to transition probabilities and intensities. In agreement with the classical result, the π component is found to be twice as strong as each σ component, in transverse observation.

The second-order perturbation by a magnetic field gives rise to terms proportional to $\mathscr{H}^2$. They are noticeable at extremely high field strengths and have been observed in other elements (see p. 215).

4. The linear Stark effect

a. Outline of the observed facts

The effect of electric fields on the Balmer lines was discovered independently by Stark[(12)] and Lo Surdo[(13)] and is known as *Stark effect*; this term is also applied to the effect on lines of other elements. In order to observe the light from excited atoms in strong fields, of the order of 100,000 volts/cm, Stark used positive rays from a hydrogen discharge tube. Owing to charge exchange, they contain a sufficient number of neutral atoms to make the Balmer lines visible. These *canal rays* were made to pass into the space between the plates of a highly charged condenser. Lo Surdo used the light from the cathode dark space of a glow discharge.

The main features of the phenomena are the following:

(i) all hydrogen lines form symmetrical patterns, but the pattern depends markedly on the quantum numbers n of the terms involved; the number of lines and the total width of the pattern increases with n.

(ii) the wave-number differences are integral multiples of a unit which is proportional to the field strength F and is the same for all hydrogen lines.

(iii) Certain components appear only in transverse observation and are polarised with the electric vector parallel to the field (π-components), the other components (σ-components) appear polarised with the electric vector in the x, y plane in transverse observation, and unpolarised in longitudinal observation. Except for the absence of circular polarisation, the polarisation properties thus resemble those in the Zeeman effect. But in contrast to the latter, the π-components show greater shifts than the σ-components.

It must be stressed that this *linear* Stark effect is, on the whole, restricted to hydrogen-like spectra, and occurs in these only in fields large enough for the fine structure to become negligible. Spectra of other elements show a *quadratic* effect which is generally very small, especially in the heavier elements. In helium, linear effects occur in strong fields and have been thoroughly studied (see p. 218).

b. The old quantum theory[14,15]

Though the treatment based on the Bohr–Sommerfeld form of quantum theory was able to account for all details of the linear Stark effect, with the exception of the intensities, it is rather involved, and only some of its features may be mentioned, partly as introduction to the wave mechanical treatment. The Stark effect was the first problem to which Schrödinger applied his wave mechanical perturbation theory.[16]

The effect of a homogeneous electric field of intensity F, parallel to the z axis, on an electron moving in a Kepler ellipse appears, at first sight, similar to that of a magnetic field. For the ordinary Kepler motion, in the absence of external fields, the centre of gravity of the electron, averaged over its orbit, does not coincide with the nucleus and thus forms an electric dipole along the major axis of the ellipse. One will therefore expect a linear effect, described in first order by the energy resulting from the action of the field on the unperturbed motion. The torque on the orbital dipole will cause a precession of the orbit about the z axis. But the calculation shows that, owing to the degeneracy, the motion is more complicated. The angle of azimuth, φ, is still a cyclic variable, for obvious reasons of symmetry, L_z is still a constant of the motion and the quantum number m associated with it retains its strict meaning. But $|\mathbf{L}|$ itself is no longer constant and the orbital quantum number

k (= $l+1$) is no longer a "good" quantum number. If the potential energy $V' = eFz$ is added to the central potential $V(r) = -Ze^2/r$, the classical equation is no longer separable in polar co-ordinates, but in parabolic coordinates. The orbits are not closed and, in infinite time, completely fill the space between two pairs of paraboloids. The calculation gives the energy values

$$E = E_0 - \frac{3a_0 e}{2Z} Fn\,(n_2 - n_1) \tag{35}$$

where a_0 is the Bohr radius (5) and n_1 and n_2 are two new quantum numbers whose sum is restricted by the values of n and m. In order to agree with modern nomenclature, we have to write this restriction as $n = n_1 + n_2 + |m| + 1$. It was on the basis of the Stark effect that Sommerfeld excluded the value $m = 0$, because orbits with zero z-component of the angular momentum can approach the nucleus infinitesimally closely; by analogy, the value $m = 0$ was also excluded in magnetic fields, and only this produced agreement with the results of the Stern Gerlach effect showing two deflected beams for the ground state of silver.

The selection rules are again, by the correspondence principle, $\Delta m = 0, \pm 1$, but since different senses of precession must cause the same energy, the term values depend only on $|m|$ and the sense of rotation in the σ-components is lost.

c. Quantum-mechanical treatment[16,17]

In the wave mechanical perturbation theory, the perturbation operator is $eFz = eFr \cos \vartheta$, and the matrix element

$$H'_{n'l'm'\,nlm} = eF \int_0^\infty R_{n'l'} R_{nl} r r^2 \,\mathrm{d}r \times \times \int_0^\pi P_{l'}^{m'} P_l^m \times \cos\vartheta \sin\vartheta \,\mathrm{d}\vartheta \int_0^{2\pi} \exp[i(m'-m)\varphi]\,\mathrm{d}\varphi. \tag{36}$$

Apart from the factor F, eq. (36) is identical with (II.123) and leads to similar conclusions. As $\cos\vartheta$ is an odd function, the integral vanishes unless the two functions Y_{lm} and $Y_{l'm'}$ have opposite parity. The first-order perturbation energy therefore vanishes unless the term is degenerate; it vanishes in hydrogen-like atoms for the ground term $n = 1$. But since only the parity property has been used, the result holds quite generally for all atoms, that *only degenerate terms can have a first-order Stark effect.*

If the term is degenerate, only those matrix elements of H' for which $l' = l \pm 1$ have finite values: in contrast to the Zeeman effect, the Stark effect perturbation connects states of different l. Owing to the last factor in (36) H' also vanishes unless $m' = m$. This is the result of the fact that the perturbation potential is independent of φ and does not act on the azimuthal integral or, in the language of matrix mechanics, that L_z commutes with z. The terms of different m can then be regarded quite separately in the perturbation calculation, the operator L_z remains diagonal in the perturbed system.

In accordance with the rules of perturbation theory (II.C.3a) the eigenvalues are found as the roots of the secular equation including those unperturbed states which are connected by the perturbation, i.e. for which m is equal and l differs by ± 1.

This may be illustrated by the example of the term $n = 2$ of hydrogen. Of the four degenerate states

$$\overbrace{l = 0, m = 0}^{2S} \quad \overbrace{l = 1, m = 0; \quad l = 1, m = 1; \quad l = 1, m = -1,}^{2P}$$

only the first two have equal m and will show first-order perturbation. The elements of the sub-matrix H' can be numbered $H'_{ll'}$ with the two values 0 and 1 of the suffix and can be calculated by substituting the appropriate eigenfunctions from table III, 4. in (36):

$$H'_{10} = H'_{01} = \frac{eF}{16 \times 2\pi a_0^3} \int_0^\infty \rho(2-\rho) \exp(-\rho/2) r^3 \, dr \times$$

$$\times \int_0^\pi \cos^2\vartheta \sin\vartheta \, d\vartheta \times 2\pi$$

$$= 3eFa_0,$$

$$H'_{11} = H'_{00} = 0.$$

The eigenvalues E' of the matrix H' are found from the determinant equation

$$\begin{vmatrix} H'_{00}-E' & H'_{01} \\ H'_{10} & H'_{11}-E' \end{vmatrix} = \begin{vmatrix} -E' & 3eFa_0 \\ 3eFa_0 & -E' \end{vmatrix} = 0$$

which has the roots

$$E' = \pm 3eFa_0. \tag{37}$$

The eigenfunctions of the first order are linear combinations of the unperturbed functions having the same m and values of l differing by ± 1. With regard to properties such as line strength, such states will get mixed as the result of the perturbation by the electric field.

For the example of $n = 2$, one finds by methods of matrix transformation the following functions $|nlm\rangle$:

$$\text{without field}\left\{\begin{array}{l} \left.\begin{array}{l}|2\ \ 0\ \ 0\rangle \\ |2\ \ 1\ \ 0\rangle\end{array}\right\} \\ |2\ \ 1\ \ 1\rangle \\ |2\ \ 1 -1\rangle \end{array}\right. \quad \left.\begin{array}{c} \left\{\begin{array}{l}\sqrt{2}(|2\ \ 0\ \ 0\rangle + |2\ \ 1\ \ 0\rangle) \\ \sqrt{2}(|2\ \ 0\ \ 0\rangle + |2\ \ 1\ \ 0\rangle)\end{array}\right. \\ |2\ \ 1\ \ 1\rangle \\ |2\ \ 1 -1\rangle \end{array}\right\}\text{with field.}$$

The quantum-mechanical result (37) agrees with (35) for the states $n = 2$, $n_2 - n_1 = 1$. The four states can be interpreted by three orientations, parallel, anti-parallel and normal to the field, of a permanent dipole of moment $3ea_0$, but it would be wrong to associate a definite orientation of an angular momentum vector with any one of the states.

The result of the quantum-mechanical perturbation calculation of the first order agrees, for all terms, with that of the old theory, but it differs in the higher orders. The energy of the atom, up to the second-order perturbation term, is found, according to quantum mechanics, as

$$E = -\frac{RZ^2ch}{n^2} + F\frac{3a_0e}{2Z}n(n_2-n_1) - $$
$$-F^2\frac{a_0^3}{16Z^4}n^4[17n^2 - 3(n_2-n_1)^2 - 9m^2 + 19]. \tag{38}$$

This formula agrees well with experiments.

The absolute value of the linear effect was tested very accurately by Kassner[18]. With the numerical values of the constants, the linear term shift is found theoretically to be

$$\Delta\tilde{\nu} = F_V\frac{n(n_2-n_1)}{Z} \times 6{\cdot}40 \times 10^{-5}\ \text{cm}^{-1}, \tag{39}$$

where F_V is the value of the field strength in volts per cm. The experiments gave an average value of the numerical constant of 6·44 for measurements on H_β, H_γ and H_δ.

Measurements at very high fields, up to 10^6 volts/cm,[19,20,21] have confirmed the correctness of the second-order term and even of the third-order term; but for the very large term shifts involved in the tests of the third-order effect, the perturbation theory begins to lose its validity.

Figure III, 5 shows the term diagram of the Stark effect of H_α and fig. III, 6[21] a photometer tracing of the pattern. The positions of the components agree with the theory, but the relative intensities depend markedly on the conditions of excitation.

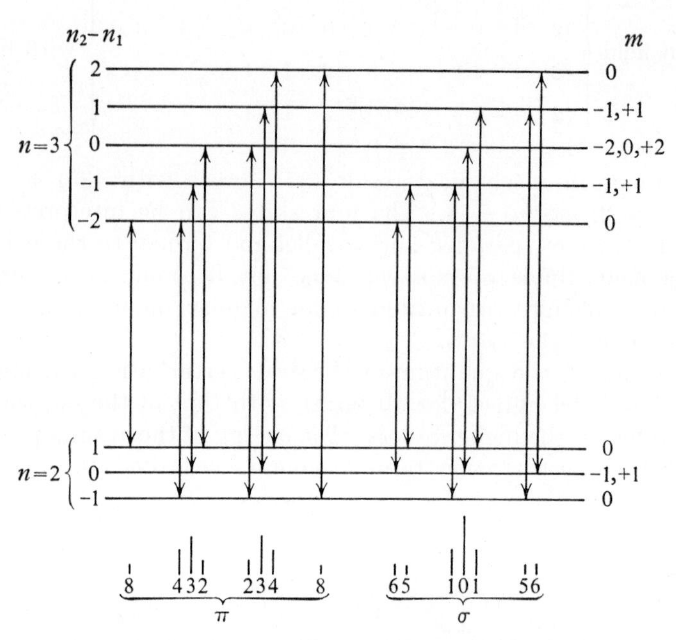

FIG. III, 5. Stark effect of H_α, term diagram (from H. G. Kuhn, *Atomspektren*, Akad. Verl. Ges. Leipzig 1934).

d. Tunnel effects

The influence of an electric field on an atom raises interesting questions if either the field is very strong, or very high terms are involved. In fig. III, 7 the potential energy $-e^2Z/r$ of the electron in a hydrogen-like atom is plotted as a dotted curve. The superposition of the potential $-eFz$ (thin line) of an external field causes a potential maximum to appear on the "downfield" side, defined by the condition $-eF = e^2Z/r^2$. The ionisation potential is lowered, in the presence of the field, by

$$E' = 2e\sqrt{(eZF)}.$$

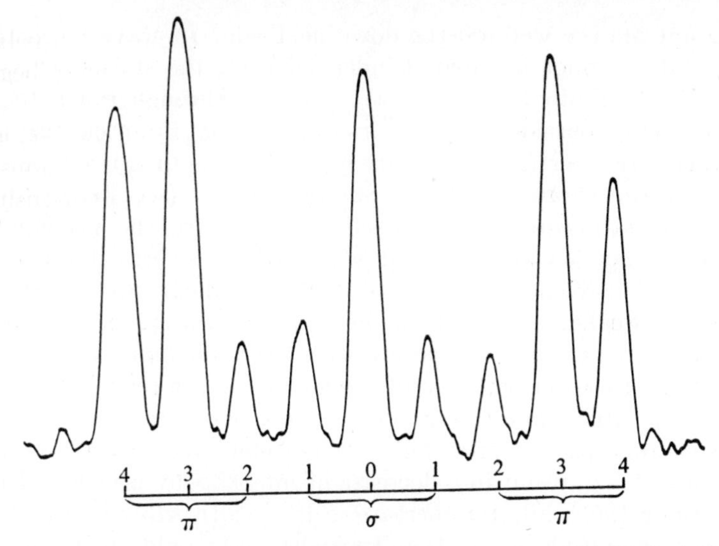

FIG. III, 6. Stark effect of H_α, photometer tracing. (H. Mark & R. Wierl, Ref. 21.)

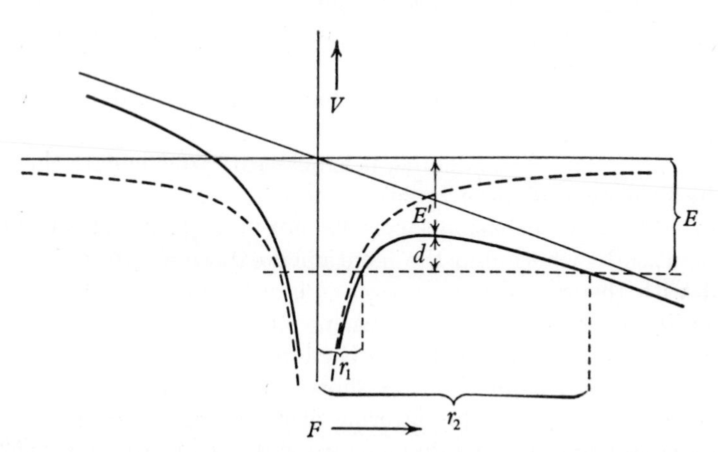

FIG. III, 7. Potential energy curve for hydrogen in an electric field (tunnel effect). The arrow gives the direction of the force.

All quantum states above this limit will be continuous, corresponding to non-periodic orbits of the classical motion.

But also the quantum states below this maximum potential undergo an important change: for any energy level E inside the potential well, there is an equal energy which the electron can assume while being outside the well, on the down-field side. In wave mechanics, such states cannot be treated independently; the stationary wave inside has a finite probability of "leaking" through the potential barrier to the outside. The electron of an atom in an electric field has therefore a certain probability of suddenly finding itself outside. This causes a shortening of the *natural life time* (see VII.B) of the quantum state which shows itself, according to the principle of indeterminacy, as an increased width of the energy level. This theory, first given by Oppenheimer[22], predicts two observable effects; a weakening of certain lines due to the process of *auto-ionisation* (see p. 297) competing with the probability of radiative transitions to lower states, and a widening if the reciprocal of the life time is larger than the smallest frequency difference resolved by the spectrograph. A probability of auto-ionisation of 10^9 or 10^{10} per sec will, e.g. result in a decrease of intensity by a factor of perhaps 10 or 100, while the increase in line width will be only of the order of magnitude of the Doppler width and would not be noticed in an ordinary spectrograph.

The probability of auto-ionisation by the field (*tunnel effect*) increases with decreasing width and height of the potential peak, relative to the term considered, and is appreciable only for terms very close to the maximum.

Traubenberg and Gebauer[20,23] have demonstrated these effects in some very striking experiments on the lines H_β, H_γ and H_δ. For an interpretation of the details, it is necessary to calculate the effect of the field on the wave functions and terms which are so highly distorted that perturbation theory cannot be applied. This was done by Lanczos[24] who found the mentioned experiment in good agreement with theoretical expectation. Plate 2 shows the "dying out" of the hydrogen lines with increasing field strength, and it is also apparent that the components on the long-wavelength side are more strongly affected than those of shorter wavelength. This is to be expected because both the lowering of the energy level (red shift of components) and the weakening are due to the same cause, the lowering of the potential energy by the field. The wave functions extending preferentially in the field direction will therefore show both effects most markedly. The experiments of Traubenberg and Gebauer showed clearly the weakening of the lines below the point of the potential maximum, thus verifying the quantum-mechanical tunnel effect.

They also found some evidence of a broadening effect, but its cause is not certain.

5. The fine structure

a. The basic facts and Sommerfeld's theory

The study of the fine structure of hydrogen-like spectra has played a great part in the development of the fundamental theories of atomic physics. Owing to the simplicity of the model consisting of a single electron moving in the Coulomb field of the nucleus the fundamental equations can be solved more rigorously than for other atoms and can be subjected to crucial tests by comparison with experimental results. Unfortunately, however, the spectroscopic measurements encounter one great difficulty; the small mass of H and He^+ causes the lines to have a large Doppler width. The discovery of the heavy isotope H^2 improved the conditions to some extent, and the use of liquid hydrogen as coolant for discharge tubes has resulted in further advance.

The main steps in the exploration of the fine structure can best be sketched with reference to the H_α line: (1) discovery of a doublet structure by Michelson and Morley (1887), its interpretation by Sommerfeld as relativistic effect; (2) growing evidence for, and finally discovery of a third component between the two main components; interpretation of the new component by the introduction of the electron spin; (3) experiments with deuterium, and with improved technique, showing discrepancies between the observed doublet splitting and its theoretical value, and finally leading to the assumption that the $2S$ term is displaced upward, compared with the theoretical value; (4) definite confirmation and accurate measurement of this displacement by the micro-wave experiments of Lamb and Retherford; (5) new formulation of the radiation theory leads to a revision of the theory, giving agreement with the experiments (Bethe); (6) improved spectroscopic technique confirms the results and establishes a similar shift in the $3S$ term.

To this brief summary of the historical development must be added a mention of the important part played by the line $n = 4 \rightarrow 3$ of He^+ (4686A) the structure of which has recently been determined with great accuracy.

Only a brief outline of Sommerfeld's theory will be given here. It was pointed out before that the velocity of the electron in the lower quantum orbits is not negligible compared with the velocity of light. For the first Bohr orbit in hydrogen, (6) gives the ratio

$$\alpha = v_1/c = e^2/\hbar c = 7{\cdot}2973 \times 10^{-3} = 1/137{\cdot}038. \tag{40}$$

For an elliptic orbit, the velocity is specially large when the electron is near the perihelion, and one must expect the relativistic corrections to differ for orbits of different k (or l), for the same n. By applying the equations of relativistic mechanics, instead of Newton's mechanics, to hydrogen-like atoms, Sommerfeld[11] found that all stationary energy levels were lowered, due to an additional term

$$\Delta E_{\text{rel}} = -\frac{Rch\alpha^2 Z^4}{n^3}[1/k - 3/(4n)] \tag{41}$$

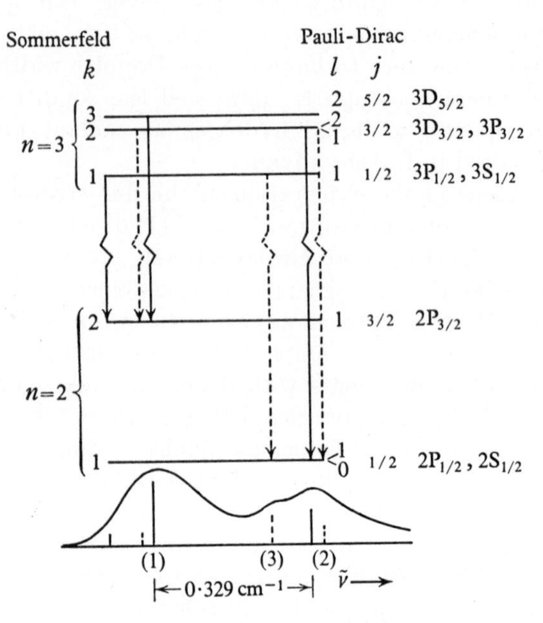

FIG. III, 8. Fine structure of H_α, according to the Pauli–Dirac theory.

where α is the *fine structure constant* defined by (40). According to this theory, the H_α line would consist of a doublet (fig. III, 8) of spacing $0{\cdot}329\ \text{cm}^{-1}$, with a further component too faint and too close to one of the other lines to allow observation. The dotted transition (3) is forbidden. Within the accuracy achieved at the time, this result appeared well confirmed.

Better resolution, mainly due to the use of deuterium, revealed the existence of a third component in the approximate position of the forbidden transition (3). Also the spin and magnetic moment of the electron had been established and their influence on the fine structure had to be taken into account; finally the development of wave mechanics called for a treatment of the problem by this new method, with the inclusion of the spin.

b. The Pauli–Darwin theory

The quantum-mechanical treatment of the spinning electron is based on the fundamental assumption that the absolute value of the spin moment, or $\mathbf{S}^2$, is constant. In the free atom, and even in weak external fields, also the absolute value of the total angular momentum, or $\mathbf{J}^2$, is constant. The magnetic moment associated with the spin interacts with the magnetic field due to the orbital motion, tending to change the angle between $\mathbf{S}$ and $\mathbf{L}$. Provided the interaction is small, this causes a gyroscopic precession of $\mathbf{S}$ and $\mathbf{L}$ about the vector resultant $\mathbf{J}$ while $\mathbf{L}^2$ remains constant.

Applying the quantum-mechanical theory of angular momenta (II.C.2c) to this situation, we find that the operators $\mathbf{J}^2$, $\mathbf{L}^2$ and $\mathbf{S}^2$ commute with one another and with J_x, J_y, J_z, but the three latter do not commute with one another. A commuting set of operators is then $\mathbf{L}^2$, $\mathbf{J}^2$, J_z, where the z-component has been chosen arbitrarily and $\mathbf{S}^2$ has been omitted as trivial since it has only one eigenvalue $s(s+1)\hbar^2 = \frac{3}{4}\hbar^2$ and can always be treated as a constant. A representation can therefore be chosen in which the three operators are diagonal and, since they are constant, in which also the energy is diagonal. States of constant energy can then be labelled by the quantum numbers l, j, m.

As a next step, the operands on which these operators act have to be defined, which can be done in several ways. The so-called Pauli functions can be written as two-component ψ-functions or two-row symbols whose elements are functions of the coordinates and are, in fact, identical with Schrödinger functions:

$$\psi = \begin{pmatrix} \psi_1(x, y, z) \\ \psi_2(x, y, z) \end{pmatrix} = (+)\psi_1 + (-)\psi_2. \tag{42}$$

The probability density is given by $\psi_1^*\psi_1 + \psi_2^*\psi_2$, and the functions are normalised so that the integral of this density over the whole space has the value 1. The matrices act on the two-row symbol functions according to the rules described in II.C. Differential operators, such as L_z, are written as matrices as follows:

$$L_z \cdot \begin{pmatrix} 1 & 0 \\ 0 & 1 \end{pmatrix} = \begin{pmatrix} L_z & 0 \\ 0 & L_z \end{pmatrix} = \begin{pmatrix} -i\hbar\partial/\partial\varphi & 0 \\ 0 & -i\hbar\partial/\partial\varphi \end{pmatrix}.$$

The eigenvalues of J_z can easily be found from the equation

$$J_z\psi = \text{const.} \times \psi = m\hbar\psi$$

where m is a yet undefined numerical constant, so that the assumption $m\hbar$ for the eigenvalues merely ensures the correct dimension. With $J_z = L_z + S_z$, this equation becomes

$$\hbar\begin{pmatrix} -i\partial/\partial\varphi + \frac{1}{2} & 0 \\ 0 & -i\partial/\partial\varphi - \frac{1}{2} \end{pmatrix}\begin{pmatrix} \psi_1 \\ \psi_2 \end{pmatrix} = m\hbar\begin{pmatrix} \psi_1 \\ \psi_2 \end{pmatrix},$$

giving

$$(-i\partial/\partial\varphi + \tfrac{1}{2})\psi_1 = m\psi_1$$
$$(-i\partial/\partial\varphi - \tfrac{1}{2})\psi_2 = m\psi_2,$$

the solutions of which are

$$\begin{aligned} \psi_1 &= A_1(r, \vartheta)\exp[i(m-\tfrac{1}{2})\varphi] \\ \psi_2 &= A_2(r, \vartheta)\exp[i(m+\tfrac{1}{2})\varphi]. \end{aligned} \tag{43}$$

ψ_1 and ψ_2 must have the same values for angles φ differing by multiples of 2π, therefore m is restricted to the values

$$m = \pm\tfrac{1}{2}, \pm\tfrac{3}{2} \ldots$$

and for each value of m, the eigenfunction is

$$\psi_m = \begin{pmatrix} A_1(r, \vartheta)\exp[i(m-\frac{1}{2})\varphi] \\ A_2(r, \vartheta)\exp[i(m+\frac{1}{2})\varphi] \end{pmatrix}. \tag{44}$$

Since $m \pm \frac{1}{2}$ is integral, the exponential functions are identical with the Schrödinger functions $\Phi(\varphi)$.

In the same way, the requirements of ψ to be an eigenfunction of $\mathbf{L}^2$, whose eigenvalues are known to be $l(l+1)\hbar^2$ (where $l = 0, 1, 2 \ldots$) leads to the result that the component functions ψ_1 and ψ_2 must depend on ϑ and φ in the form of spherical harmonics Y_{l,m_l}, where m_l is an integer:

$$\begin{aligned} \psi_1 &= R_1(r)Y_{l,\, m-\frac{1}{2}} \\ \psi_2 &= R_2(r)Y_{l,\, m+\frac{1}{2}} \end{aligned}. \tag{45}$$

The functions also have to be eigenfunctions of $\mathbf{J}^2$, and from general commutation properties of this operator it is known (see p. 49) that its eigenvalues are $j(j+1)\hbar^2$ where j is either integral or half integral. This leads to a matrix equation, and finally to the results

$$\psi_{l,\,j=l+\frac{1}{2},\,m} = (+)\left(\frac{l+\frac{1}{2}+m}{2l+1}\right)^{\frac{1}{2}} R(r)Y_{l,\,m-\frac{1}{2}}+(-)\left(\frac{l+\frac{1}{2}-m}{2l+1}\right)^{\frac{1}{2}} R(r)Y_{l,\,m+\frac{1}{2}}$$

$$\psi_{l,\,j=l-\frac{1}{2},\,m} = (+)\left(\frac{l+\frac{1}{2}-m}{2l+1}\right)^{\frac{1}{2}} R(r)Y_{l,\,m-\frac{1}{2}}-(-)\left(\frac{l+\frac{1}{2}+m}{2l+1}\right)^{\frac{1}{2}} R(r)Y_{l,\,m+\frac{1}{2}} \qquad (46)$$

where j is restricted to the two values $l\pm\frac{1}{2}$ while the radial function $R(r)$ is normalised but otherwise undetermined. The spin functions have been written in the form explained in II.C.1. In the states $m = l+\frac{1}{2}$ and $m = -(l+\frac{1}{2})$ one of the components of the Pauli functions is found to vanish, and S_z has the definite value $\frac{1}{2}\hbar$ or $-\frac{1}{2}\hbar$; L_z has then the definite value $+l\hbar$ or $-l\hbar$. In the other states they are indeterminate and the squares of the coefficients of the component functions give the probabilities of m_s having the values $+\frac{1}{2}$ and $-\frac{1}{2}$, and of m_l having the values $m-\frac{1}{2}$ and $m+\frac{1}{2}$.

As an example, and for later use, we write down the eigenfunctions (46) for a p-state: $l = 1$; the common factor $R(r)$ has been omitted.

$$\left.\begin{array}{lll} j = \frac{3}{2}, m = \frac{3}{2} & (+)Y_{1,1} & \\ \quad\;\; m = \frac{1}{2} & 3^{-\frac{1}{2}}[2^{\frac{1}{2}}(+)Y_{1,0}+(-)Y_{1,1}] & \\ \quad\;\; m = -\frac{1}{2} & 3^{-\frac{1}{2}}[(+)Y_{1,-1}+2^{\frac{1}{2}}(-)Y_{1,0}] & {}^2P_{\frac{3}{2}} \\ \quad\;\; m = -\frac{3}{2} & (-)Y_{1,-1} & \\ j = \frac{1}{2}, m = \frac{1}{2} & 3^{-\frac{1}{2}}[(+)Y_{1,0}-2^{\frac{1}{2}}(-)Y_{1,1}] & \\ \quad\;\; m = -\frac{1}{2} & 3^{-\frac{1}{2}}[-2^{\frac{1}{2}}(+)Y_{1,-1}+(-)Y_{1,0}] & {}^2P_{\frac{1}{2}}. \end{array}\right. \qquad (47)$$

The eigenfunctions allow the spatial distribution of the electron to be calculated and lead to an important result for the symmetry of the charge distribution. The probability density is given by $\psi_1{}^*\psi_1+\psi_2{}^*\psi_2$ as a superposition of the probability densities of the Schrödinger functions involved, with the weight factors $(l+\frac{1}{2}+m)/(2l+1)$ and $(l+\frac{1}{2}-m)/(2l+1)$ respectively. For a state $l = 1$, $j = \frac{1}{2}$, $m = \frac{1}{2}$, the probability density is found from (46) and table II, 2 to be $R^2(r)\,(\cos^2\vartheta+\sin^2\vartheta)/4\pi = R(r)^2/4\pi$, and the same result is found for $m = -\frac{1}{2}$. The charge distribution of an electron in a $P_{\frac{1}{2}}$ level is thus spherically symmetrical. The same applies obviously to the $S_{\frac{1}{2}}$ levels whose eigenfunctions contain only $Y_{0,0}$, but it can be proved also for any state of $j = \frac{1}{2}$, so that any state with total angular momentum quantum number $j = \frac{1}{2}$ has a spherically symmetrical charge distribution. The radial distribution $R^2(r)$ is the same as in Schrödinger functions.

c. The spin-orbit coupling (see also p. 165).

We consider at first the classical interaction between the magnetic moment $\boldsymbol{\mu}_s = -\mathbf{S}e/m_0c$ and the magnetic field due to the orbital motion whose angular momentum is given by

$$\mathbf{L} = m_0\mathbf{r} \times \mathbf{v}. \tag{48}$$

With a view to later application to other atoms we assume at first more generally that the force is derived from some central force potential $V(r)$. The orbital motion of velocity $\mathbf{v}$ in the electric field of strength

$$\mathbf{F} = -\mathbf{r}\frac{1}{r}\frac{\mathrm{d}V}{\mathrm{d}r}$$

causes a magnetic field at the point of the electron of strength

$$\mathscr{H} = \frac{1}{c}\mathbf{F} \times \mathbf{v} = -\frac{1}{cr}\frac{\mathrm{d}V}{\mathrm{d}r}\mathbf{r} \times \mathbf{v} \tag{49}$$

or, after substitution from (48),

$$\mathscr{H} = -\frac{1}{cm_0}\frac{1}{r}\frac{\mathrm{d}V}{\mathrm{d}r}\mathbf{L}, \tag{50}$$

giving rise to a magnetic energy

$$E' = -\mathscr{H} \cdot \boldsymbol{\mu}_s = -\frac{e}{c^2m_0^2}\frac{1}{r}\frac{\mathrm{d}V}{\mathrm{d}r}\mathbf{L} \cdot \mathbf{S}. \tag{51}$$

This calculation would be accurate if the electron were at rest and the nucleus were revolving around it. One can, e.g., derive (50) as the field produced by the circular current $i = Ze/\tau = Zev/2\pi r$ in the centre: $\mathscr{H} = 2\pi i/rc = Zev/cr^2$ which agrees with (50) for $V = Ze/r$. Thomas[25] and Frenkel[26] have shown that the relativistic transformation from the frame moving with the electron to the frame at rest introduces a further factor $\frac{1}{2}$ in the equation, so that

$$E' = -\frac{e}{2c^2m_0^2}\frac{1}{r}\frac{\mathrm{d}V}{\mathrm{d}r}\mathbf{L} \cdot \mathbf{S} = \xi(r)\mathbf{L} \cdot \mathbf{S} \tag{52}$$

where

$$\xi(r) = -\frac{e}{2c^2m_0^2}\frac{1}{r}\frac{\mathrm{d}V}{\mathrm{d}r}. \tag{53}$$

For the special case of the Coulomb field $V(r) = Ze/r$,

$$\xi(r) = \frac{Ze^2}{2c^2m_0{}^2}\frac{1}{r^3}. \tag{54}$$

While $\mathbf{L}\cdot\mathbf{S}$ is constant, the value of r and therefore of $\xi(r)$ generally varies during the motion. In the classical or Bohr–Sommerfeld treatment, the average $\overline{1/r^3}$ has to be calculated for the unperturbed orbital motion to give a first order approximation to the energy. In quantum mechanics[27] the corresponding step consists in applying $\xi(r)\mathbf{L}\cdot\mathbf{S}$ as an operator to be added as a perturbing potential to $V(r)$ in the Hamiltonian (II.60):

$$H\psi = \left(\frac{1}{2m_0}p_r{}^2+\frac{1}{2m_0r^2}\mathbf{L}^2+V(r)+\xi(r)\mathbf{L}\cdot\mathbf{S}\right)\psi = E\psi \tag{55}$$

where ψ stands for a two-component Pauli wave function.

The operator $\mathbf{L}\cdot\mathbf{S} = L_xS_x+L_yS_y+L_zS_z$ can be transformed by means of the following operator equations:

$$\begin{aligned}\mathbf{J}^2 &= (\mathbf{L}+\mathbf{S})^2 = \mathbf{L}^2+\mathbf{S}^2+2\mathbf{L}\cdot\mathbf{S},\\ \mathbf{L}\cdot\mathbf{S} &= \tfrac{1}{2}(\mathbf{J}^2-\mathbf{L}^2-\tfrac{3}{4}\hbar^2).\end{aligned} \tag{56}$$

Since H, $\mathbf{L}^2$, $\mathbf{J}^2$ and J_z are a commuting set of operators, eq. (56) shows that $\mathbf{L}\cdot\mathbf{S}$ commutes with all of them and is diagonal in the representation defined by the quantum numbers l, j, m. Consequently the matrix equation (56) can be replaced by a set of equations containing the eigenvalues instead of the operators: $\mathbf{L}^2$ is replaced by $l(l+1)\hbar^2$ and $\mathbf{L}\cdot\mathbf{S}$ becomes

$$\mathbf{L}\cdot\mathbf{S} = \tfrac{1}{2}\hbar^2[j(j+1)-l(l+1)-s(s+1)]. \tag{57}$$

With this substitution, (55) can be written separately for each of the values $j = l\pm\frac{1}{2}$. In each of these equations, the ψ is one or the other of the two space functions ψ_1, ψ_2 appearing in the two-component Pauli function (42). The magnetic interaction term in these equations can now be treated as a first-order perturbation, with the ordinary Schrödinger functions ψ_{nlm} for the point electron serving as zero-order functions.

The first-order perturbation energy (see p. 57) is then

$$H' = \tfrac{1}{2}\hbar^2[j(j+1)-l(l+1)-s(s+1)]\int\psi_{nlm}{}^*\xi(r)\psi_{nlm}\,\mathrm{d}\tau. \tag{58}$$

When $\xi(r)$ is substituted from (54), it is obvious that the integration only involves the radial Schrödinger functions $R(r)_{nl}$. The result is

$$E' = \frac{Ze^2}{2c^2m_0^2}\overline{\frac{1}{r^3}}\frac{\hbar^2}{2}[j(j+1)-l(l+1)-s(s+1)], \tag{59}$$

$$\overline{\frac{1}{r^3}} = \int \psi_{nlm}^*\frac{1}{r^3}\psi_{nlm}\,\mathrm{d}\tau = \frac{1}{a_0^3}\frac{Z^3}{n^3l(l+\frac{1}{2})(l+1)}. \tag{60}$$

Inserting R and α from (7) and (40) we find from (59) and (5):

$$E' = \frac{Rch\alpha^2Z^4}{n^3}\times\frac{j(j+1)-l(l+1)-s(s+1)}{2l(l+\frac{1}{2})(l+1)}. \tag{61}$$

The second factor has the values $1/(l+1)(2l+1)$ and $-1/l(2l+1)$ for $j = l+\frac{1}{2}$ and $j = l-\frac{1}{2}$ respectively. The magnetic interaction has caused each term of given l to split into two terms displaced upwards and downwards respectively. The term difference is found to be

$$\Delta T = \frac{R\alpha^2Z^4}{n^3l(l+1)} = 5{\cdot}84\frac{Z^4}{n^3l(l+1)}\ \mathrm{cm}^{-1}. \tag{62}$$

The displacements are in the ratio $l/(l+1)$. It is interesting to note that this is equal to the inverse ratio of the statistical weights of the terms $(2j'+1)/(2j+1)$, so that the sum of the magnetic energies of all states of a given level n, l is zero. This is an example of a general sum rule to be discussed in V.A.8.

It is instructive to compare (59) with the energy calculated by applying the classical formulae (52) and (54) to the Bohr–Sommerfeld orbits with the vector model for which $|\mathbf{L}| = \hbar l$, $|\mathbf{S}| = \hbar s$,

$$E^* = \frac{Ze^2}{2c^2m_0^2}\overline{\frac{1}{r^3}}\hbar^2ls\cos\vartheta = \frac{Ze^2}{2c^2m_0^2}\overline{\frac{1}{r^3}}\hbar^2ls\frac{j^2-l^2-s^2}{2ls}.$$

One can thus obtain the correct dependence (59) on the quantum number j from the vector model by the following change in $\cos\vartheta$:

$$\frac{j^2-l^2-s^2}{2ls} \to \frac{j(j+1)-l(l+1)-s(s+1)}{2ls}. \tag{63}$$

The right-hand side of (63) is sometimes called the quantum-mechanical cosine of ϑ. Analogous relations hold more generally and are often useful for remembering quantum-mechanical formulae, though no accurate geometrical meaning should be attached to them. The value $\overline{1/r^3}$ for Bohr–Sommerfeld orbits agrees with the wave mechanical value only in the limit of infinitely large quantum numbers l.

d. The relativity correction

Neglecting, for the moment, again the magnetic splitting, one can modify the Schrödinger equation by inserting the relativistic expression for the energy. This procedure does not, in fact, make the equation relativistically invariant and is open to some objections, but it can be regarded as an approximation somewhat similar to the relativistic form of the old quantum theory. The result of this calculation is the appearance of an additional term in the energy,

$$E'_{\text{rel}} = \frac{E_n \alpha^2 Z^4}{n}\left(\frac{1}{l+\frac{1}{2}} - \frac{3}{4n}\right). \tag{64}$$

Since this correction, as well as the magnetic spin–orbit energy, is small, we can simply add both corrections, with the result

$$E_{n,l,j} = -\frac{RchZ^2}{n^2}\Bigg[1 + \underbrace{\frac{\alpha^2 Z^2}{n}\left(\frac{1}{l+\frac{1}{2}} - \frac{3}{4n}\right)}_{\substack{\text{relativity} \\ \text{correction}}} - \underbrace{\frac{\alpha^2 Z^2}{n}\,\frac{j(j+1)-l(l+1)-s(s+1)}{l(2l+1)(l+1)}}_{\substack{\text{spin–orbit} \\ \text{interaction}}}\Bigg]. \tag{65}$$

It is found that the coefficients of these two corrections which appear to be quite different in origin are the same. This has the result that any two terms having the same n and j, and thus differing in l by one unit, coincide: by substituting $l = j+\frac{1}{2}$ and $j-\frac{1}{2}$ in turn into (65), one finds the same result:

$$E_{n,j} = -\frac{RchZ^2}{n^2}\left[1 + \frac{\alpha^2 Z^2}{n}\left(\frac{1}{j+\frac{1}{2}} - \frac{3}{4n}\right)\right]. \tag{66}$$

The resulting term values agree with those from Sommerfeld's formula if $j+\frac{1}{2}$ is replaced by k. On account of this agreement, fig. III, 8 can again be used to represent the term diagram of hydrogen according to these quantum-mechanical results. The levels merely have to be re-named as shown on the right of the figure. Following the usual spectroscopic nomenclature, the j values are given as suffix and the values of $l = 0, 1, 2 \ldots$ are denoted by the symbols S, P, D, ... In this new term diagram, more transitions are allowed, especially that marked (3) which was naturally identified with the partly resolved third component.

It is significant that the relativistic effect and the magnetic interaction appear as terms with the same constant factor in (65). This fact alone indicates that the origins of both effects must be closely related. Dirac[28] succeeded in proving that the electron spin arises, in fact, as a necessary result of a relativistic formulation of quantum mechanics. In contrast to the relativistically corrected theory of Schrödinger and Pauli, Dirac's wave equations are truly invariant for Lorentz transformations. The wave functions of this theory have four components, but two of them are mainly connected with the states of negative energy which this theory postulates, and can be neglected in most applications to atomic structure. The remaining two "large" components differ little from Pauli functions except in the immediate vicinity of the nucleus. This difference is, however, important for calculations of hyperfine structures and isotope shifts.

In Dirac's theory, not only the spin quantum number $s = \frac{1}{2}$, but also the value of the magnetic moment is derived without any specific assumptions apart from the values of m and e.

Dirac's theory removes one serious doubt left by Pauli's theory when it is applied to S terms. The magnetic interaction term should, according to (55), vanish for $\mathbf{L} = 0$, while equation (66) gives a finite interaction for $l = 0$; the latter result is confirmed by Dirac's theory. All energy levels of hydrogen-like atoms as derived from Dirac's theory agree with those of formula (66). For most applications to atomic spectra, it is therefore not necessary to use the complicated Dirac functions, and no account of the theory will be given, though its results will be used in chapter VI. The energy values (66) can be considered as the results of the most consistent and accurate form of quantum theory of the free atom with one electron and are often referred to as results of Dirac's theory. Comparisons with experiments are naturally of considerable interest as direct tests of the validity of the theory.

e. The radiation corrections (*Lamb shift*)

Many spectroscopists had found the separation between the main components of the line H_α to be slightly smaller than the value predicted by theory,[29] and Pasternack[30] had suggested that this discrepancy could be explained by assuming the term $2S_{\frac{1}{2}}$ to be about $0{\cdot}03$ cm^{-1} higher than $2P_{\frac{1}{2}}$, in contradiction to Dirac's theory. Lamb and Retherford[31] proved the correctness of this assumption by detecting radio-frequency resonance for the transition between the two terms. This was confirmed spectroscopically by the use of discharge tubes cooled with liquid hydrogen; the Doppler width was then sufficiently reduced to allow the component (3) to be resolved and thereby the term difference $2P_{\frac{1}{2}} - 2S_{\frac{1}{2}}$ to be measured.[32] Further radio-frequency measurements established this difference, sometimes called Lamb shift, to a very high accuracy as $0{\cdot}03528$ cm^{-1} for light hydrogen.

Soon after the discovery of this shift, Bethe[33] showed that a revised form of the theory of interaction between matter and radiation causes all S terms to be raised by an amount which agreed well with the experimental value for hydrogen. Later, refined calculations[34] gave the result $2P_{\frac{1}{2}} - 2S_{\frac{1}{2}} = 0{\cdot}03526$ cm^{-1}. The differences between theory and experiment are now within the limits of accuracy of higher-order terms of the theory.[35]

A quantitative account of the quantum electrodynamic derivation of the radiation shift is outside the scope of this book; only the approximate treatment due to Welton[36] may be briefly outlined, because it shows the nature of the effect in a comparatively simple form. In quantum electrodynamics, the lowest state of energy of the radiation field is not one of vanishing field strengths but one which has zero point energy. The fluctuating electric field due to this zero point energy causes the electron to carry out rapid, irregular trembling movements which have the same effect as if the electron were spread over a larger space; the greater spread reduces the value of ψ^2 near the origin. For S terms for which ψ^2 has large values close to the nucleus the potential energy is noticeably increased (the attraction reduced) and the term value raised. For other terms, for which $\psi_{(r=0)}$ vanishes, the effect is negligibly small.

Figure III, 9 shows the revised term diagram of H_α and the positions and intensities of the components according to theory. Plate 3 gives photometer tracings of spectrograms obtained[32] with a Fabry Perot etalon; the light was taken from a discharge in a helium-deuterium mixture cooled with liquid hydrogen. The shortest exposure (a) shows the main components (1) and (2) and the longer exposure (b) shows component (3) resolved. In the overexposure (c), a weak component (5) is visible. The latter was later completely resolved,[37] by the use of a double etalon, and its measurement established an upward shift of the term $3S_{\frac{1}{2}}$ by approximately the amount predicted by the radiation theory. The influence of the unresolved components, most of which are very weak, on the measured positions of the lines can be estimated; when this is done, the results are in good agreement with the radiation theory.

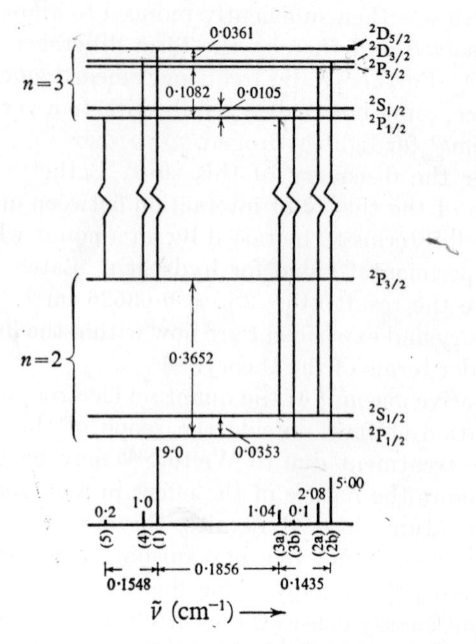

FIG. III, 9. Fine structure of H_α, with Lamb shift. (H. G. Kuhn & G. W. Series, ref. 32).

The only permitted transition for hydrogen-like atoms in the state $2S_{\frac{1}{2}}$ is that to $2P_{\frac{1}{2}}$, but the frequency is so low that the transition probability for spontaneous emission is negligible. The term $2S_{\frac{1}{2}}$ is therefore metastable. The method of Lamb and Retherford is based on this property and cannot be applied to higher S terms. It has, however, been applied in a modified form to the $2S_{\frac{1}{2}}$ term of He^+. The latest measurements[38] gave the value 14040 Mc/sec (0·46833 cm^{-1}) for the difference $2P_{\frac{1}{2}} - 2S_{\frac{1}{2}}$. The latest theoretical value of 14057 Mc/sec is slightly in excess of this, apparently owing to a higher-order relativistic effect not included in the theory.[35]

The Lamb shift of the level 1S of deuterium has been studied by an accurate determination of the wavelength of the Lyman-α line[39].

The difficulty of wavelength standards in the vacuum ultra-violet was overcome by the use of the combination principle applied to several reference lines. The shift of the term was found to be 0·26 cm^{-1}, in good agreement with the theoretical value 0·273 cm^{-1}.

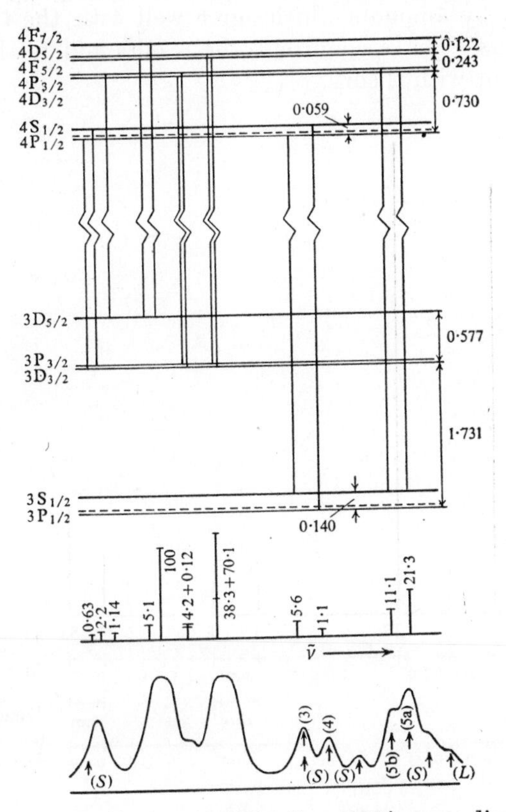

FIG. III, 10. Fine structure of He$^+$ line 4686A, term diagram and photometer tracing (G. W. Series, ref. 44).

Spectroscopic investigations of the fine structure in He$^+$ have been mainly restricted to the line $n = 4 \rightarrow 3$ (4686A) which was first studied by Paschen(40). Later observers(41,42,43) proved the existence of the shifts of the levels $3S_{\frac{1}{2}}$ and $4S_{\frac{1}{2}}$. With the use of a source cooled with liquid hydrogen and of a double etalon, the relative positions of all fine structure terms of the levels $n = 3$ and $n = 4$ could be measured with considerable accuracy.(44) Figure III, 10 shows the term diagram and photometer tracing and fig. III, 11 the comparison of measured and theoretical values of the positions of the lines. According to Dirac's theory, the doublets (3) + (4) and (5a) and (5b) would be single lines; the splitting demonstrates directly the existence of radiation shifts. The measurements gave an upward displacement, compared with the terms of Dirac's theory of the terms $3S_{\frac{1}{2}}$ and $4S_{\frac{1}{2}}$ by amounts which agree well with the theory. Later measurements[45] have confirmed the results and further improved the agreement with theory.

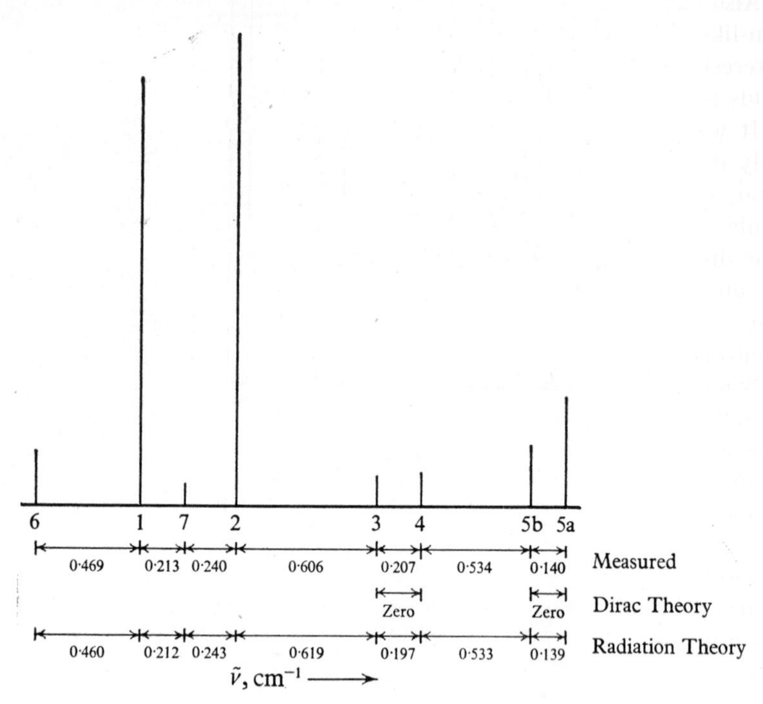

FIG. III, 11. Comparison of observed and calculated components of the He^+ line 4686 A. (G. W. Series, ref. 44).

6. Zeeman- and Stark-effect in weak fields

If the external magnetic field is so weak that the energy of interaction with the atom is small compared with the energy difference between fine structure levels, we have to consider the influence of the field on each level separately. The coupling between **L** and **S**, being stronger than the influence of the field, will still cause the absolute value of the total angular momentum, or J^2, to be constant and j a good quantum number. The states of different J_z, or different m, which are degenerate in the absence of a field pass individually into the states of the perturbed system. Each fine structure level splits into $2j+1$ equidistant energy states with the values $m = j, j-1, \ldots -j$, but the splitting factor is different for the different levels. This *anomalous Zeeman effect* will be treated later in more detail (p. 199). For the optical spectra of the hydrogen-like atoms it is of little importance because the smallness of the splitting does not allow the pattern to be properly resolved.

Also the Stark effect in small fields is difficult to observe in hydrogen-like spectra, for the same reason, but its knowledge is of some interest for an estimate of disturbances of atomic spectra by ionic fields in discharges.

It was shown in A.4 that the first-order perturbation is present only if terms of different l have the same energy. All terms of this kind, such as $3P_{\frac{3}{2}}$, $3D_{\frac{3}{2}}$, are expected to show linear Stark effect, while single terms such as $3D_{\frac{5}{2}}$ can only show quadratic Stark effect. The displacements and intensities of the lines have been calculated by quantum-mechanical methods(46,47,48) and the calculations have also been extended to the levels in which the radiation shift has removed the degeneracy.(49)

Some experiments in intermediate fields have given results which do not agree with the theory,(48) but it is possible that the discrepancies are caused by anomalies in the intensities of the unresolved components. Discrepancies between observed and theoretical intensities of Stark components have often been reported. In a review of the subject, Foster(50) comes to the conclusion that the theoretical intensities are best realised by means of a Lo Surdo source containing rare gas with only a trace of hydrogen. This is in accordance with observations of fine structure in the absence of fields (see section 7) and the view that collisions with foreign gas atoms tend to establish thermal equilibrium among states of closely similar energies.

The Stark effect in the line 4686A of He^+ in medium and weak fields have been calculated by Kullenberg(51) and compared with experiments. Some discrepancies found may well be due to the limited resolution.

7. Intensities and f-values

a. Calculations

The measurement of the strength of lines in hydrogen and the hydrogen-like spectra meets with a peculiar difficulty connected with the l-degeneracy. In most experiments on the atomic spectrum of hydrogen, the pressure is kept low, because collisions with other hydrogen atoms, or atoms of an admixed rare gas favour recombination and weaken the atomic spectrum. But in the absence of collisions during the life time, the distribution of the atoms among the excited states is often far from an equilibrium distribution, and terms of the same n and different l are by no means populated according to their statistical weight.

The theoretical calculations of transition probabilities, on the other hand, can be carried out accurately, since the wave functions are well known, and these results will be described first. The essential step is the calculation of the integrals

$$\int R_{nl} R_{n'l'} r^3 \, dr$$

which was carried out by Gordon[52] and others[53]. A few typical results may be quoted; for a full discussion the reader may be referred to Bethe's article[54].

Table III, 5 gives the oscillator strengths (f-values) of the transitions from the terms listed in the first line, to various other discrete states and to the continuous range of energies beyond the series limit. The values show a rapid decrease with increasing n. The sum of the f-values of all transitions from one term, including the continuous spectrum and the negative values for downward transitions, is found to be exactly 1, in accordance with the f sum rule.

Also the transition probabilities, which depend on the frequency as well as on the f-values, decrease with increasing n within a series. For the Lyman series, the transition probabilities are given by the formula

$$A_{n,1\to 1,0} = 8\times 10^9 \frac{2^8 n(n-1)^{2n-2}}{9(n+1)^{2n+2}} \sec^{-1}. \tag{67}$$

For large n, this value decreases according to

$$A_n \sim n^{-3}, \tag{68}$$

a result which applies also to other series of hydrogen-like spectra.

Table III, 6 gives some calculated transition probabilities and life times of states with specified l. Remarkable is the long life time of all S states; this is due to the fact that all spontaneous transitions involve increase of l and decrease of n, and are therefore weak (see p. 68). P terms are especially short-lived because transitions to the ground state are always extremely strong.

TABLE 5

f-values for hydrogen (*) (**)

Initial state		1S	2S	2P		3S	3P		3D	
Final state		nP	nP	nS	nD	nP	nS	nD	nP	nF
$n =$	1	—	—	−0·139	—	—	−0·026	—	—	—
	2	0·416				−0·041	−0·142	—	−0·417	—
	3	0·079	0·425	0·014	0·694	—	—	—	—	—
	4	0·029	0·102	0·003	0·122	0·484	0·032	0·619	0·011	1·016
	5	0·014	0·042	0·0012	0·044	0·121	0·007	0·139	0·0022	0·156
	6	0·0078	0·022	0·0006	0·020	0·052	0·003	0·056	0·0009	0·053
$\Sigma f(n = 1 \to \infty)$		0·564	0·638	−0·119	0·923	0·707	−0·121	0·904	−0·402	1·302
contin. spectr.		0·436	0·362	0·008	0·188	0·293	0·010	0·207	0·002	0·098
total sum		1·000	1·000	−0·111	1·111	1·000	−0·111	1·111	−0·400	1·400

(*) H. Bethe, *Handb. d. Physik* Bd. **24**, Tl. 1 (Springer 1933)

(**) see also J. M. Harriman, *Phys. Rev.* **101**, 594, 1956.

The mean life times of the levels of given n, defined as the reciprocal of the transition probabilities averaged over all different l, with due regard to their statistical weights, are

$n =$	2	3	4	5	6	
	0·21	1·02	3·35	8·8	19·6	10^{-8} sec.

The life time of states of any fixed value of l also increases with n (table III, 6).

TABLE 6

Transition probability (A) and life-time (τ) in hydrogen

	A (in 10^8 sec^{-1})	τ (in 10^{-8} sec)		A (in 10^8 sec^{-1})	τ (in 10^{-8} sec)
2P → 1S	6·25	0·16	4P → 1S	0·68	1·24
3S → 2P	0·063	16	4P → 2S	0·095	
3P → 1S	1·64	0·54	4P → 3S	0·030	
3P → 2S	0·22		4P → 3D	0·003	
3D → 2P	0·64	1·54	4D → 2P	0·204	3·65
4S → 2P	0·025	23	4D → 3P	0·070	
4S → 3P	0·018		4F → 3D	0·137	7·3

For thermal equilibrium at infinite temperature, the intensities of the Balmer lines with the upper state n are, in arbitrary units:

$n =$	3	4	5	6	7	8	9–∞
	11·8	5·38	2·87	1·50	1·04	0·73	2·4

For the opposite extreme assumption that the number of excitation processes per sec is equal for all states, the calculated intensities of all lines in a series are equal; this rule probably applies to series of all atoms in an approximate form.

b. Measurements

Owing to the difficulty of realising well-defined conditions of population of levels of different l, fully satisfactory, quantitative confirmations of the theoretical intensity relations are still lacking. The strength of absorption in the Balmer lines in electrically excited hydrogen has been measured[55] and the results are compared with theory in table III, 7. The two values in the first two columns refer to the two main components of the fine structure; the difference between them is in agreement with the expectation that the P levels are excited preferentially (see p. 81). Within the rather large experimental uncertainties, the results agree with the theory.

TABLE 7

Relative *f*-values in the Balmer series

	H_α/H_β	H_β/H_γ	H_γ/H_δ	H_δ/H_ϵ
exper.	4·9, 5·5	1·6, 2·46	1·78	1·7
theor.	5·37	2·66	1·94	1·74

The measurement of anomalous dispersion in excited hydrogen[56] led to values of the ratio of the strengths of H_α and H_β between 4·66/1 and 5·91/1, compared with the theoretical values 4·16/1 and 5·68/1 for the S levels alone and the P levels alone.

Comparisons of the intensities of different lines originating from the same level, namely of the line 4 → 2 of the Balmer series and 4 → 3 of the Paschen series have led to inconclusive results.[57]

In investigations of the fine structure of H_α, the theoretical intensity ratios calculated with the assumption of thermal distribution, were the more closely approached the higher the total pressure, either of hydrogen itself or of the admixed rare gas.[29,32] The residual differences indicated over-population of the P states.

Also the intensities of the components of the He^+ line 4686A show some deviations from the theoretical values indicating non-equilibrium populations. Lines originating from the same level appear to have the correct intensity ratio, though no accurate measurements have been made.

The life time of the excited states of He^+ was measured directly[58] and found to be in good agreement with theory.

B. HELIUM AND THE HELIUM-LIKE IONS

Next to the one-electron atoms one would expect the atoms with two electrons, He, Li^+, Be^{++} . ., to show the simplest spectra. In fact, these spectra are in many respects more complex than the alkali spectra which are therefore usually treated first. The more systematic procedure of dealing with two-electron spectra first will, however, be adopted here; it is hoped that this will give a more realistic idea of the approximations involved in the theories of atomic spectra.

1. Some facts and empirical relations

The term analysis of the spectrum of helium leads to two systems of terms which are practically independent since transitions between the two systems are scarcely observable. The two spectra used to be known as Par- and Ortho-helium spectra and were at one time believed to belong to different elements; they are now known as *singlet*- and *triplet*-systems owing to the fact that all terms of one system, except the S-terms, are triple.

Plate 5 shows the spectrum of helium taken with a quartz spectrograph, in the visible and near ultra-violet, and fig. III, 12 the term diagram; the break in the scale of the latter, at the bottom part, is to be noted. Most of the stronger lines in the spectrogram are marked, the singlets below, the triplets above the spectrum. Several series can be seen and can be compared with the term diagram. The singlet spectrum is generally weaker than the triplet spectrum. The position of the line 7065A has been marked, though it is not visible, owing to lack of sensitivity of the plate for this wavelength. The aluminium lines are caused by the electrodes of the discharge tube.

The triplet structure is far too small to be resolved in an ordinary spectrograph; we shall neglect its presence at first and use the expressions *singlet* and *triplet* merely as distinguishing labels for the two non-combining term systems.

A striking feature which is not obvious from fig. III, 12 is the comparatively high energy of the first excited level which is about 4/5 of the ionisation energy, so that all the excited levels occupy only a small fraction of the ionisation energy. The absorption series lies in the far ultra-violet, with the resonance line at 584·4A. The lowest triplet term is highly metastable, owing to its parity which agrees with that of the ground state.

For comparison, the terms of hydrogen are indicated on the right in fig. 16. In the manner described in the introduction, the terms are divided into S. ., P. ., D. . terms according to the way in which they combine. The close relation to the hydrogen terms is obvious from the diagram. For each hydrogen term there are several helium terms: for $n = 3$, e.g. there are three terms 3S, 3P and 3D in the singlet system, written more specifically as $3\,^1S$, $3\,^1P$, $3\,^1D$, and three triplet terms $3\,^3S$, $3\,^3P$, $3\,^3D$.

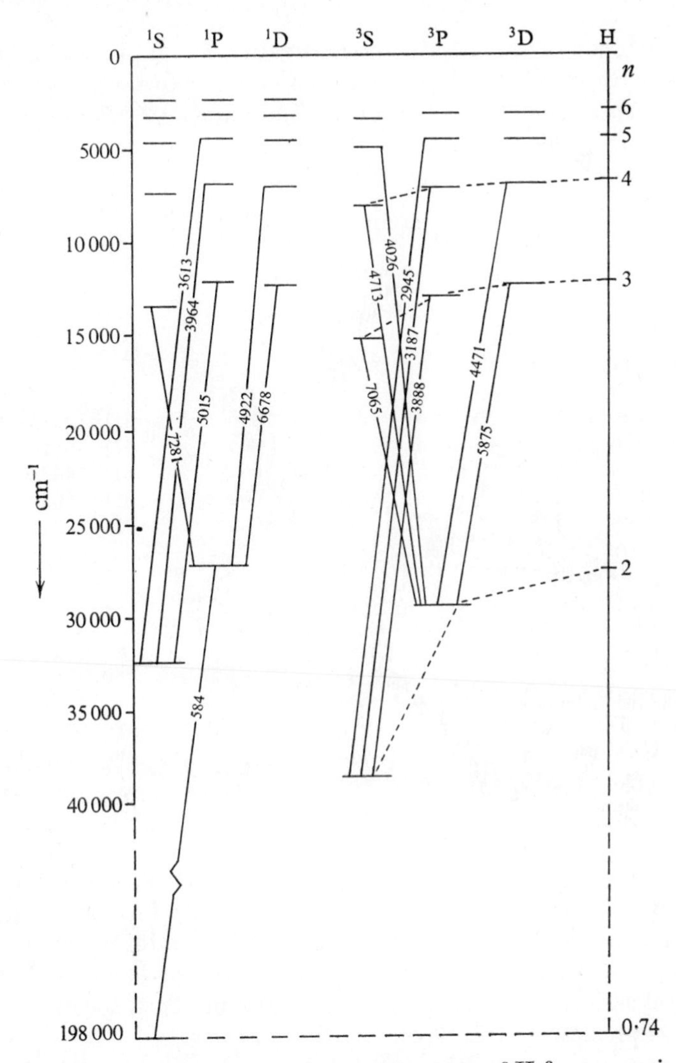

FIG. III, 12. Term diagram of He, with terms of H for comparison and for assignment of n-values.

The relation to the hydrogen terms can be shown more quantitatively by the following method: in hydrogen-like spectra, the term values T_n are connected with the quantum number n and the charge number Z by the relation $n = Z\sqrt{(R/T_n)}$. If we define a number n^* for each series of terms in helium or helium-like ions by

$$n^* = Z_a\sqrt{(R/T_n)}, \tag{69}$$

a comparison of n^* with the integral values n will provide a measure of the hydrogen-likeness of the terms. n^* is known as *effective quantum number*, though it is not a quantum number in the proper sense.

TABLE 8

Terms of helium

a. Singlet terms

^{1}S-terms		n^*	^{1}P-terms		n^*	^{1}D-terms		n^*
1S	198311	0·7438						
2S	32039	1·8505	2P	27182	2·009			
3S	13452	2·8557	3P	12107	3·010	3D	12212	2·997
4S	7376	3·8563	4P	6824	4·009	4D	6870	3·996
5S	4653	4·855	5P	4374	5·007	5D	4398	4·994
6S	3202	5·852	6P	3042	6·004	6D	3056	5·990
7S	2338	6·848	7P	2238	6·999	7D	2247	6·985

b. Triplet terms

^{3}S-terms		n^*	^{3}P-terms		n^*	^{3}D-terms		n^*
2S	38461	1·689	2P	29230	1·937			
3S	15080	2·697	3P	12752	2·933	3D	12215	2·997
4S	8019	3·699	4P	7100	3·930	4D	6872	3·994
5S	4970	4·698	5P	4516	4·928	5D	4400	4·993

Z_a has been written instead of Z to indicate that it is to be chosen as 1 for He, 2 for Li^+ . .. Table III, 8 shows that for the terms n^1D, n^3D, n^1P and n^3P the values n^* are very close to integers and increase in steps of very nearly one. Similar regularities occur in most spectral series. They were first discovered in alkali spectra and are often expressed by the Rydberg–Ritz formula

$$T_n = \frac{Z_a{}^2 R}{n^{*2}} = \frac{Z_a{}^2 R}{(n-\alpha(l)-\beta(l)/n^2)^2}. \tag{70}$$

α and β are usually positive and are constant within one series of terms, and β is usually small, often quite negligible.

Another empirical adaptation of the Balmer formula to non-hydrogen-like spectra can be made by the addition of a constant in the numerator, instead of the denominator:

$$T_n = \frac{(Z-\sigma)^2 R}{n^2} = \frac{Z^{*2} R}{n^2}. \tag{71}$$

It is found useful mainly for the comparison of corresponding terms in isoelectronic spectra where (71) describes the variation of T as function of Z fairly well. If $\sqrt{(T/R)}$ is plotted as function of Z the resulting curve is often close to a straight line of slope $1/n$; the point of intersection with the Z-axis has the abscissa σ. Such plots are known as *Moseley graphs*, from their original use in X-ray spectra.

TABLE 9

Ground terms of helium-like atoms

	$1s^2\ {}^1S_0$ (cm^{-1})	$Z^* = \sqrt{({}^1S_0/R)}$	ΔZ^*
He	198311	1·3445	
			1·0140
Li^+	610080	2·3585	
			1·0055
Be^{++}	1241225	3·3640	
			1·0033
B^{3+}	2091960	4·3673	
			1·0025
C^{4+}	3162450	5·3698	
			1·0019
N^{5+}	4452800	6·3717	
			1·0016
O^{6+}	5963000	7·3733	
			1·0019
F^{7+}	7693400	8·3752	

Table III, 9 shows how well the relation applies to the ground terms of two-electron spectra. The nearly constant differences in the last column show the near linearity of $\sqrt{(T/R)}$, with a slope factor very close to 1. Since the numbering of Z in (71) does not affect the linearity of the relation, it can be chosen arbitrarily, but for later reference it is convenient to choose the true nuclear charge number. If this is done a value $\sigma = 0{\cdot}63$ fits all values of higher Z very well and is a fair approximation for the lighter ions and helium.

Each triplet term is lower than the singlet term of the same n. The lowest S–, P– . . terms of the singlet system are 1S, 2P, 3D . .; the same applies to triplet terms except for the lowest term $1\,^3S$ which is missing.

The structures of the triplet lines of helium are very narrow and liable to be distorted by self absorption. Accurate and reliable experimental results have only recently been obtained by means of the most refined interferometric methods. The structure of a line $2\,^3P$–$n\,^3S$ is shown diagrammatically in fig. III, 13a where the heights of the lines indicate the intensities. For very small current densities where self absorption becomes progressively smaller, the intensity ratios approach the values 1 : 3 : 5. The same appears to apply to the even narrower structures of the lines arising from combinations with $3\,^3P$. Fig. III, 13b(59) is a direct recording showing the fine structure of the line 7065 A, $2\,^3P - 3^3S$, with an intensity ratio 1 : 3 ; 5. The peak P_0–S_1 belongs to a different order of the Fabry–Perot interferometer, so that its position with regard to the other lines cannot be directly inferred from the figure.

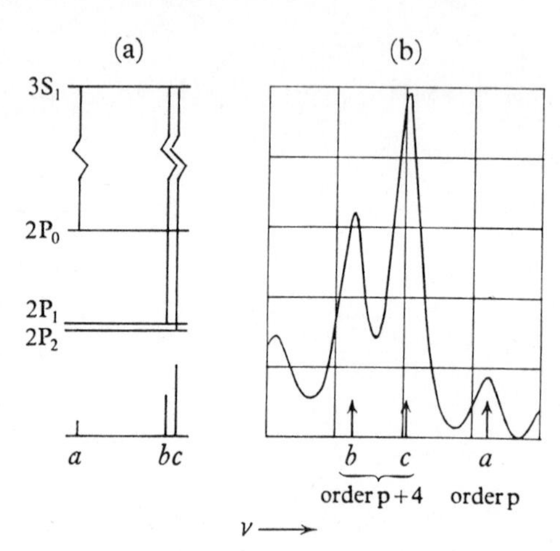

FIG. III, 13. Triplet structure of He line $2\,^3P - 3\,^3S$ (7065 A); (a) term diagram, (b) photometer tracing of Fabry–Perot fringes (J. Brochard, R. Chabbal, H. Chantrel & P. Jacquinot, ref. 59).

The observed term structure in helium suggests a purely formal interpretation. The triplet splitting in helium is of similar order of magnitude as the doublet splitting in hydrogen; as the latter could be explained by magnetic coupling of the orbital motion to an angular spin momentum with the quantum number $s = \frac{1}{2}$, a triplet can be similarly explained by the assumption of an angular momentum whose absolute value is characterised by a quantum number $S = 1$. If we assume that, for some reason, the two electron spins are strongly coupled together while their interaction with the orbital angular momentum is much weaker, we come to a qualitative explanation of the facts by applying the quantum rules of addition of angular momenta. The quantum number S is determined by

$$S = s_1 \pm s_2 = 1 \quad \text{or} \quad 0$$

and is associated with a resultant angular spin momentum $\mathbf{S}$ by

$$\mathbf{S}^2 = \hbar^2 S(S+1).$$

The weaker coupling with the orbital momentum gives

for $S = 1$, $J = L+1, L, L-1$ (three levels)
for $S = 0$, $J = L$ (one level).

The triplet and singlet terms are thus formally explained. This coupling scheme is known as *Russell–Saunders coupling*, or L,S coupling. This concept will be generalised presently by giving a wider meaning to the quantum number L.

The classical motion has to be imagined as a rapid precession of two vectors $\mathbf{S}_1$ and $\mathbf{S}_2$ about the resultant $\mathbf{S}$, which itself precesses more slowly, together with $\mathbf{L}$, about the overall resultant $\mathbf{J}$ (fig. III, 14).

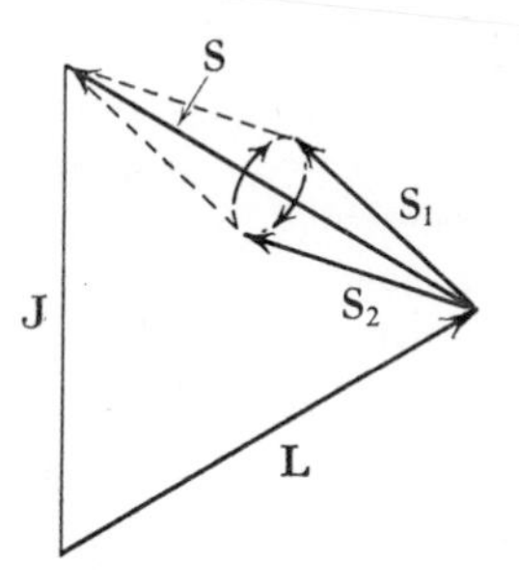

FIG. III, 14. Vector diagram for two electrons in Russell–Saunders-coupling; sp $^3P_1$-level.

The origin of the strong interaction coupling the spins together is the electrostatic repulsion between the electrons, a fact which is far from obvious and can only be understood by means of quantum mechanics (see sections 4 and 5).

For a ^{3}P term, the values of J are 2, 1 and 0, forming levels of statistical weights 5, 3 and 1. These weights are in the same ratio as the observed intensities of the components of the line arising from combinations with a single term. This is in accordance with a more general rule to be discussed below (p. 170). The strong coupling between $\mathbf{S}_1$ and $\mathbf{S}_2$ is demonstrated by the large term difference $2\,^3\text{P} - 2\,^1\text{P} = 2048\ \text{cm}^{-1}$ as compared with the difference $2\,^3\text{P}_2 - 2\,^3\text{P}_0 = 1{\cdot}06\ \text{cm}^{-1}$.

The spectra of helium-like ions are qualitatively similar to that of helium. Though their stronger lines lie in the far ultra-violet the multiplet structures are much wider and have been extensively studied.(60)

2. Screening effects

We consider at first only electrostatic forces.

The classical motions of two charged particles attracted by a fixed, charged centre are highly complicated. Attempts at applying Bohr–Sommerfeld's theory to the helium atom not only led to involved mathematics but even failed to account qualitatively for many of the observed facts. The interpretation of the helium spectrum by means of wave mechanics was one of its first and most striking successes.

If we assume, as a zero-approximation, the two electrons to be entirely independent of one another, the Schrödinger equation (II, 36) separates into two independent equations, one for each electron. The solutions are formed by products of ψ functions each of which depends only on the coordinates of one electron. Each quantum state is then described by the quantum numbers of the individual electrons. The symbol 1s2p thus means that one electron is in the state 1s, the other in the state 2p. Small letters s, p . . are used to indicate a quantum state of a single electron, while capital letters are used in a more general sense to be explained below. Either can be used in one-electron spectra.

When the mutual repulsion of the electrons is included, this description by quantum numbers of single particles is only a rough approximation for the energies of the terms. One can, however, imagine the forces of interaction to be brought into play by a gradual, "adiabatic" process which preserves the identity of each term. A term symbol such as 1s2p will then mean that for gradually vanishing interaction between the electrons the term passes into a state with one 1s electron and one 2p electron.

In this sense all terms of the ordinary helium spectrum (fig. III, 12) arise from the excitation of only one electron and are described by the symbol 1snx where x stands for s, p, d, . . and $n = 1, 2, \ldots$ The first symbol is often omitted as obvious and helium terms are described as, e.g. 3p or 3P instead of 1s 3p. The description by quantum numbers of the individual electrons only, such as 1s 3p, is known as *configuration*.

We can see qualitatively why the helium terms with $l \geqq 1$ are very nearly hydrogen-like. The graphs of the hydrogen wave functions show that the 1s electron forms a spherically symmetrical cloud of negative charge closely surrounding the nucleus. For a nuclear charge $2e$ the field outside this cloud—strictly speakingly outside a sphere enclosing most of the cloud—will be almost identical with a field due to a single charge e at the origin. If we assume that for the second electron, for which $l \geqq 1$, the wave function is hydrogen-like, with $Z = 1$, we find that its charge is distributed almost wholly outside the charge cloud of the first electron (fig. III, 3). This argument, however crude, suggests that our assumption on the wave function of the second electron was justified: the assumed wave functions of the two electrons produce charge distributions and thus potential fields which make the assumed wave functions possible. This is a highly simplified example of a *self-consistent field* (see p. 229).

In this extreme case, the nuclear charge acting on the second, external electron is $(Z-\sigma)e \approx (2-1)e$. The constant σ (see p. 133) which is here almost unity, is called the *screening constant*.

The hydrogen functions ns are only partly outside the charge cloud of 1s, leading to a screening constant of the 1sns terms between 0 and 1. The fact that σ is found to be almost independent of Z (table III, 9) can be made plausible from the similarity of the wave functions for different Z.

3. The Calculation of the ground state

For a more accurate calculation of the term values and for even a qualitative interpretation of the term structure of helium, the crude concept of screening has to be replaced by more detailed methods. With the assumption of only electrostatic interactions, the wave equation for helium-like atoms can be written

$$\left(-\frac{\hbar^2}{2m_0}\nabla_1{}^2 - \frac{\hbar^2}{2m_0}\nabla_2{}^2 - \frac{Ze^2}{r_1} - \frac{Ze^2}{r_2} + \frac{e^2}{r_{12}}\right)\psi = E\psi \qquad (72)$$

where r_{12} is the distance between the electrons. This equation cannot be solved rigorously, but numerous approximation methods have been applied to it. In the simplest way, the ground term can be calculated by means of the perturbation methods outlined in II, where the term e^2/r_{12} is regarded as perturbing potential. The unperturbed state is the state 1s1s without interaction; this is non-degenerate and its eigenfunction is the product of hydrogen-like wave functions:

$$\psi = \frac{Z^3}{\pi a_0^3} \exp(-\rho_1/2) \exp(-\rho_2/2). \tag{73}$$

The first-order perturbation energy, for non-degenerate states, is always equal to the perturbing potential applied to the unperturbed wave function. In the present case this is simply the electrostatic energy of repulsion between two equal, spherically symmetrical charge clouds. The charge density of each of these is equal to euu^* where u is the wave function of the ground state of hydrogen, but with $Z = 2$. A straightforward integration gives $E' = -\frac{5}{4}ZE_{\mathrm{H}}$ where E_{H} is the energy of the ground state of hydrogen. The total energy of helium, which is the energy required to bring both electrons from infinity, is $E = 2Z^2E_{\mathrm{H}}-\frac{5}{4}ZE_{\mathrm{H}}$, and the energy of single ionisation

$$E_i = Z^2E_{\mathrm{H}}-\tfrac{5}{4}ZE_{\mathrm{H}}. \tag{74}$$

Expressed in electron-volts, one finds $E_{\mathrm{H}} = -13{\cdot}53$ eV, and the total energy of helium ($Z = 2$) $E_{\mathrm{He}} = -74{\cdot}42$ eV, as compared with the experimental value of 78·62 eV. For higher values of Z the perturbing potential is smaller in relation to the central potential, so that the first-order approximation can be expected to give more accurate results. This is, in fact, found: for C^{4+}, the theoretical value of 872·7 eV agrees very closely with the experimental value of 876·2 eV.

The interaction between the two electrons in helium is not so small compared with the attraction by the nucleus that the first-order perturbation treatment could be expected to give very accurate results; the perturbation terms of second order are difficult to calculate.

The variation method (see II.C.3b) is much better suited for accurate calculations of the ground term of a two-electron atom. In the simplest form, the function (73) can be used with the constant Z replaced by the variable parameter $Z' = Z-\sigma$. The minimum condition for the variation integral can easily be solved and leads to the value $\sigma = 5/16$. The perturbation integral, which is an upper limit for the energy, is then found to exceed the observed value by only 2 per cent.

By introducing other than central force parameters into the variation method, Hylleraas[61] achieved a great advance in the accurate calculation of the ground term of helium. The precision was further increased considerably by other authors using electronic computers. The most recent value derived by means of a formula with 203 parameters[62] gives the value, for ^{4}He:

$$1\,^1S = 198312{\cdot}01 \text{ cm}^{-1} \text{ (excluding radiation field effects)}$$

with an accuracy of probably about $0{\cdot}1$ cm^{-1}.

In this work, the method of *perimetric coordinates*[63] was employed. Other workers, using different forms of the variation method, have found very similar results.[64]

It was difficult to match the accuracy of this theoretical work by experiment. The accurate measurement of the limits of the visible and near ultra-violet series originating from the terms with $n = 2$ was not difficult, but all combinations with the ground term $1\,^1S$ fall into the far ultra-violet and their measurement is technically difficult and suffers from lack of standards. A thorough study of this problem was made by Herzberg[65] who succeeded in measuring the wavelengths of the lines $1\,^1S-2\,^1P_1$, $1\,^1S-3\,^1P_1$ and the weak intercombination line $1\,^1S-2\,^3P_1$ with an accuracy of $0{\cdot}0005$ A. The required wavelength standards were built up from various lines of longer wavelengths by the use of the combination principle. The resulting value for ^{4}He was

$$1\,^1S = 198310{\cdot}82 \pm 0{\cdot}15 \text{ cm}^{-1}.$$

The difference between this and the theoretical value, of $1{\cdot}2$ cm^{-1}, agrees well with the calculated value $1{\cdot}3$ cm^{-1} of the Lamb shift which was not included in the theoretical value quoted. Similarly excellent agreement was found for ^{3}He.

Table III, 10[62] summarises the results for other hydrogen-like ions. Within the much larger limits of error—they are greater than the Lamb shift—the agreement is satisfactory.

Hylleraas has also derived an approximation formula for the energies of the ground states of all helium-like ions. The most recent form[66] which gives reliable values for all values of Z including $Z = 1(\text{H}^-)$, is

$$T = R_M(Z^2-5/4Z+0{\cdot}315311-0{\cdot}01707/Z+0{\cdot}00068/Z^2-0{\cdot}00489/Z^4) \tag{75}$$

where T is the term value of the ground state, i.e. the first ionisation potential in cm^{-1} and R_M the Rydberg constant for the appropriate nuclear mass. Some relativistic corrections have to be applied and also the Lamb shift has to be considered for accurate comparison with experimental values.

TABLE 10

Ground levels of He-like ions (cm^{-1})

Z		Lamb shift	$1\,^1S$ theor.(*)	$1\,^1S$ exper.
1	H-	0·0037	6083·08	—
2	He	1·339	198310·67	198310·82 ± 0·15
3	Li II	7·83	610079·61	610079 ± 25
4	Be III	27·1	1241259·4	1241225 ± 100
5	B IV	65·7	2092003·3	2091960 ± 200
6	C V	132	3162441	3162450 ± 300
7	N VI	235	4452758	4452800 ± 500
8	O VII	383	5963135	5963000 ± 600
9	F VIII	584	7693810	7693400 ± 800

(*) including Lamb shift.

Approximate wave functions for the ground state of helium have been calculated[67,68] by the method of expanding Hylleraas wave functions in terms of central field functions. This procedure includes configuration intraction (see p. 258).

4. The theory of the excited states

In applying perturbation methods to the excited states we assume that one electron, regarded singly, is in a definite excited state which we specify at first as the non-degenerate state 2s. We thus consider the *configuration* 1s2s. Again, we start by disregarding all magnetic interactions.

One would be inclined to use as zero-order wave function a simple product $\psi_{1s}(1)\psi_{2s}(2)$. But this would violate the postulate that electrons are indistinguishable from one another. As explained in II.C the two degenerate functions $\psi_{1s}(1)\psi_{2s}(2)$ and $\psi_{1s}(2)\psi_{2s}(1)$ have to be replaced by the linear combinations ψ_s and ψ_a. The electrostatic interactions between the electrons will then remove the degeneracy and result in two states of different energies, one described by a symmetrical, the other by an antisymmetrical wave function. To include the electron spin, though still neglecting magnetic interactions, we must multiply each Schrödinger function by a spin function of appropriate symmetry so as to make the product always antisymmetrical, in agreement with the Pauli principle. This gives only one possible product for ψ_s, namely by multiplication by χ_a (see p. 36) and three possible products for ψ_a by multiplication by χ_s, χ_s', χ_s''. This results in one single and one triply degenerate term.

In the functions χ_a and χ_s the spin function is not defined for any one of the electrons, but the sum $S_z = S_z(1)+S_z(2)$ has definitely the value 0 in either of these functions. This can be seen from the fact that the signs of the spin functions are opposite to each other in any one term of the sum.

The values $S_z = 0, 0, +\hbar, -\hbar$ for the four χ functions (with the arbitrary choice of the z-axis for the diagonal representation) lead to ascribing χ_a to a resultant spin quantum number $S = 0$ and χ_s, χ_s', χ_s'' to the resultant spin $S = 1$, in accordance with the vector diagram.

The energy splitting between the singlet and the triplet term is entirely due to electrostatic forces and was first explained by Heisenberg[(69)]. It is known as *resonance* splitting, owing to its close analogy to effects in classical mechanics. If two exactly similar, linear harmonic oscillators have no mutual interactions they represent a system of two degrees of freedom whose two frequencies ν_0 coincide (degeneracy). If a small interaction is introduced so that the equation of motion for the coordinate x_1 of one oscillator depends on that of the other, x_2, the amplitude of each oscillator is no longer constant but varies periodically with frequency $\delta(\ll \nu_0)$; the energy passes to and fro between the two oscillators. If *normal coordinates* are introduced by the substitution $\xi = x_1+x_2$, $\xi' = x_1-x_2$, it is found that ξ and ξ' carry out simple harmonic motions independently, with constant energies and frequencies $\nu_0+\delta$ and $\nu_0-\delta$ where δ is the larger the stronger the coupling force. The interaction has thus removed the degeneracy.

In quantum theory the existence of two electrons in states of different energy, 1s and 2s, is equivalent to two classical oscillators, and the splitting of the excited states of helium into singlet and triplet states can be interpreted by a fluctuation of the excitation energy between the two electron oscillators,with frequency $(E_{\text{singl.}}-E_{\text{tripl.}})/h$.

For the configuration 1s 2p we have to take account of the fact that 2p is itself degenerate, consisting of the three states, $2p_1$, $2p_0$

and $2p_{-1}$, where the suffix indicates the magnetic quantum number m. There are thus 8 degenerate states in the zero approximation, namely the four products $1s(1)2s(2)$, $1s(1)2p_1(2)$, $1s(1)2p_0(2)$, $1s(1)2p_{-1}(2)$ and the four products arising from the interchange of the numbers in brackets. If the electrostatic repulsion is now introduced as perturbation, the perturbation matrix written with the use of these product functions has to be diagonalised. While the diagonalisation of a matrix with 8 rows and columns is, in general, equivalent to the solution of a determinant equation of the 8th order, it can be shown from parity considerations that the matrix, in the present problem, breaks up into four sub-matrices of two rows and columns; the perturbation considered does not "mix" any two products unless they contain the same two single-electron functions. The configuration 1s2s thus leads to only two kinds of matrix elements:

$$\begin{aligned} J_s &= \int \psi_{1s}^*(1)\psi_{1s}^*(2)\frac{e^2}{r_{12}}\psi_{1s}(1)\psi_{2s}(2)\,d\tau_1 d\tau_2 \\ K_s &= \int \psi_{1s}^*(1)\psi_{2s}^*(2)\frac{e^2}{r_{12}}\psi_{1s}(2)\psi_{2s}(1)\,d\tau_1 d\tau_2. \end{aligned} \tag{76}$$

If the function $\psi_{2s}(1)$ in the second integral is replaced by one of the three functions $\psi_{2p_m}(1)$, the integral can be shown to vanish for reasons of parity: simultaneous reflection of the coordinates of both particles at the origin leaves the distance r_{12} unchanged, so that e^2/r_{12} is even. In the integration one can imagine the elements $d\tau_1$ and $d\tau_2$ to be taken in pairs (x, y, z), $(-x, -y, -z)$. The integral then vanishes if one of the product functions involved has the opposite parity of the other; perturbations by an even perturbing potential do not mix terms of opposite parities. Similar arguments show that products containing functions ψ_{2p_m} with different m do not mix either, in spite of their equal parity.

Apart from the two matrix elements J_s and K_s there are thus three equal integrals J_p and three equal integrals K_p. Of the 8 roots of the four quadratic determinant equations, two are J_s+K_s and J_s-K_s; of the remaining six, any three, with $m = 0, \pm 1$ coincide, forming the triple roots J_p+K_p and J_p-K_p. The wave functions belonging to the diagonal form of the perturbation matrix, forming the unperturbed eigenfunctions into which the perturbed functions pass in the limit of vanishing perturbation, are listed in table III, 11, together with their symmetry properties. For brevity the symbols ψ_{1s}, etc., have been replaced by 1s etc.

The capital symbols S and P indicate that the quantum number $L = l$ is 0 and 1 respectively. Since one electron is always an s-electron, this symbol is identical with the small-letter symbol of the other electron. The superscripts ^{3} and ^{1} express the fact that the wave-function of the coordinates is odd (even) and is to be multiplied by the three even (one odd) spin functions leading to a further three-fold (no further) degeneracy. The ^{3}P term is then 9-fold degenerate while ^{1}P remains triply degenerate.

TABLE 11

States for two electrons, $n_1 = 1, n_2 = 2$

coordinate function	symmetry	spin-function	E'	term symbol
1s(1)2s(2) + 1s(2)2s(1)	s	χ_a	J_s+K_s	$^1S_0$
1s(1)2s(2) − 1s(2)2s(1)	a	$\chi_s, \chi_s', \chi_s''$	J_s-K_s	$^3S_1$
$1s(1)2p_1(2) + 1s(2)2p_1(1)$	s	χ_a	J_p+K_p	$^1P_1$
$1s(1)2p_0(2) + 1s(2)2p_0(1)$	s	χ_a	J_p+K_p	
$1s(1)2p_{-1}(2) + 1s(2)2p_{-1}(1)$	s	χ_a	J_p+K_p	
$1s(1)2p_1(2) - 1s(2)2p_1(1)$	a	$\chi_s, \chi_s', \chi_s''$	J_p-K_p	$^3P_{2,1,0}$
$1s(1)2p_0(2) - 1s(2)2p_0(1)$	a	$\chi_s, \chi_s', \chi_s''$	J_p-K_p	
$1s(1)2p_{-1}(2) - 1s(2)2p_{-1}(1)$	a	$\chi_s, \chi_s', \chi_s''$	J_p-K_p	

These results can be summarised thus: if we regard as zero-order approximations the uncoupled electrons moving in the Coulomb field of the nuclear charge, the energy depends only on the two values of n of the electrons, and the state is highly degenerate. For a perturbation due to the electrostatic repulsion, those linear combinations of the product functions which have definite symmetry properties form the appropriate zero-order wave functions. The electrostatic interaction removes firstly the degeneracy due to the difference of the l-values, secondly the exchange degeneracy by splitting each term into two having wave functions of opposite symmetry in the coordinates. Two kinds of integrals occur in the expressions of the perturbation energy; the J integrals describing the electrostatic interaction between the two electrons each of which is in a definite quantum state; they are known as *Coulomb integrals*, and the K integrals involving two quantum states of the unperturbed system differing by exchange of the electrons. The latter are known as *exchange integrals* or resonance integrals. Their numerical value is positive and they appear with positive and negative signs in the perturbation energy of singlet and triplet terms respectively. The exchange effect thus causes the *triplet* terms to be *lower* than the corresponding singlet terms, in agreement with the observed facts.

The configuration 1s1s, also written as $1s^2$, has only a single, symmetric product ψ-function. The absence of exchange degeneracy leads to one state only, described by the product of this ψ-function with χ_a; there is no triplet term for the configuration $1s^2$. This is a direct result of the Pauli exclusion principle: the two electrons agree in all quantum numbers of the space coordinates and must therefore have different spin quantum numbers $m_{s_1} = +\frac{1}{2}$, $m_{s_2} = -\frac{1}{2}$, or $S = 0$. The absence of the lowest ^{3}S term in helium and similar spectra was one of the facts leading Pauli to the formulation of his principle. Electrons of equal n and l are called *equivalent*.

A study of table III, 8 shows that the resonance effect decreases with increasing l, for any given value of n. The greater width of the singlet–triplet spacing of P terms compared with D terms leads to an apparent anomaly in the order of the singlet terms: the terms 3 ^{1}P, 4 ^{1}P, . . are higher than 3 ^{1}D, 4 ^{1}D . . respectively. If one compares the centres of gravity of the pairs n ^{1}P, n ^{3}P and n ^{1}D, n ^{3}D, counting each triplet term with three times the statistical weight, the terms are in the expected order P, D, F . ..

5. The triplet structure

All magnetic interactions had been neglected in the preceeding section. In fact we must expect three kinds of magnetic interactions: that between one spin and its own orbit, (s_1, l_1), that between the other spin and the orbit of the first electron (s_2, l_1) and that between the two spins. We have to investigate how far the degeneracy giving rise to three terms ^{1}P and nine terms ^{3}P is removed by these interactions. The structure of the spin function χ_a (see p. 36) to be multiplied by each of the three space functions ^{1}P expresses the fact that the spin of one electron has always a direction opposite to that of the other electron. As a result of this, the interactions of the spins with the orbital motion are found to cancel out in the perturbation terms. The spin–spin interaction is the same for the three states, leaving the term triply degenerate and merely causing a small shift; only external magnetic fields can split the term. This result agrees with the vector diagram attributing to the term ^{1}P a quantum number $J = L = 1$ which, in an external field, forms the three magnetic levels with $m = 1, 0, -1$.

In order to discuss the nine states ^{3}P of the last three lines of table III, 11 we tabulate the values of m_l, the magnetic quantum number of the orbital angular momentum, and of $m_s = m_{s_1} + m_{s_2}$ (table III, 12). The first column contains the suffix of the coordinate wave function in table III, 11, the second column the sum of the suffices of the spin functions (p. 36). Since the table contains all possible wave functions it must account for all possible states.

TABLE 12

m_l	m_s	$m_l+m_s=m$	j
1	1	2	
1	0	1	
1	−1	0	2
0	1	1	
0	0	0	1
0	−1	−1	
−1	1	0	0
−1	0	−1	
−1	−1	−2	

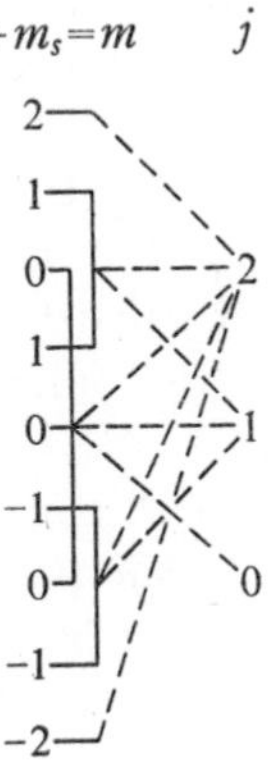

We can imagine an external magnetic field to exist and the interactions between the magnetic moments of orbit and spins to be negligible (strong field case). Classically each of the vectors $\mathbf{S}_1$, $\mathbf{S}_2$, $\mathbf{L}_1$ precesses about the direction of the field axis, which may be chosen as z-axis, and the z-components of the three momenta are constant. In the language of quantum mechanics, the operators $S_z = S_{1z}+S_{2z}$ and L_{1z} of the three uncoupled angular momenta commute and the representation in which they are diagonal has the eigenvalues $m_s\hbar$ and $m_l\hbar$ respectively.

Alternatively we can assume that the external field is infinitesimally weak but that the spin- and orbit-vectors are coupled with one another. Classically the absolute value of the total angular momentum and its z-component will be constants of the motion. In quantum mechanics a representation exists in which the operator $\mathbf{J}^2 = (\mathbf{L}+\mathbf{S}_1+\mathbf{S}_2)^2$ is diagonal simultaneously with the operator $J_z = L_z+S_{1z}+S_{2z}$. The rows and columns of this representation are numbered by pairs of the eigenvalues of these two quantities, or of

j and m. The procedure is analogous to that described in II.C.2. The energy will, in the absence of an external field, depend only on j.

Both representations must contain the same number of states and each state must have the same value of m in both representations since $L_z + S_z$ is diagonal in both. The rule that each value of j occurs with all the values of m from $+j$ to $-j$, with unit intervals, leads to the correlation with the three values $j = 2, 1, 0$ indicated in table III, 12. The coupling between the vectors $\mathbf{S}_1$, $\mathbf{S}_2$, $\mathbf{L}_1$ thus leads to three energy levels with $j = 2, 1, 0$, justifying the expression triplet term. A weak external magnetic field will split the first two of these terms into 5 and 3 levels respectively.

Table III, 12 is another example of a transformation from the representation $|m_l m_s\rangle$ into the representation $|jm\rangle$ as described in II.C.2. Again the state of the largest value $m_l + m_s (= 2)$ passes into one single state $|jm\rangle$, and the same applies to the state of smallest $m_l + m_s (= -2)$. For all other values of $m_l + m_s$, several states in one representation are connected with several in the other scheme by linear relations.

For the calculation of the energy differences between the triplet levels $^3P_2$, $^3P_1$ and $^3P_0$ the perturbation term in the Hamiltonian has to include the magnetic interactions (S_1, L_1), (S_2, L_1), (S_1, S_2) and relativistic corrections similar to those which contribute to the fine structure in hydrogen-like spectra. In helium and helium-like ions, all these effects are of similar order of magnitude, and this fact makes the calculation complicated. By means of some simplifying assumptions, Heisenberg[69] derived the following formula for the energies of the triplet terms $2\,^3P_{2,1,0}$:

$$\Delta E(^3P_{2,1,0}) = a[(Z-3)(1, -1, -2) + 1/4(1, -5,, 10)], \qquad (77)$$

where the constant a is proportional to $(Z-1)^3$; it is essentially the average value $\overline{1/r^3}$ calculated with the use of the effective charge number $Z-1$ for the p electron. Comparison of the observed structures of the term $2\,^3P$ in the isoelectronic spectra from He ($Z = 2$) to CV ($Z = 6$) has shown that the general structure and its dependence on Z agrees with Heisenberg's formula though the agreement in the absolute value of a has been shown to be partly fortuitous.[70]

The first term in (77) is due to the spin–orbit interactions (S_1, L_1) and (S_2, L_1). For very large values of Z this first term predominates and becomes in the limit proportional to Z^4. This suggests its relation with the term E' due to the magnetic spin–orbit interaction in hydrogen (p. 118). It is due to the perturbation term const. $\times \mathbf{L} \cdot \mathbf{S}$, where **S** is equal to the vector sum $\mathbf{S}_1+\mathbf{S}_2$ and has, by arguments similar to those given on p. 117, the values

$$\text{const.}\times[j(j+1)-l(l+1)-S(S+1)].$$

S now stands for the quantum number $S = 1$.

The first term alone of Heisenberg's formula (77) would give the ratio 2 : 1 for the two intervals (see III.D.3). But in helium the presence of the second term, due to spin–spin interaction, completely alters the positions of the terms, not only reversing their order but producing an entirely different splitting ratio as shown in fig. III, 13a. In the spectrum of CV, the larger Z increases the relative influence of the first term in (77), as the result of which the term $^3P_2$ is the highest, but $^3P_1$ and $^3P_0$ are still reversed.

TABLE 13

Fine structure in He (in cm^{-1})

	$\Delta\tilde{\nu}_{opt.}$ (59)	$\Delta\tilde{\nu}_{r.f.}$ (73)	theory (72)
$2P_0-2P_1$	−0·988		−0·997
$2P_1-2P_2$	−0·076	−0·076441	−0·076
$3P_0-3P_1$	−0·265	−0·27065	−0·218
$3P_1-3P_2$	−0·025	−0·02196	−0·0176
$3D_1-3D_2$	−0·0445		−0·0426
$3D_2-3D_3$	−0·004		−0·0025

The triplet structures in helium have been calculated somewhat more rigorously by Breit(70), Inglis(71) and Araki(72), with the use of Dirac wave functions. In table III, 13 the results of Araki's work are compared with the most reliable optical values(59) and recent radio-frequency measurements.(73) It is likely that the remaining discrepancies are mostly due to the inaccuracy of the theoretical calculations.

6. Intercombination lines and metastable levels

In perfect Russell–Saunders coupling the influence of the magnetic interactions on the space part of the eigenfunctions is negligible. The operator of the electric dipole moment (and also that of the electric quadrupole moment) therefore commutes with **S** and does not connect states of different S. This leads to the selection rule for both dipole and quadrupole radiation:

$$\Delta S = 0.$$

The rule expresses the almost trivial fact that the electric vector of the light wave can only act on the orbital motion of the electrons and, if the coupling of the latter with the electron spins is neglected, cannot change the orientation of the two spins relative to one another. The smallness of the multiplet structure in helium and the helium-like ions, compared with the singlet–triplet difference, shows that the assumption of Russell–Saunders coupling is a very good approximation, and the above selection rule can be expected to hold very well.

From the wave functions derived by the various approximation methods, the strength of allowed transitions can be calculated. Bates and Damgaard[(74)] give a critical survey of results.

If the effect of the magnetic interaction on the wave function is taken into account as a second-order perturbation, it has the effect of "admixing" some of the singlet function into the triplet function. An estimate of the resulting transition probability of the line $2s^2\ {}^1S_0 - 2s2p\ {}^3P_1$ of CV[(75)] gives a strength of about 10^{-5} of that of the corresponding singlet–singlet transition.

In fact, intercombination lines in the spectrum of helium are extremely weak. In the helium-like ions from CV upwards, however, the line $2s^2\ {}^1S_0 - 2s2p\ {}^3P_1$ is easily observable. Since the magnetic interaction increases more rapidly with Z than the term value itself, the deviation from Russell-Saunders coupling, though very small, increases with increasing Z and causes an increase in the transition probability of the intercombination. Furthermore, the ratio of the intensities of intercombination to allowed lines is by no means equal to the ratio of the transition probabilities[(60)]; the very high frequency makes the life time, even of intercombination lines, comparable with the time between collisions. One then begins to approach the conditions found in highly rarified gases where the intensities are independent of the transition probabilities (at least for single transitions), as explained above (p. 82).

The terms $2\ {}^1S$ and $2\ {}^3S$ of helium cannot lose their excitation energy by any radiactive process, under conditions of a laboratory experiment, and are therefore *metastable*. It was, in fact, for these terms that the concept of metastability was first developed. Even in weak glow discharges in helium, the infra-red lines 20582A ($2\ {}^1S - 2\ {}^1P$) and 10830A($2\ {}^3S - 2\ {}^3P$) were found to be strongly absorbed.

While the atoms in the state $2\,^1P$ lose their energy preferentially by emission of the line $1\,^1S - 2\,^1P$, on account of its great strength and very high frequency, the light absorbed in the triplet line $2\,^3S - 2\,^3P$ was found to be almost completely re-emitted at the same frequency. This observation first demonstrated the non-existence of the term $1\,^3S$.

In discharge tubes, the life of any of the metastable states of helium will be terminated by collisions with the wall or, at higher densities, with other atoms. The same applies to helium-like ions.

In the vacuum ultra-violet the lines of the principal series of helium $1\,^1S_0 - n\,^1P_1$, can be observed as strong absorption lines. At the series limit, 504A, a continuous absorption starts abruptly and decreases regularly with decreasing wavelength. Recent measurements(76) of the absorption coefficient are in good agreement with calculated values.

7. The negative hydrogen ion

In the series of helium-like ions, the negative hydrogen ion H^- has to be included. If an electron is brought near a hydrogen atom, an attractive force must exist for two reasons: (i) the charge of the electron polarises the atom, inducing an electric dipole which attracts the electron; (ii) owing to the penetration of the second electron into the charge cloud of the bound electron the nuclear charge is not completely screened by the latter. Both effects lead to a force whose potential decreases with a high power of the distance from the nucleus. Under the influence of such a force the electron can be shown to have only a finite number of discrete quantum states. Detailed calculations suggest that H^- has only one quantum state, so that the spectrum of this ion is expected to consist only of a continuum. The ionisation energy of H^-, which is also called the electron affinity of H, has been calculated by Hylleraas and others(77) and was found to be 0·750 eV.

The continuous emission spectrum caused by transitions

$$\text{H} + \text{electron} + \text{kin. energy} \rightarrow \text{H}^-$$

is responsible for a considerable part of the light from the sun.(78, 79) It has also been observed in laboratory sources.(80)

The continuous absorption of light by H^- ions has recently been observed and measured.(81) The results agree well with the calculated transition probability.(82, 83)

C. ALKALI-LIKE SPECTRA

1. General features

In proceeding to the third element in the periodic table, we can follow Bohr's "building-up principle" imagining the charge of the helium nucleus to be increased by one unit and subsequently a third electron to be added. If this electron could assume the same quantum numbers $n = 1$, $l = 0$ as the other two, the ionisation energy of the third element would be greater than that of helium, continuing the trend from hydrogen to helium, since the screening effect of the added electron could only partly compensate the increase in nuclear charge. In fact, the very much lower ionisation energy of lithium (app. table 3) shows that the third electron is in a higher quantum state. This is explained by Pauli's principle according to which the term $n = 1$ has only two states $m_s = +\frac{1}{2}$ and $m_s = -\frac{1}{2}$ so that two electrons "fill" the shell $n = 1$. The lowest quantum state available to the third electron is 2s.

The two electrons with $n = 1$ form a helium-like state $1s^2\,{}^1S_0$. It has zero angular momentum and a charge distribution of perfectly spherical symmetry and is described as the *core* of the lithium atom. It contributes nothing to the angular momentum of the atom which can be regarded as consisting of a single electron moving in a central force field.

The complete term symbol of the ground state is $1s^22s\,{}^2S_{\frac{1}{2}}$, and of the excited states $1s^23s\,{}^2S_{\frac{1}{2}}$, $1s^22p\,{}^2P_{\frac{1}{2},\frac{3}{2}}$, $1s^23d\,{}^2D_{\frac{3}{2},\frac{5}{2}}\ldots$ In shortened symbols in which the core is omitted as self-evident the first excited state is written as $2p\,{}^2P_{\frac{1}{2},\frac{3}{2}}$ or $2\,{}^2P_{\frac{1}{2},\frac{3}{2}}$. The suffix denotes the quantum number j.

For large values of l the wave function of the valence electron will be almost entirely restricted to the space outside the charge distribution of the two core electrons. By the same arguments as those used in the discussion of helium, these terms of the lithium atom will be very closely hydrogen-like.

For lower values of l, the energy of the term becomes progressively lower compared with the hydrogen term. Smaller value of l implies closer approach of the outer electron to the nucleus, either in Bohr–Sommerfeld's theory or in wave mechanics. This results in two effects: (i) the greater penetration into the core causes the screening to become less, the electron moves part of the time in a field of a greater charge, (ii) the proximity of the outer electron distorts the core in a way which can be imagined as an electrostatic polarisation. Both effects increase the force of attraction and lower the energy; they can be regarded as first- and second-order perturbation effects.

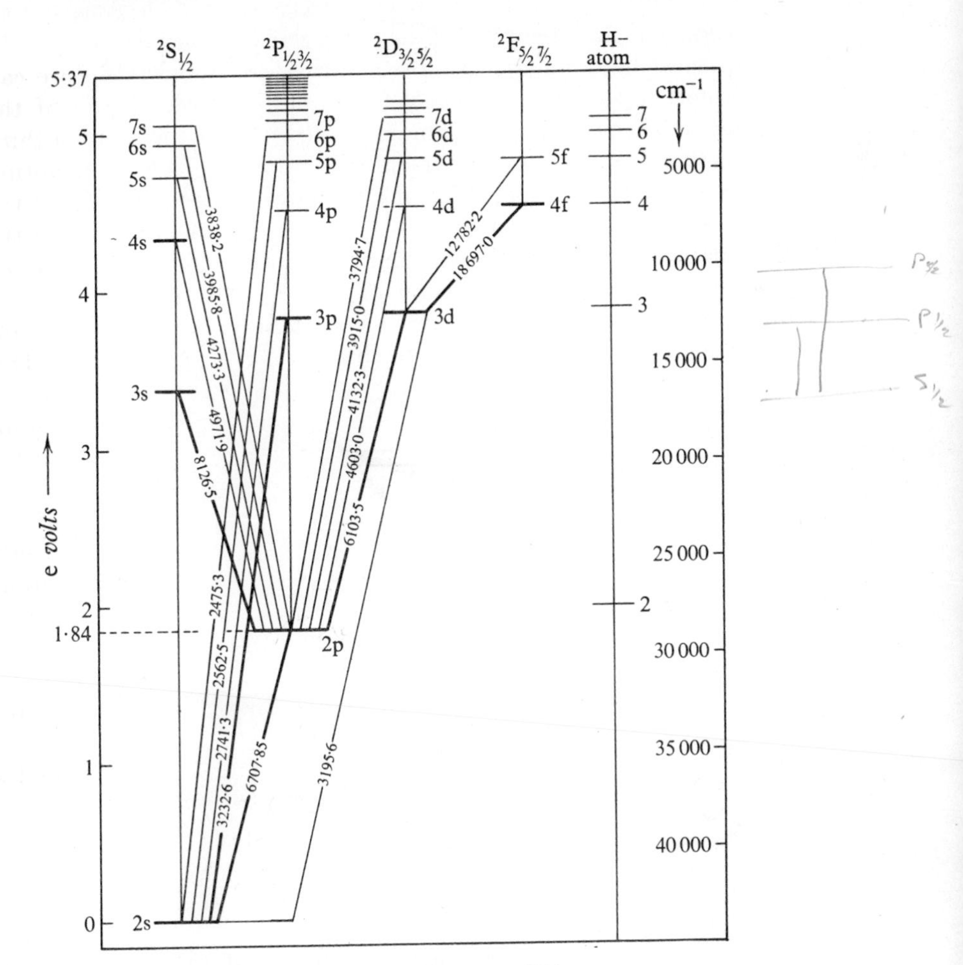

FIG. III, 15. Term diagram of Li.

By similar arguments, we can understand why the other elements in the first column of the periodic table, Na, K, Rb, Cs, have similar spectra which can be ascribed to the motion of a single electron. In every case the atom contains one electron outside closed shells ns^2np^6, any of which forms a system of complete spherical symmetry without angular momentum (see p. 96). In contrast to the hydrogen spectrum, the deviation of the potential from the Coulomb potential causes the term value to depend on l, to an extent far exceeding relativity effects. Also the doublet splitting due to spin–orbit interaction is much larger in the alkali atoms than in hydrogen (see under 3). The fine structure of the hydrogen atom has become partly a difference between S–, P–, D– . . terms, partly a *multiplet structure*, more specifically a *doublet* structure. Both these effects, the dependence of the term value on l and on j, are the more marked the smaller l and the greater the atomic number of the element.

The correlation of the terms of lithium with those of hydrogen is shown in fig. III, 15; the ground term and the lowest P-term have the quantum number $n = 2$. Similar comparisons show that the ground term and lowest P-term of Na, K, Rb and Cs have the quantum numbers $n = 3, 4, 5$ and 6 respectively. The shell structure of these atoms will be discussed in chapter IV.

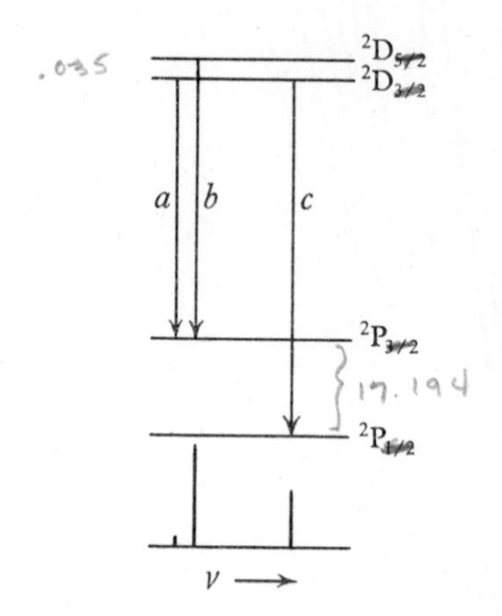

FIG. III, 16. Composite doublet $^2P - {}^2D$.

Owing to the low ionisation potentials of the alkali atoms their spectra do not extend far into the ultra-violet. Since the vapour of alkali metals is mainly monatomic the absorption spectra of the atoms can be produced easily. At moderate temperature practically all atoms are in the ground state and cause only a single series of doublets, $n_0\,{}^2S_{\frac{1}{2}} - n\,{}^2P_{\frac{1}{2},\frac{3}{2}}$, to appear. Plate 4 shows the absorption spectrum of sodium vapour; the doublets are not resolved in this spectrogram. With spectrographs of larger resolving power, over seventy members of the absorption- or principal series of Na, K, Rb and Cs[84, 85, 86] have been measured. The extrapolation of the series

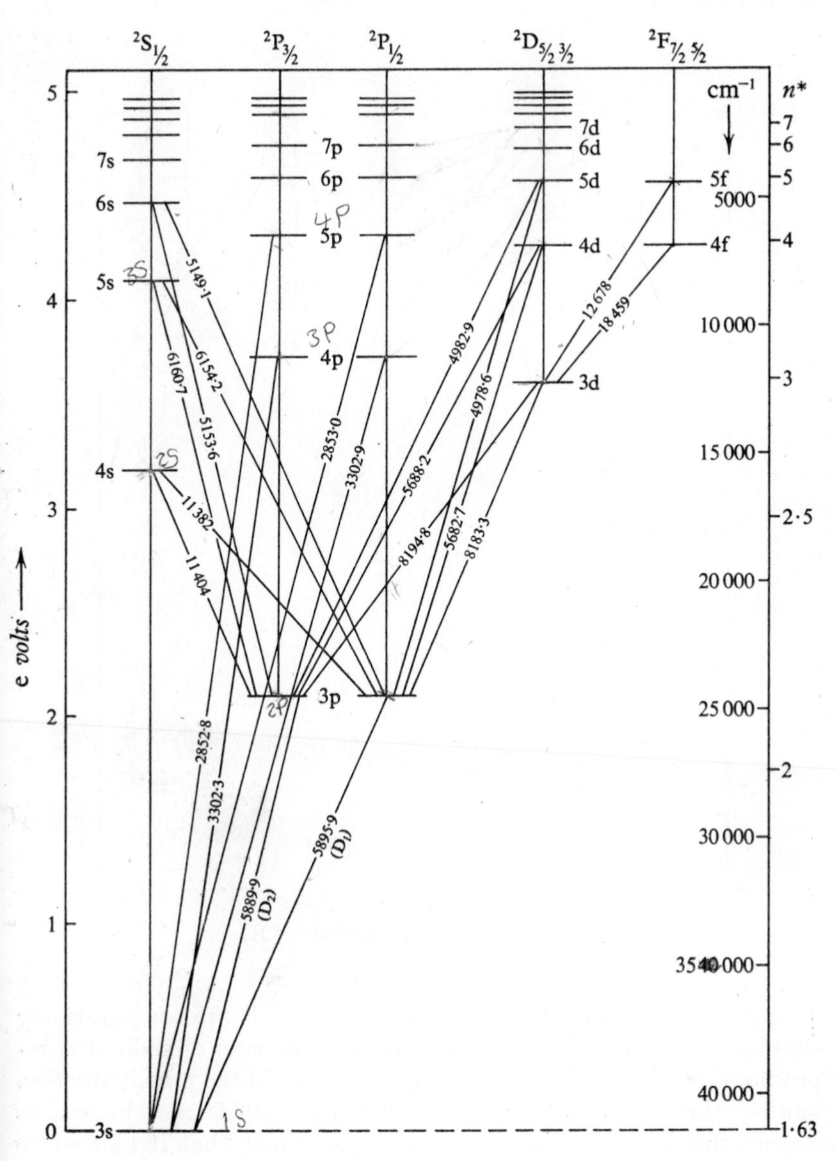

FIG. III, 17. Term diagram of Na.

limits have led to very accurate values of the ionisation potentials of these elements.

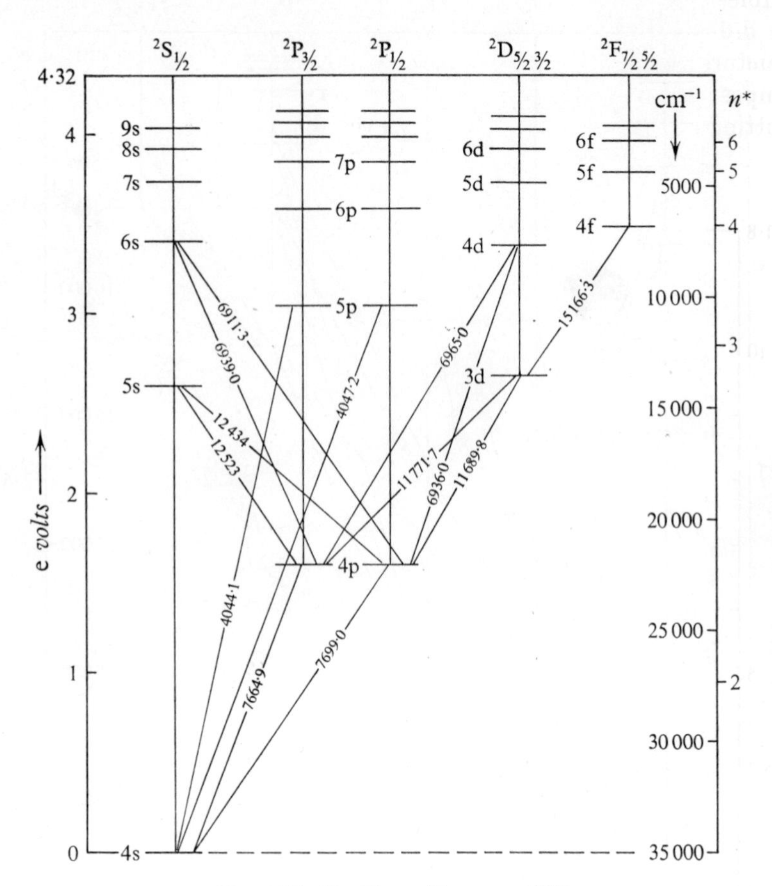

Fig. III, 18. Term diagram of K.

Even the emission spectra of the alkali metals are comparatively simple. The visible part of the emission spectrum of sodium is reproduced in plate 6; it shows only one member of the principal series, namely the resonance doublet 5889·96, 5895·93A, also known as Fraunhofer D-lines. Their intensity is so great that it had to be reduced in the photograph by means of a colour filter having an absorption band in the yellow, in order to avoid blurring by over-exposure. The lines $3P - nD$ are known as *diffuse series* and the somewhat weaker lines $3P - nS$ as *sharp series*. The doublet splitting, in wave-numbers, has the same width for all lines of both series and is equal to the splitting of the resonance lines. This fact shows that the doublet splitting is due to the common term 3 $^2P_{\frac{1}{2},\frac{3}{2}}$. The lines of the diffuse series, however, have actually a slightly more complex structure; higher resolution reveals that the stronger component is composed of two lines one of which is very weak. This is due to the splitting of the D-term as shown in the diagram fig. III, 16.

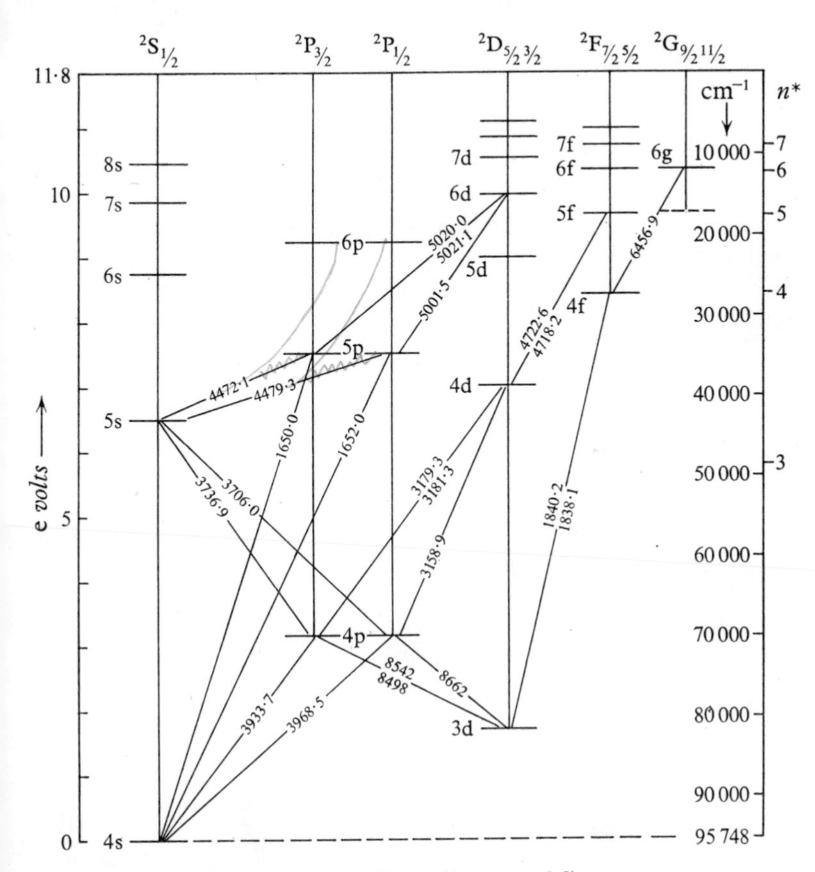

FIG. III, 19. Term diagram of Ca^+.

Some very accurate measurements of small splittings of P- and D-terms have been made by the use of atomic beam light sources.[87]

The diffuse and sharp series converge to a common limit whose wave-number is equal to the value of the lowest P-term (strictly speaking, the doublet components converge to two limits $^2P_{\frac{1}{2}}$ and

$^2P_{\frac{3}{2}}$). The fact that the difference of the wave-numbers of this common limit and of the limit of the principal series is equal to the wave-number of the first member of the latter series in all alkali spectra formed one of the starting points of the term analysis of atomic spectra.

Figure III, 17 shows the term diagram of Na. The structure of the term diagrams of K (fig. III, 18) and the iso-electronic CaII (fig. III, 19) is similar to that of Na, but the doublet splitting of the lowest ^{2}P-term is now large enough to be visible. Also the lowest D-term is somewhat lower in K and very much lower in CaII. In the next member of the iso-electronic sequence, ScIII, the D-term has become the ground term, so that this spectrum can no longer be described as alkali-like. In contrast to Na, the lowest D-term has a lower quantum number n in K than the lowest S- or P- terms. The bearing of these facts on the theory of the shell structure will be discussed in IV.

Table III, 14 lists some data for the resonance lines and resonance potentials.

TABLE 14

Resonance lines of alkali atoms

	λ(A)	$\tilde{\nu}$ (cm^{-1})	$V_{res.}$(ev)
Li	6707·85	14903·8	1·85
Na	5895·92 5889·96	16956·1 16973·4	2·10
K	7698·98 7664·91	12985·1 13042·9	1·61
Rb	7947·64 7800·23	12578·9 12816·5	1·56
Cs	8943·46 8521·12	11178·3 11732·3	1·45

2. Quantitative relations

A number of empirical formulae have been proposed for representing term values of alkali spectra as functions of n and l. Most widely used are those which are based on the hydrogen formula. If the term value is expressed by $T = -R/n^{*2}$ the effective quantum number n^* can be written to a first approximation as

$$n^* = n-\alpha(l) \quad \text{(Rydberg formula)} \quad (78)$$

or more accurately

$$n^* = n-\alpha(l)-\beta(l)/n^2$$

or

$$n^* = n-\alpha(l)+\beta'(l)T \quad \text{(Rydberg–Ritz formula).} \quad (79)$$

The usefulness of these formulae is shown by the values in tables III, 15: the *term defect* $n-n^*$ is almost constant within one term series, but slightly larger for the lowest terms owing to the Ritz correction term β/n^2. The tables show the increase of the term defect with decreasing l and increasing atomic number.

The Rydberg–Ritz formula can be derived by Bohr–Sommerfeld's theory on either of two different assumptions. For nearly hydrogen-like terms the attractive force between valence electron and core due to the polarisation of the latter was calculated[88, 89] and gave not only the correct form of the function of n but also a good estimate of the absolute values of the term defects. For greater term defects it was assumed[90] that the orbit consisted of elliptical segments outside the core and others lying inside the closed shell, assumed to consist of a negatively charged spherical surface. The calculation of this composite orbit again gave a dependence of the type of Rydberg–Ritz's formula. In fact, the distinction between penetrating and non-penetrating orbits is by no means as definite as these simplified models made it appear.

Wave-mechanical approximation methods led to expansions whose first three terms have again the form of the Rydberg–Ritz formula.[91] For a calculation of absolute values of α and β, approximate wave-functions for the core electrons are required. In a semi-empirical way these can be derived from X-ray term values, and an "effective" field can be calculated for the motion of the valence electron. In this way[92] very good agreement with the observed values has been achieved; if the wave functions of the core electrons are computed by the self-consistent field method, the agreement is much less good.

Many attempts have been made to correlate term values and Ritz- and Rydberg-corrections with polarisabilities of the core, i.e. of the positive ion. Sternheimer[93] and Bockasten[94], in recent contributions to this subject, give references to earlier papers.

TABLE 15 (*a*)

Li

S terms	n^*	P terms	n^*	D terms	n^*	F terms	n^*
2S 43487	1·588	2P 28583	1·966				
3S 16281	2·596	3P 12561	2·956	3D 12204	2·999		
4S 8476	3·598	4P 7018	3·95	4D 6864	3·998	4F 6857	4·00
5S 5188	4·60	5P 4474	4·95	5D 4390	5·00	5F 4382	5·00
6S 3500	5·60	6P 3100	5·95	6D 3047	6·00		

TABLE 15 (*b*)

Na

S terms	n^*	P terms	$\Delta\tilde{\nu}$	n^*	D terms	n^*	F terms	n^*
3S 41449·7	1·626	$3P_{1/2}$ 24493	17·18	2·116	3D 12277	2·99		
		$3P_{3/2}$ 24476		2·117				
4S 15710·2	2·643	$4P_{1/2}$ 11182	5·49	3·133	4D 6901	3·99	4F 6861	4·00
		$4P_{3/2}$ 11177		3·134				
5S 8249·0	3·647	$5P_{1/2}$ 6409	2·49	4·138	5D 4413	4·99	5F 4391	5·00
		$5P_{3/2}$ 6407		4·139				
6S 5078·0	4·649	$6P_{1/2}$ 4153	1·50	5·140	6D 3062	5·99	6F 3042	6·00
		$6P_{3/2}$ 4152		5·141				
7S 3438·0	5·650	$7P_{1/2}$ 2909	1·47	6·141				
		$7P_{3/2}$ 2908		6·143				

TABLE 15 (*c*)

K

S terms	n^*	P terms	$\Delta\tilde{\nu}$	n^*	D terms	$\Delta\tilde{\nu}$	n^*	F terms	n^*
					$3D_{3/2}$ 13471	$-2\cdot3$	$2\cdot854$		
					$3D_{5/2}$ 13474				
4S 35009·78	1·771	$4P_{1/2}$ 22025	57·7	2·232	$4D_{3/2}$ 7612	$-1\cdot1$	3·797	4F 6882	3·994
		$4P_{3/2}$ 21967		2·235	$4D_{5/2}$ 7612				
5S 13984	2·802	$5P_{1/2}$ 10308	18·7	3·263	5D 4826		4·77	5F 4408	4·991
		$5P_{3/2}$ 10290		3·266					
6S 7559	3·800	$6P_{1/2}$ 6013	8·1	4·272	6D 3314		5·76	6F 3060	5·99
		$6P_{3/2}$ 6005		4·275					
7S 4737	4·81	$7P_{1/2}$ 3940	4·4	5·28					
		$7P_{3/2}$ 3935							

TABLE **15** (*d*)

Rb

S terms	n^*	P terms	$\Delta\tilde{\nu}$	n^*	D terms	$\Delta\tilde{\nu}$	n^*	F terms	n^*
					4D 14336	−0·44	2·767	4F 6899	3·989
5S 33691	1·805	$5P_{1/2}$ 21112	237·6	2·280	$5D_{3/2}$ 7991	3·0	3·706	5F 4420	4·984
		$5P_{3/2}$ 20874		2·293	$5D_{5/2}$ 7988		3·707		
6S 13560	2·846	$6P_{1/2}$ 9976	77·5	3·317	$6D_{3/2}$ 5004	2·3	4·68	6F 3070	5·98
		$6P_{3/2}$ 9899		3·331	$6D_{5/2}$ 5002				
7S 7380	3·856	$7P_{1/2}$ 5856	35·0	4·329	$7D_{3/2}$ 3411	1·5	5·67		
		$7P_{3/2}$ 5821		4·342	$7D_{5/2}$ 3409				

TABLE 15 (*e*)

Cs

S terms	n^*	P terms	$\Delta\tilde{\nu}$	n^*	D terms	$\Delta\tilde{\nu}$	n^*	F terms	$\Delta\tilde{\nu}$	n^*
								4F 6936	−0·2	3·977
					$5D_{3/2}$ 16907	97·6	2·548	5F 4437	−0·2	4·975
					$5D_{5/2}$ 16809		2·555			
6S 31406·7	1·869	$6P_{1/2}$ 20228	554·1	2·392	$6D_{3/2}$ 8817	42·9	3·528	6F 3079	−0·1	5·97
		$6P_{3/2}$ 19674		2·362	$6D_{5/2}$ 8775		3·537			
7S 12870·9	2·920	$7P_{1/2}$ 9641	181·0	3·374	$7D_{3/2}$ 5358	20·9	4·53			
		$7P_{3/2}$ 9460		3·406	$7D_{5/2}$ 5337					

The ground state of the lithium atom has been calculated very accurately by means of variation methods similar to those used by Hylleraas for helium. The calculation of the ground term agrees to within 1/1000 with the measured value.(95, 96) Various modifications and improvements of the self-consistent field method have been applied to alkali-like atoms(97, 98, 99).

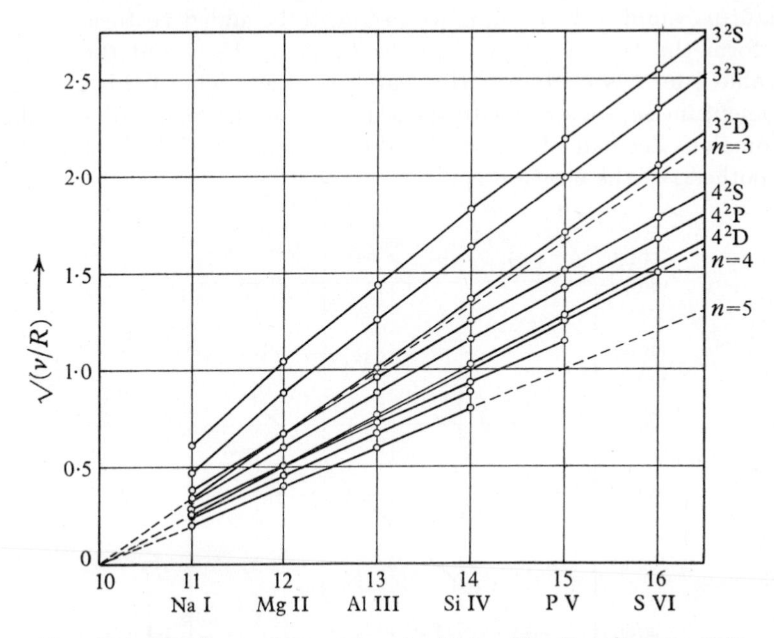

FIG. III, 20. Law of screening doublets; Moseley diagram (W. Grotrian, *Graph. Darst. d. Spektren*, II, J. Springer, 1928).

In comparing iso-electronic alkali-like spectra one can again use the definition of the effective charge number and the screening constant σ according to (71). It is found that the value of σ remains almost constant within an iso-electronic series. This can be shown by means of a *Moseley graph*, in which $\sqrt{(T/R)}$ is plotted against Z (fig. III, 20). Points of the same n but different l, such as 3 ^{2}P and 3 ^{2}S, lie very closely on parallel lines; this fact is known as the law of *screening doublets*. The corresponding term differences, plotted against Z, lie very nearly on a straight line (fig. III, 21). It will be noted that the slopes of the lines in the Moseley plot are equal to the values of $1/n$.

3. The doublet structure

The motion of a point particle in a central force field forms a mechanical system of three degrees of freedom, singly degenerate owing to the isotropy of space, and is fully described by two quantum numbers n and l. The occurrence of doublet structure in alkali spectra was thus difficult to reconcile with the assumption of a point electron and was, for many years, described merely formally by means of a quantum number $\frac{1}{2}$, of unknown origin, to be added vectorially to l to form the two resultant j-values $l+\frac{1}{2}$ and $l-\frac{1}{2}$, and the single j-value $\frac{1}{2}$ for $l = 0$. It was the search for the origin of this angular momentum $\frac{1}{2}\hbar$, together with the search for an understanding of the anomalous Zeeman effect, which led Goudsmit and Uhlenbeck to the hypothesis of the electron spin.

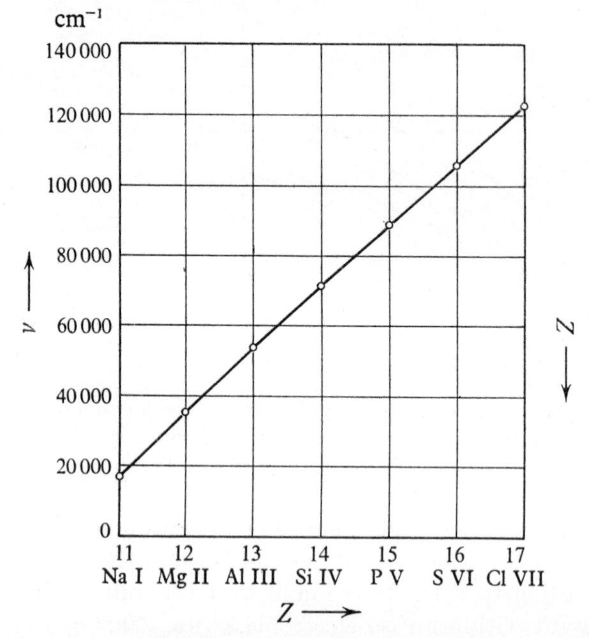

FIG. III, 21. Law of screening doublets; plot of differences $3\,^2S - 3\,^2P$ (W. Grotrian, *Graph. Darst. d. Spektren*, II, J. Springer 1928).

With this hypothesis, they explained the doublet splitting as due to the magnetic interaction between the spin and the orbital motion of the valence electron. The theory follows the same lines as that for the corresponding splitting of hydrogen terms as contained in eq. (62).

Owing to the penetration effect, the alkali electron moves partly in a field due to an effective charge number $Z_{eff} > 1$ which causes the splitting to be considerably greater than that in hydrogen, and the more so the less hydrogen-like the term.

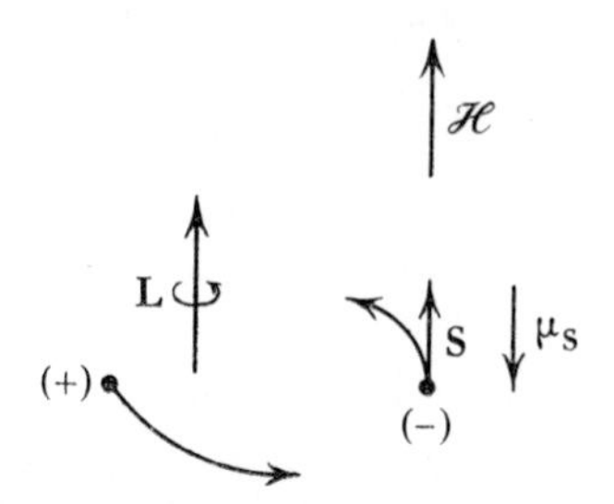

FIG. III, 22. Diagram illustrating spin–orbit interaction and order of doublet levels.

The increase of the width of splitting with decreasing l, in a given atom, and with increasing atomic number for different atoms, is analogous to the dependence of the term defect on the same variables and can be ascribed to the same cause. As a qualitatively useful picture, we can imagine the electron to be stationary and the nucleus, together with the core electrons, to move around the valence electron. This picture allows us most easily to find the order of the levels: the revolution of the positive charge $Z_{eff} \times e$ causes a magnetic field of the same sign as **L** at the point of the electron. The latter acts as a magnetic dipole, opposite in sign to the angular momentum **S**, owing to the negative electronic charge (fig. III, 22). For $j = l+\frac{1}{2}$ the angular momenta are in the same direction, the magnetic spin moment is opposed to $\mathcal{H}$, leading to a higher energy for the greater value of j in agreement with (61).

This *normal* order of the terms is found in all terms of Li and in other alkali atoms for all terms of comparatively large splitting. The fact that ^{2}D terms of Na and K and most of the ^{2}D and ^{2}F terms in the other alkali spectra are inverted shows that the simple assumption of l,s-interaction of a single electron moving in a central force field is not always justified. This has to be kept in mind in the following quantitative discussion based on this assumption.

The model of the penetrating orbit, consisting of segments of elliptical orbits of effective charge number Z_a (= 1 for neutral atoms, = 2 for singly ionised atoms, etc.) and orbits of effective charge number $Z_i = Z - \sigma$ can be used for deriving a formula for the width of the doublet splitting.[100]

The spin–orbit interaction of an electron moving in a Coulomb field of charge number Z at a distance r was found to be proportional to Z/r^3 (eq. 54) or for a non-spherical orbit proportional to $\overline{Z/r^3}$. In Bohr–Sommerfeld's theory, integration over the elliptical orbit gives the result

$$\overline{\frac{1}{r^3}} = \frac{Z^3}{a_0{}^3 n^3 l^3}$$

which passes into the wave-mechanical result (60) by substitution of $l(l+\frac{1}{2})(l+1)$ for l^3. If the classical formula is applied to the motion in segments of ellipses and the wave-mechanical substitution subsequently made, the result is, for $Z_i \gg Z_a$

$$\overline{\frac{1}{r^3}} = \frac{Z_a{}^2 Z_i}{a_0{}^3 n^{*3} l(l+\frac{1}{2})(l+1)} \tag{80}$$

where n^* can be derived from the term value in the usual way. This leads to the *Landé formula*

$$\Delta T = R\alpha^2 \frac{Z_a{}^2 Z_i{}^2}{n^{*3} l(l+1)} = 5{\cdot}84 \frac{Z_a{}^2 Z_i{}^2}{n^{*3} l(l+1)} \tag{81}$$

which can be regarded as a generalisation of (62).

It is not possible to calculate the absolute value of the doublet splitting from this formula which contains the unknown quantity Z_i. But various indirect tests have shown that the Landé formula is a useful approximation. If the values of $Z_i = Z - \sigma$ are calculated for an iso-electronic series from the observed doublet splittings and the experimental values of n^* by means of (69), a plot of Z_i against Z is found to give nearly a straight line, at any rate for the lighter elements. Landé found that the assumption $Z_i = Z - 2$ for the first and $Z - 4$ for other periods represents the doublet splitting of the P series to within 10 to 20 per cent.

The dependence of ΔT on n^* could be tested on the large number of doublets $n_0\,{}^2\mathrm{S} - n\,{}^2\mathrm{P}_{\frac{1}{2},\frac{3}{2}}$ which were resolved in the absorption spectra of K, Rb and Cs. The term splitting ΔT was found to be proportional to $1/n^{*3}$ to a fairly high degree of accuracy in all three atoms.[86]

In the quantum-mechanical treatment the width of the doublet splitting can be calculated as in the hydrogen-like atoms, as a perturbation due to the magnetic spin–orbit interaction term $\xi\mathbf{L}\cdot\mathbf{S}$. The factor ξ can be expressed in terms of $\mathrm{d}V/\mathrm{d}r$ as in (53). The energy of interaction can then be written

$$W' = \int \psi^*\xi\mathbf{L}\cdot\mathbf{S}\psi\,\mathrm{d}\tau = \tfrac{1}{2}a[j(j+1)-l(l+1)-s(s+1)] \qquad (82)$$

where a is defined by

$$a = \hbar^2 \int R(r)^2\xi r^2\,\mathrm{d}r. \qquad (83)$$

The calculation of a requires the knowledge of the radial wave functions which are of a complex nature for all except hydrogen-like atoms. Casimir[101] considers the ψ-function as composed of a hydrogen-like function for very small radii, with effective quantum number $Z_i \approx Z$, and a hydrogen-like function with Z_a ($= 1$ for neutral atoms) for large radii; the function for intermediate radii is interpolated by approximation methods. The result is a modified Landé formula

$$\Delta T = H_r\frac{\mathrm{d}n^*}{\mathrm{d}n}R\alpha^2\frac{Z_i{}^2Z_a{}^2}{n^*l(l+1)}$$

where the correction factor $\mathrm{d}n^*/\mathrm{d}n$ is due to the deviation from the Rydberg formula and is equal to $1+2\beta/n^3$, with the definitions of eq. (70). H_r contains a relativity correction and differs appreciably from unity only for heavy elements (see VI, fig. 6). Barnes and Smith[102] compare the formula with experimental values of term splittings and find that the lowest P term of alkali-like atoms is well represented by the assumption $Z_i = Z-n$, where n is the principal quantum number of the p-electron; these results do not differ much from those found by Landé.

If one considers the electron as moving in a Coulomb field due to a charge Z_{eff}, the energy of interaction can be written in analogy to (59):

$$E' = \frac{\hbar^2e^2}{4c^2m^2}Z_{\mathrm{eff}}\overline{\frac{1}{r^3}}[j(j+1)-l(l+1)-s(s+1)]. \qquad (84)$$

This expression can be used for calculating E' from a value of $Z_{\mathrm{eff}}\overline{(1/r^3)}$ derived from approximate wave functions. Conversely, the observed doublet splitting can be used for deriving a value of $Z_{\mathrm{eff}}\overline{(1/r^3)}$ which is often required in hyperfine structure research.

For many purposes this simple model is too crude and it is necessary to take exchange effects into account. In two-electron spectra, the possibility of exchanging the symbols of the two electrons in the product-wave function led to exchange degeneracy and to corresponding terms in the perturbation energy. Similar effects occur in all atoms with more than one electron. If only interactions of each spin with its own orbit are considered, it can be rigorously shown that only the L, S-interaction of the valence electron appears in the final result, so that the simple formula (52) remains valid.

If spin–spin interaction and spin–other orbit interaction are included, it is found that the exchange between valence electron and any of the core electrons affects the two doublet levels differently. This effect can not only reduce the width of the splitting appreciably but might even reverse the order of the terms. The calculation of these interactions is very difficult; they are specially important in the lighter atoms. In heavier atoms, they are usually small, but the calculation of the *L,S*-interaction depends critically on the wave functions which are not known very accurately. The calculation of the doublet splitting is therefore difficult in all atoms.

With the use of Hartree–Fock functions, the following values of the doublet splitting of the lowest ^2P terms have been derived[103,104] (table III, 16):

TABLE 16

Doublet splitting in lowest P-terms (cm^{-1})

	calc.	obs.
Li	0·26	0·337
Na	9·6	17·2
K	65	57·7

For lithium, the omission of the exchange integral which contains the spin–spin interactions would lead to the entirely wrong value of 0·71 cm^{-1}. The difference between the calculated value of 0·26 and the observed value is probably due to the inaccuracy of the wave function.

The main cause of the inversion of the higher alkali terms appears to be still uncertain. Some authors ascribed it to perturbations,[105,106,107] others to exchange effects.[103] Araki[108] explained them by the influence of the spin–orbit interaction on the screening constant σ, but the agreement which he found for alkali-like terms of copper has been regarded by others[109] as accidental.

4. The intensity ratio

It has long been known that the intensity ratios of the components of doublets, and also of triplets and higher multiplets, tend to have certain rational "normal" values which depend on the angular quantum numbers but not on n. It would be difficult to state rigorously the conditions for this normal intensity ratio to hold, but the main requirement is this: the width of the doublet- (or multiplet-) splitting of any term must be small compared with the distance from any other term. Strictly speaking, it is the ratio of the strengths (see p. 70) which obeys simple rules, but if the splitting is small, the Boltzmann factor and the factor ν^3 in (II.121) can be regarded as identical and the intensities are in the same ratio as the strengths. It is most important that self absorption should be negligible.

Most of the doublets in the spectrum of Na fulfil these conditions well enough. The doublet components of the lines of the sharp series show the intensity ratio 2 : 1, and the same is true for the diffuse series if the splitting of the ^{2}D term is unresolved, and for the principal series if self absorption is avoided. Absorption is an easier method, in the latter series, for determining the ratio of the strengths which is again found to be 2 : 1. The most accurate measurements of this ratio were made by means of the method of anomalous dispersion.

These striking facts can easily be understood in terms of the old quantum theory and the correspondence principle (see II.D.1). Consider the two levels $n\ ^2P_{\frac{1}{2}}$ and $n\ ^2P_{\frac{3}{2}}$; owing to the smallness of the energy difference between the doublet levels we can assume the influence of the electron spin on the orbital motion to consist merely in causing different orientations of the orbital axis with regard to the spin for the two levels, while the motion in the orbital plane remains essentially unchanged. The correspondence principle links the transition probability with the classical rate of loss of energy by radiation, which is naturally independent of the orientation and must be the same for the two doublet terms $^2P_{\frac{1}{2}}$ and $^2P_{\frac{3}{2}}$. There are, however, under normal conditions of excitation, twice as many atoms in the state $^2P_{\frac{3}{2}}$ than in $^2P_{\frac{1}{2}}$, according to the ratio of the statistical weights $2j+1$ for the two terms. Since the emission leads to the same single state $^2S_{\frac{1}{2}}$ for both, the lines appear in the intensity ratio 2 : 1.

If both terms are multiple, as in the composite doublets $^2P-{}^2D$ or in higher multiplets to be discussed later, these considerations have to be generalised and lead to the intensity sum rule of Ornstein, Burger and Dorgelo[110,111]:

For the transition between two term multiplets, the two sums of the intensities of all transitions from any one level and from any other level are in the ratio of the statistical weights of these levels.

Applying this rule to the composite doublet of fig. III, 16 we find

$$\frac{I_b}{I_a+I_c} = \frac{2\times\frac{5}{2}+1}{2\times\frac{3}{2}+1} = \frac{3}{2} \quad \text{and} \quad \frac{I_c}{I_a+I_b} = \frac{2\times\frac{1}{2}+1}{2\times\frac{3}{2}+1} = \frac{1}{2}$$

giving the result $I_a : I_b : I_c = 1 : 9 : 5$.

For a doublet $^2D-{}^2F$ the same procedure gives the intensity ratio 1: 20 : 14.

For a triplet transition $^3S_1-{}^3P_{0,1,2}$ the sum rule gives directly the intensity ratio

$$I_0 : I_1 : I_2 = (2\times 0+1) : (2\times 1+1) : (2\times 2+1) = 1 : 3 : 5$$

which has in fact been observed in the helium line 7065 A (see fig. III, 13).

The extreme narrowness of the triplet splitting shows that the interactions connected with different orientations of the spin with regard to the orbital axis are very weak indeed. The conditions for the validity of the intensity sum rule are therefore singularly well fulfilled, regardless of the nature of the interactions.

On the basis of the quantum-mechanical results, the intensity sum rule follows immediately from the independence of the sum (II.125) of m and j. The sum taken for each m has the same value, and there are $2j+1$ possible values of m for each j. For normal excitation conditions the equal population of the states of different m and j then implies the sum rule.

If the line strength S which is symmetrical with regard to initial and final state, is substituted for the intensity, the sum rule applies to absorption as well as emission.

It is interesting to note that the centre of gravity of any doublet line agrees with the wave-number which the line would have in the absence of magnetic interaction. This is easy to see if only one term is split, e.g. for a transition $^2P_{\frac{1}{2},\frac{3}{2}}-{}^2S_{\frac{1}{2}}$. It was pointed out on p. 118 that the magnetic interaction (82) causes shifts of the levels 3/2 and $\frac{1}{2}$ in the ratio $l/(l+1)$, which is equal to 1 : 2, while the ratio of the strengths of the transitions is 2 : 1.

If the L, S-interaction is strong, as indicated by a large width of the doublet splitting, it acts as a perturbation of the Schrödinger functions $\psi_{n,l,m}$, making the radial function dependent on j. This causes the intensity rule to break down. An example of this is found in the higher doublets of the principal series of the alkali atoms, especially of caesium. While the first doublet of Cs shows the normal intensity ratio 2 : 1, this ratio is 3·8 in the second doublet and appears to rise steadily towards higher members where values of 25 : 1 have been reported.(86, 112, 113)

Fermi(114) has given an interesting interpretation of these anomalies. The effect of the strong magnetic L, S-interaction is treated as a perturbation as mentioned before. As a result of this, the wave functions of the perturbed, lowest states $6p\ ^2P_{\frac{1}{2},\frac{3}{2}}$ of caesium contain some admixture of the zero-order functions 7P, 8P . . . . and vice versa. Since the term 6P has a far greater strength of transition to the ground term 6S than any of the higher P terms, a small admixture of the latter will have little effect on the strength of the transition 6P – 6S. On the other hand, the strengths of the transitions from higher P-levels will be very sensitive to a small admixture of 6P. Since the perturbation connects only levels of the same j (see p. 295), a level $n\ P_{\frac{3}{2}}$ will only be perturbed by $6\ P_{\frac{3}{2}}$, and $n\ P_{\frac{1}{2}}$ only by $6\ P_{\frac{1}{2}}$. But $6\ P_{\frac{3}{2}}$ is considerably higher than $6\ P_{\frac{1}{2}}$ and therefore closer to the perturbed term n P. The level $n\ P_{\frac{3}{2}}$ is thus perturbed more strongly than $n\ P_{\frac{1}{2}}$ and acquires a greater increase in transition probability. Fermi estimated the magnitude of the effect on the second member of the principal series and found it sufficient to account for the observed anomaly.

5. Absolute strength of lines and continuous spectra

The absolute strength of spectral lines is difficult to measure and difficult to calculate. The same applies to the relative strength of lines of appreciably differing frequencies such as the different lines of a series. In consequence of this, accurate comparisons between experiment and theory are not very numerous. The calculation involves the radial wave functions (see p. 93) which have to be computed by laborious approximation methods. Such calculations have been carried out for Li and Na(74,115–118) and for K.(74, 119, 120) Hylleraas(121) and Bates and Damgaard(74) have made use of the fact that the radial integral of the dipole moment depends predominantly on the outer part of the wave function where the force is not very different from a Coulomb force. On account of this, comparatively simple approximations lead to results which are almost as accurate as those gained by highly involved calculations

Measurements by the method of anomalous dispersion are capable of great accuracy, but the value of the vapour pressure contains an important source of error.[122]

If an excited state gives rise to only one emission line, the transition probability or strength can be immediately calculated from the life time of this state which can be measured either directly or from the contour of the line[123, 124, 125] or by methods of magnetic rotation.[126] The latter method involves some difficulties due to hyperfine structure and has apparently not yet been used very successfully. Table III, 17 shows experimental *f*-values of the first four lines of the principal series of Na, from absolute measurements[122–127] for the first line and relative measurements for the other lines.[128] They are compared with calculations by Trumpy.[115] Similarly good agreement is found in lithium, though only relative experimental values appear to be available.[129]

TABLE 17

f values in the principal series of Na

	f exper.	*f* theor.
3S – 3P	0·95 to 1·05	0·975
3S – 4P	0·0136	0·0144
3S – 5P	0·0020	0·0024
3S – 6P	0·00062	0·00098

According to the *f*-sum rule (p. 67) the sum of the *f*-values for the principal series, including the continuum, should be 1. The bulk of the total strength of all transitions from the ground state of Na is found to be in the resonance doublet. The sodium atom is thus seen to approach very closely the model of a single harmonic oscillator if the doublet structure is disregarded. The same applies to K and Rb for which both experimental and theoretical values are close to 1, but probably to a lesser extent to Li. Also *f*-values for lines of the sharp and diffuse series have been calculated.[120]

Within each series, the strength of the lines decreases towards the limit, apparently in a uniform manner. The rate of decrease is greater than in hydrogen, and it increases markedly from Li to Cs.

The series limit is followed by a comparatively weak continuous spectrum. Absorption of light in the continuous range causes photo-ionisation of the atom; the electron is ejected with kinetic energy $h(\nu - \nu_{\text{lim}})$. Emission of the continuous spectrum results from the capture of a free electron by an alkali ion.

By analogy with hydrogen, one would expect a steady decrease in the strength of the continuum towards increasing frequencies, continuing the decrease of the line strengths in the series. Measurements of the continuous absorption spectrum and of the ionisation effect showed that the expected, initial drop of strength near the limit was followed by a rise towards still higher frequencies in Na and especially in K and Cs. It is experimentally very difficult to separate atomic and molecular absorption in alkali vapours, but the evidence for a rise of the truly atomic absorption with increasing frequency is fairly strong.[130]

In the calculation of the strength of the continuous spectrum it has been found essential to take exchange effects into account. If this is done the occurrence of a minimum in the absorption can be explained.[131] Calculations of this kind[118, 132] show some measure of agreement with experimental values. The exchange effects modify the validity of the f-sum rule, but the extent to which the continuous absorption appears to exceed the limits set by this rule must still be regarded as unexpected.

The continuous emission spectrum beyond the series limit of the sharp and diffuse series in Cs has been used in studies of the velocity distribution and concentration of free electrons in gas discharges.[133, 134]

In some respects the atoms Cu, Ag and Au have the characteristics of alkali atoms. Their ground states are $^2S_{\frac{1}{2}}$ terms and the strong resonance doublets $^2S_{\frac{1}{2}} - {}^2P_{\frac{1}{2},\frac{3}{2}}$ (3248, 3274A in Cu, 3281, 3383A in Ag, 2428, 2678A in Au) correspond to the resonance lines of alkali atoms. But many of the terms of these atoms have higher multiplicity and are to be regarded as belonging to a complex spectrum. Though, in these atoms, the ground term and some other terms can be ascribed to a single electron, moving in a central field due to closed shells of electrons, the electrons of one of these shells (see p. 317) can be excited easily. The core, though of spherical symmetry, must be regarded as easily deformable.

Even in alkali atoms, electrons of the core can be excited, but the energy required is so high that the spectral lines due to these processes are in the far ultra-violet; they are not included in the concept of "alkali-like spectra".

6. Forbidden lines

Absorption spectra of alkali vapours of high density show the transitions $n_0{}^2S - n\ {}^2D$.[135–137] Their strengths are of the order of 10^5 to 10^6 times smaller than those of the principal series lines. The Zeeman effect of these forbidden lines[138] proves that they are caused, at any rate in some of these experiments, by quadrupole radiation and not by dipole radiation enforced by perturbing influences of neighbouring atoms.

In the presence of electric fields, violations of the l-selection rule can also be produced, since l is not a good quantum number in Stark effects (see p. 106). For high quantum numbers n this mixing of states of different l becomes noticeable even in very weak fields: High members ($n \approx 20-30$) of the forbidden series $n_0S - nS$, $n_0S - nD$ and $n_0S - nF$ have been observed in fields of only 100 volts per cm.[139–141] Their Zeeman effect proves that they are due to electric dipole radiation.[142]

D. THE SIMPLE SPECTRA OF THE ALKALINE EARTH ELEMENTS

1. Beryllium

In treating atoms of increasing numbers of electrons we had, after reaching Li with three electrons, interrupted the progression by considering, together with lithium, the spectra of elements in the same column of the periodic table. In resuming the original sequence we come, after lithium, to beryllium with four electrons.

Disregarding, at first, weaker lines and lines of very high excitation potential, one can represent the spectrum of beryllium by two term diagrams, one in which all terms are single, and one in which they are triple except for the S-terms which are always single. Intercombinations are so weak that they have not been observed with certainty. The spectrum of Be has thus the characteristic features of the two-electron spectra (He, LiII . .).

But in comparison with helium, the ionisation potential of Be is very small; in fact, the term diagram of Be is very similar to that of He with the omission of the lowest S-terms $1\ {}^1S_0$ and $2\ {}^3S$. Principal quantum numbers n can be assigned to the terms by comparison with hydrogen in the way explained before; this leads to the numbering in the term diagram shown in fig. III, 23.

The qualitative interpretation follows the same lines as in the spectra of He and Li. It makes use of the Pauli principle and the fact that the energy of an electron orbit depends primarily on n, secondarily on l. If an electron is added to the configuration $1s^2 2s$ of Li, the configuration of lowest energy is $1s^2 2s^2$. The two 2s electrons, being equivalent (see p. 144), must have opposite values of m_s and can thus only form one term $1s^2 2s^2\ {}^1S_0$.

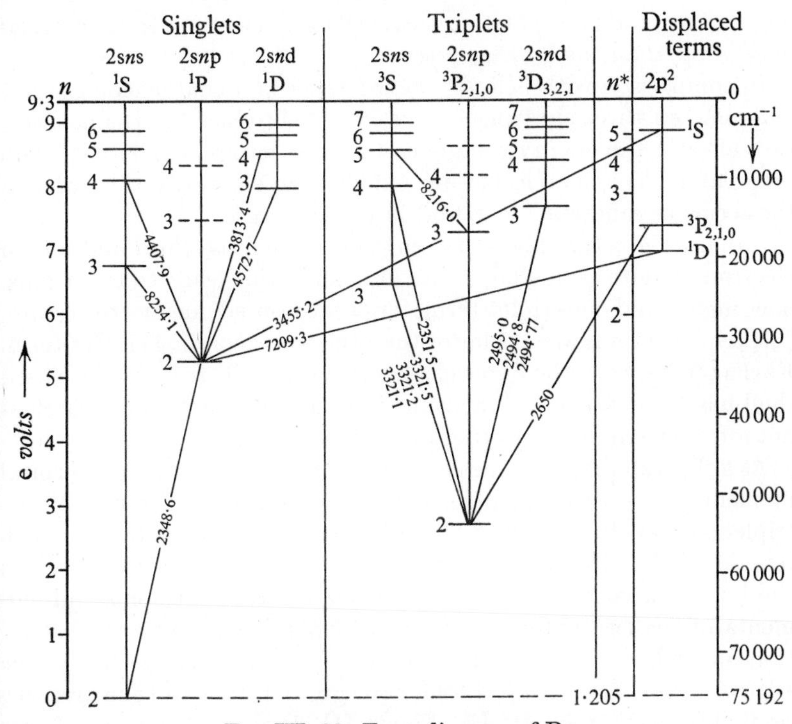

FIG. III, 23. Term diagram of Be.

The lowest excited states arise from excitation of one of the two 2s electrons, resulting in the following terms:

$1s^2 2sns\ {}^1S_0$ and $1s^2 2sns\ {}^3S_1$ $\quad (n = 3, 4, \ldots)$

$1s^2 2snp\ {}^1P_1$ and $1s^2 2snp\ {}^3P_{0,1,2}$ $\quad (n = 2, 3, 4, \ldots)$

$1s^2 2snd\ {}^1D_2$ and $1s^2 2snd\ {}^3D_{1,2,3}$ $\quad (n = 3, 4, \ldots)$, etc.

which can be written in abbreviated symbols $n\ {}^1S_0$, $n\ {}^3S_1 \ldots$ For the same n, the energy of the terms increases in the order nS, nP nD . ..

In comparison with helium, the doubly charged nucleus of the latter is replaced by the nucleus Be^{4+} together with the closed shell of two electrons $1s^2\ {}^1S_0$ forming a core of net charge $2e$ and spherical symmetry. The following arguments are closely analogous to those used for helium: If the third electron is kept in its lowest state 2s which has a charge distribution of spherical symmetry, the fourth electron can be regarded as moving in a central force field due to the core $1s^22s\ ^2S_{\frac{1}{2}}$ of single net charge. For large values of l of the fourth electron the terms will be closely hydrogen-like; for lower values of l polarisation and penetration effects will cause the term to be appreciably lower than the hydrogen term.

In comparison with helium, the existence of a charge cloud of not only one but three electrons can be expected to increase the penetration effects: the energy difference between terms of different l will be greater. This simple picture explains satisfactorily the order of the terms of different l as found in the spectrum.

As a next step the effects of exchange between the third and fourth electron have to be included. As in helium, this leads to two terms, one singlet- and one triplet-term, for every excited one-electron state. The triplet term is always lower than the corresponding singlet term. Exchange between the outer electrons and the electrons in the closed shell has to be taken into account in accurate calculations but does not affect the number or order of terms.

As a third step, the magnetic interactions between spins and orbital momenta have to be included. They cause again all terms of the triplet system to split into three levels, except the S-terms which remain single. The latter results can, as in helium, be expressed in the form of a vector diagram which now refers to the spins and momenta of the two outer electrons (see fig. III, 14). The wave functions of Be have been calculated by the Hartree–Fock method of the self-consistent field (see p. 228) for the ground state and the first two excited terms 2s2p ^1P, ^3P.[143] Figure III, 24 shows the radial charge density (charge per dr) as function of r/a_0. The resonance effect is seen to cause the charge distribution of the 2p electrons to differ markedly for the different multiplicities. The smaller average radius in the triplet term makes the electrostatic energy smaller than that of the singlet term, in accordance with the observed term order. The values of the three terms $2s^2\ ^1S$, 2s2p ^1P, ^3P as derived from these wave functions agree with the observation to within 10000, 2000 and 1000 cm^{-1} respectively, the difference $^3P - {}^1P$ to within 8 per cent.

2. The other alkaline earth elements

The elements in the same column of the periodic table, Mg, Ca, Sr, Ba and Ra show spectra of the same type as beryllium. The quantum numbers n of the lowest S and P-terms are empirically found

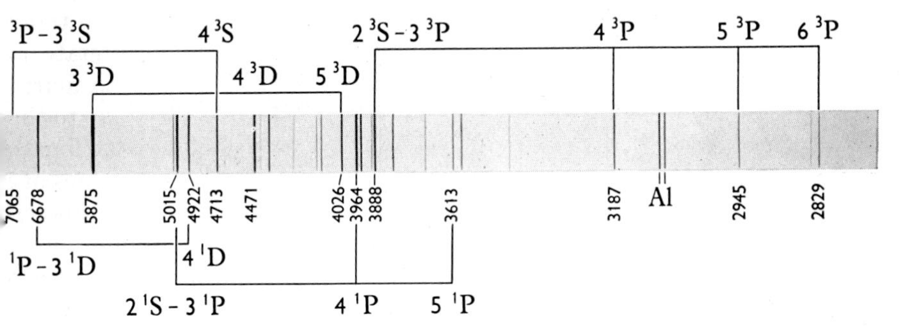

PLATE 5. Spectrum of He, quartz spectrograph. Owing to the rapid drop of sensitivity of panchromatic plates beyond 6800 A, the line 7065 is not visible in the reproduction.

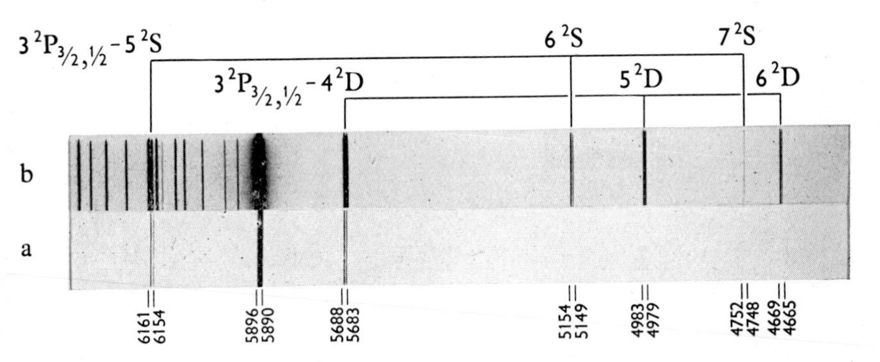

PLATE 6. Emission spectrum of Na; grating spectrograph; (a) short exposure, with argon-filled sodium lamp, (b) long exposure, with neon-filled sodium lamp; the orange and red lines on the left are due to neon.

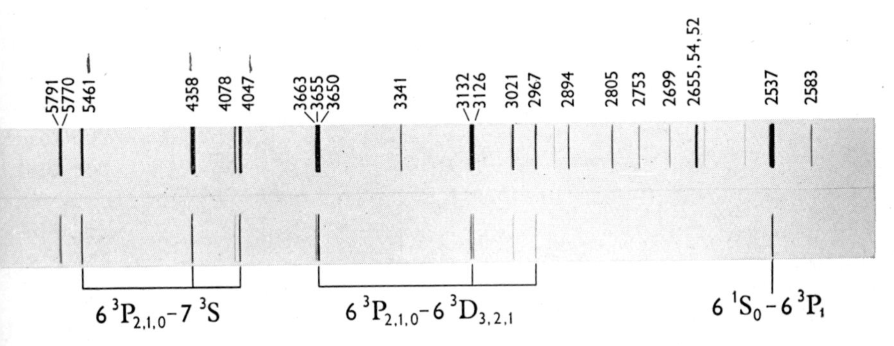

PLATE 7. Spectrum of Hg; quartz spectrograph.

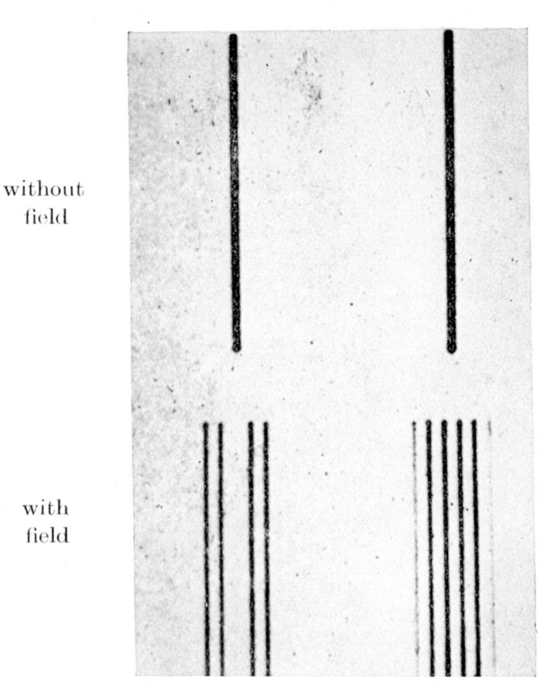

PLATE 8. Anomalous Zeeman effect of Na resonance lines 3 $^2S_{1}/_{2} - 3\ ^2P_{3}/_{2}$ (5890 A) and 3 $^2S_{1}/_{2} - 3\ ^2P_{1}/_{2}$ (5896 A). (E. Back & A. Landé, *Zeeman-effekt u. Multiplettstruktur*, Springer, 1925).

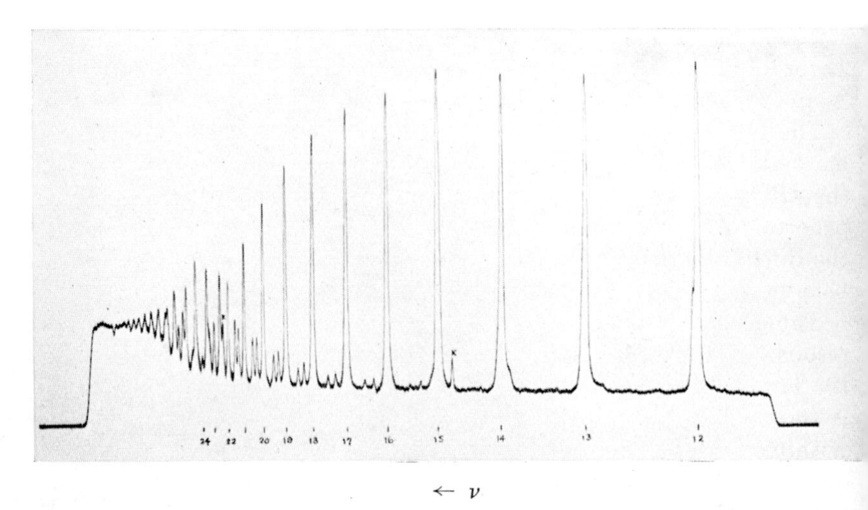

PLATE 9. Forbidden lines 5 $^2S - n\,^2S$, ^2D in absorption spectrum of Rb in electric field of 300 volts/cm. (Ny Tsi-Ze & Choong Shin-Piaw, ref. III. 175).

to be 3, 4, 5, 6 and 7 respectively for these elements, as in the adjacent alkali atoms. Again, we have to assume that the lower s- and p-states are occupied by the electrons forming the core which consists entirely of closed shells and has spherical symmetry.

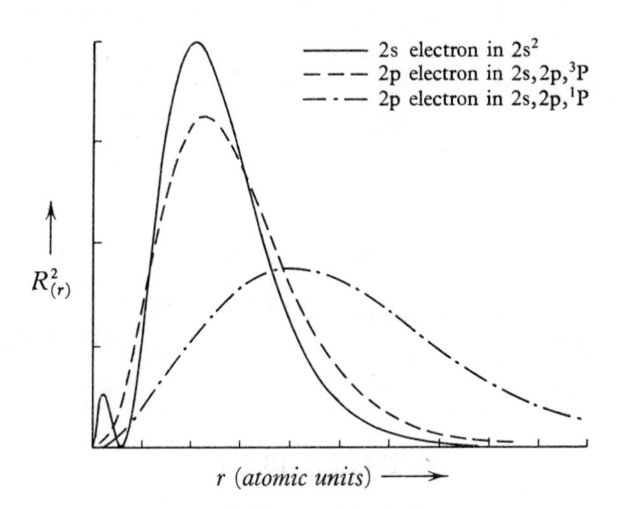

FIG. III, 24. Charge distribution in Be. (D. R. Hartree & W. Hartree, ref. 143.)

The term values within a series can generally be represented by a Rydberg–Ritz formula, and the transitions to the lowest S-term are again described as *principal series*, those to the lowest P-term from S- and D-terms as *sharp* and *diffuse series*.

The differences between terms of equal n but different l increase markedly with atomic number, and the same is true for the width of the triplet splitting. Both effects must be ascribed to increasing penetration of the outer electron into the core. In contrast to this, the difference between singlet- and triplet-terms does not show this behaviour.

Table III, 18 lists some of the term values and table III, 19 the resonance lines of the alkaline earth elements, and fig. III, 25 shows the term diagram of Mg as a typical example. The triplet splitting is too small to be shown in the figure. The arrangement of terms within each system is very similar to that in Na except for the absence of the lowest S-term in the triplet system. As in beryllium, the triplet terms are lower than the corresponding singlet terms.

TABLE 18 (*a*)

Be, Singlet system

S-terms	n^*	P-terms	n^*	D-terms	n^*
2S 75192	1·205	2P 32627	1·83		
				3D 10764	3·20
3S 20515	2·30				
				4D 6411	4·13

Be, Triplet system

S-terms	n^*	P-terms	$\Delta\tilde{\nu}$	n^*	D-terms	n^*
		$2P_0$ 53213				
			0·7			
		$2P_1$ 53212		1·436		
		$2P_2$ 53210	2·4			
3S 23110	2·179				3D 13137	2·890
4S 10685	3·204				4D 7249	3·890

TABLE 18 (*b*)

Mg, Singlet system

S-terms	n^*	P-terms	n^*	D-terms	n^*
3S 61669	1·334	3P 26618	2·030	3D 15266	2·681
4S 18166	2·457	4P 12323	2·984	4D 8525	3·586
5S 9113	3·470	5P 6970	3·966	5D 5361	4·524

Mg, Triplet system

S-terms	n^*	P-terms	$\Delta\tilde{\nu}$	n^*	D-terms	n^*
		$3P_0$ 39818		1·660		
			19·9			
		$3P_1$ 39798		1·660		
			40·9		3D 13712	2·829
		$3P_2$ 39758		1·661		
		$4P_{0,1}$ 13821				
4S 20472	2·315		4·1	2·818	4D 7477	3·830
		$4P_2$ 13817				
5S 9796	3·347					
		5P 7416		3·846	5D 4701	4·830
6S 5778	4·357					

TABLE 18 (*c*)

Ca, Singlet system

S terms	n^*	P terms	n^*	D terms	n^*	F terms	n^*
4S 49305	1·492	4P 25652	2·068	3D 27455	2·000		
5S 15988	2·620	5P 12573	2·954	4D 12006	3·023	4F 6961	3·970
6S 7518	3·820	6P 7626	3·794				

Ca, Triplet system

S terms	n^*	P terms	$\Delta\tilde{\nu}$	n^*	D terms	$\Delta\tilde{\nu}$	n^*	F terms	n^*
		$4P_0$ 34147		1·793	$3D_1$ 28970				
			52·2			13·9			
5S 17765	2·485	$4P_1$ 34095		1·794	$3D_2$ 28955		1·946		
			105·9			21·7			
		$4P_2$ 33989		1·797	$3D_3$ 28934				
6S 8830	3·525	$5P_0$ 12757			$4D_1$ 11556				
			7·0			3·7			
		$5P_1$ 12750		2·935	$4D_2$ 11553		3·083	4F 7134	3·922
			20·4			5·8			
		$5P_2$ 12730			$4D_3$ 11547				
7S 5324	4·540	$6P_0$ 6790							
			3·9						
		$6P_1$ 6786		4·022					
			7·8						
		$6P_2$ 6778							

TABLE 18 (*d*)

Sr, Singlet system

S terms	n^*	P terms	n^*	D terms	n^*	F terms	n^*
5S 45926	1·525	5P 24227	2·128	4D 25776	2·063	4F 6387	4·145
6S 15335	2·675	6P 11828	3·045	5D 11110	3·143		
7S 7482	3·830	7P 7019	3·953				

Sr, Triplet system

S terms	n^*	P terms	$\Delta\tilde{\nu}$	n^*	D terms	$\Delta\tilde{\nu}$	n^*	F terms	$\Delta\tilde{\nu}$	n^*
6S 16887	2·549	$5P_0$ 31608		1·863	$4D_1$ 27766			$4F_2$ 7175		
			187			59·6			1·7	
7S 8501	3·591	$5P_1$ 31421		1·869	$4D_2$ 27706		1·99	$4F_3$ 7173		3·91
			394			100			2·7	
8S 5163	4·610	$5P_2$ 31027		1·880	$4D_3$ 27606			$4F_4$ 7170		
		$6P_0$ 12110		3·010	$5D_1$ 10918					
			41·4			15·0				
		$6P_1$ 12068		3·015	$5D_2$ 10903		3·17			
			105			22·8				
		$6P_2$ 11964		3·024	$5D_3$ 10880					

TABLE 18 (*e*)

Ba, Singlet system

S terms	n^*	P terms	n^*	D terms	n^*	F terms	n^*
6S 42032	1·615	6P 23972	2·139	5D 30637	1·893	4F 13478	2·854
7S 16402	2·587	7P 9485	3·402	6D 13803	2·820		
		8P 5042	4·67				

Ba, Triplet system

S terms	n^*	P terms	$\Delta\tilde{\nu}$	n^*	D terms	$\Delta\tilde{\nu}$	n^*	F terms	$\Delta\tilde{\nu}$	n^*
7S 15872	2·629	$6P_0$ 29766		1·924	$5D_1$ 32998		1·824	$4F_2$ 7430		
			370			181			14	
8S 8127	3·675	$6P_1$ 29396		1·935	$5D_2$ 32817		1·829	$4F_3$ 7416		3·85
			878			381			14	
9S 4937	4·515	$6P_2$ 28518		1·961	$5D_3$ 32436		1·840	$4F_4$ 7402		
		$7P_0$ 11289		3·118	$6D_1$ 11337		3·111			
			72			55				
		$7P_1$ 11217		3·128	$6D_2$ 11282		3·119			
			172			67				
		$7P_2$ 11045		3·152	$6D_3$ 11214		3·128			

TABLE 18 (*f*)

Zn, Singlet system

S terms	n^*	P terms	n^*	D terms	n^*
4S 75767	1·204	4P 29022	1·944	4D 13309	2·871
5S 19979	2·343	5P 12858	2·921		
6S 9730	3·358				

Zn, Triplet system

S terms	n^*	P terms	$\Delta\tilde{\nu}$	n^*	D terms	$\Delta\tilde{\nu}$	n^*
5S 22094	2·228	$4P_0$ 43455		1·589	$4D_1$ 12998		2·905
			190			3·4	
6S 10334	3·259	$4P_1$ 43265		1·592	$4D_2$ 12994		2·906
			389			5·5	
		$4P_2$ 42876		1·599	$4D_3$ 12989		2·906
		$5P_0$ 14519		2·749			
			26·7				
		$5P_1$ 14493		2·752			
			56·2				
		$5P_2$ 14436		2·757			

TABLE 18 (*g*)

Cd, Singlet system

S terms	n^*	P terms	n^*	D terms	n^*
5S 72539	1·230	5P 28846	1·949	5D 13319	2·870
6S 19229	2·388	6P 12633	2·947		
7S 9452	3·407				

Cd, Triplet system

S terms	n^*	P terms	$\Delta\tilde{\nu}$	n^*	D terms	$\Delta\tilde{\nu}$	n^*
6S 21055	2·286	$5P_0$ 42425		1·608	$5D_1$ 13052		2·899
			542			11·7	
7S 9976	3·317	$5P_1$ 41883		1·618	$5D_2$ 13041		2·901
			1171			18·2	
		$5P_2$ 40712		1·642	$5D_3$ 13023		2·903
		$6P_0$ 14148		2·785			
			70·7				
		$6P_1$ 14077		2·792			
			174				
		$6P_2$ 13903		2·810			

TABLE 18 (*h*)

Hg, Singlet system

S terms	n^*	P terms	n^*	D terms	n^*
6S 84184	1·142	6P 30115	1·909	6D 12851	2·922
7S 20256	2·328	7P 12889	2·919		
8S 9779	3·350				

Hg, Triplet system

S terms	n^*	P terms	$\Delta\tilde{\nu}$	n^*	D terms	$\Delta\tilde{\nu}$	n^*
7S 21834	2·242	$6P_0$ 46539		1·535	$6D_1$ 12848		2·922
			1767			60·1	
8S 10223	3·277	$6P_1$ 44772		1·565	$6D_2$ 12788		2·930
			4631			35·1	
		$6P_2$ 40141		1·653	$6D_3$ 12753		2·934
		$7P_0$ 14667		2·735			
			145·4				
		$7P_1$ 14522		2·749			
			1546				
		$7P_2$ 12976		2·908			

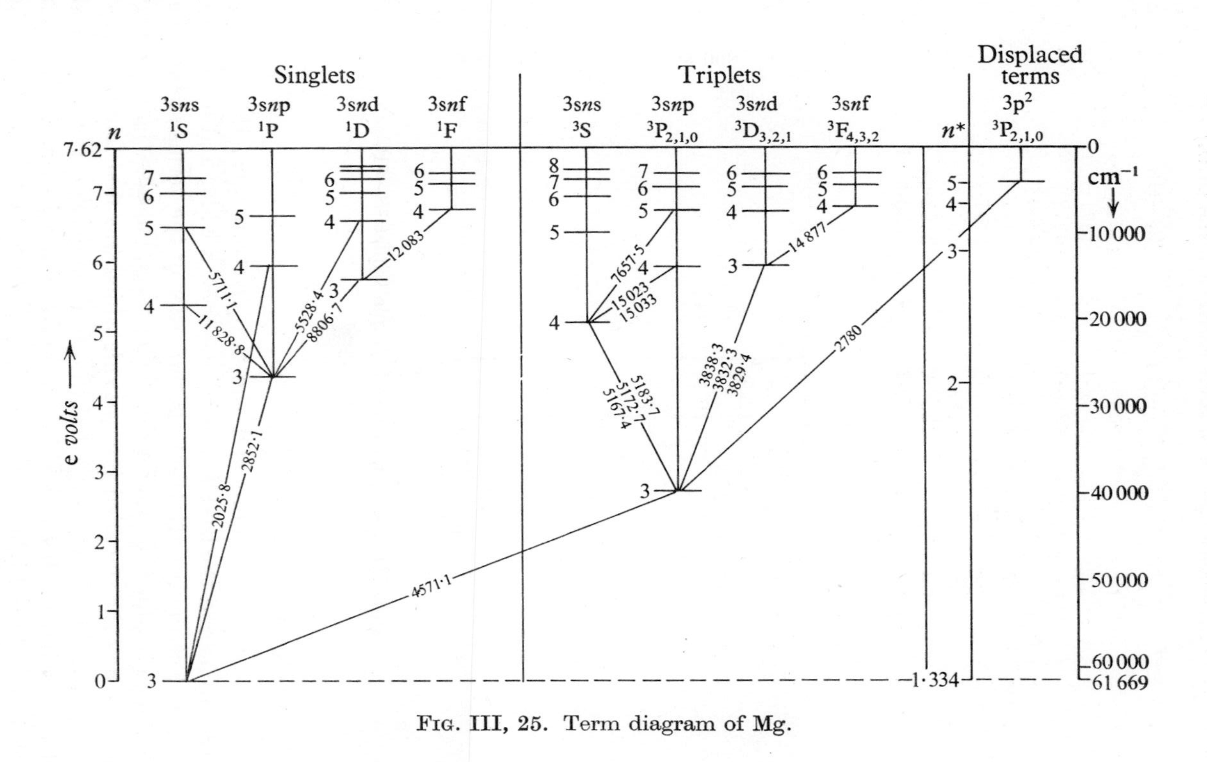

FIG. III, 25. Term diagram of Mg.

TABLE 19

Resonance lines of alkaline earth elements (A.U.)

	$^1S_0 - {}^1P_1$	$^1S_0 - {}^3P_1$
Be	2348·6	(4548·3)
Mg	2852·1	4571·1
Ca	4226·7	6572·8
Sr	4607·3	6892·6
Ba	5535·5	7911·3
Ra	4825·9	7141·2
Zn	2138·6	3075·9
Cd	2288·0	3261·0
Hg	1849·6	2536·5

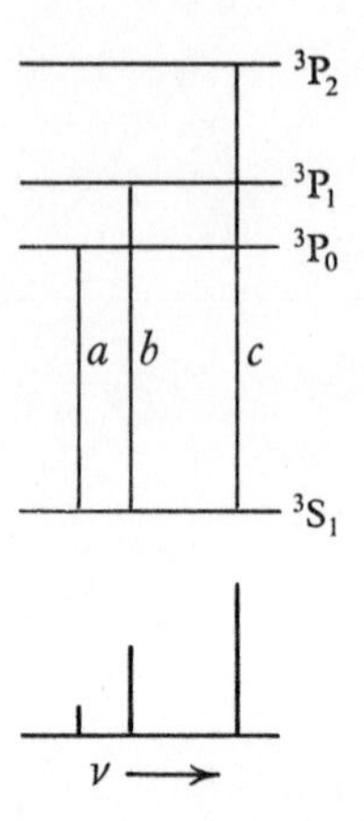

FIG. III, 26. Structure of simple triplet $^3S - {}^3P$.

3. The triplet structure

Compared with helium and the two-electron ions, the alkaline earth atoms and the iso-electronic ions show a wider and more regular triplet structure, Not only is the order of the terms *regular*, i.e. the higher j-value is found for the higher term, but also their spacings tend to obey the so-called *interval rule* which was first found empirically by Landé.[(144)] It states that in a term multiplet, the differences between adjacent levels are in the ratio of the j-values, where for each interval the higher of the two j-values is to be taken. Thus, the differences $^3P_2 - {}^3P_1$ and $^3P_1 - {}^3P_0$ are in the ratio 2:1, and the differences $^3D_3 - {}^3D_2$ and $^3D_2 - {}^3D_1$ in the ratio 3:2.

Figures III, 26 and III, 27 show the normal triplet structure of a $^3S-{}^3P$- and of a $^3P-{}^3D$-transition.

Theoretically, the interval rule arises from the magnetic interaction between two mechanical systems each of which is characterised by an angular momentum of fixed absolute value. In hydrogen these were seen to be due to the orbital motion causing a magnetic field in the direction of **L**, and the spin **S** of the same electron. In the present case of *two* electrons outside a closed shell, the strong electrostatic interaction causes the two spins to form a resultant whose value $\mathbf{S}^2$ is fixed and therefore described by a "good" quantum number $S = 1$ (or 0). This results in a magnetic dipole moment in the direction of **S**. The orbital motion causes a magnetic field in the direction of **L** (which coincides with $\mathbf{L}_2$ since $\mathbf{L}_1 = 0$ in simple spectra). We shall see later that, for more than one electron having a finite angular momentum, the same considerations apply, but **L** is then the resultant of several momenta.

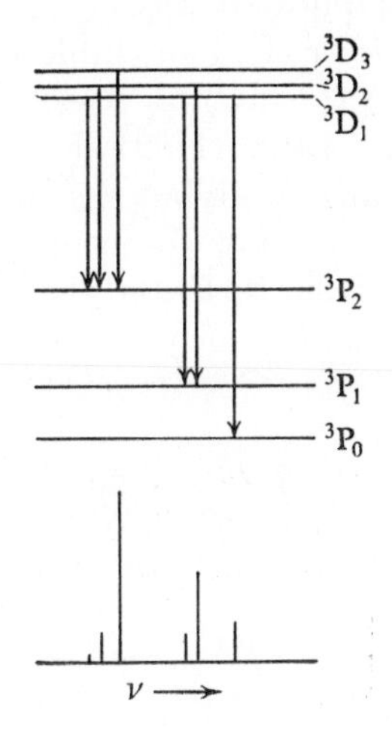

FIG. III, 27. Structure of composite triplet $^3P-{}^3D$.

The magnetic energy is, as in (51) given by

$$W' = \text{const. } \mathbf{L}\cdot\mathbf{S} = \text{const.}|L||S| \cos(L, S) \tag{85}$$

which gives the quantum-mechanical result (see p. 117)

$$W' = \text{const.}[J(J+1)-L(L+1)-S(S+1)]. \tag{86}$$

For three levels $J, J-1, J-2$ of a term multiplet (in the present case of $S = 1$ a triplet) (86) gives the ratio of the differences

$$\frac{\Delta_{J,J-1}}{\Delta_{J-1,J-2}} = \frac{J}{J-1} \tag{87}$$

which expresses the interval rule.

One condition of the validity of the interval rule is the smallness of the *magnetic* L, S-interaction compared with the *electrostatic* interaction coupling the two vectors $\mathbf{S}_1$ and $\mathbf{S}_2$ together. This is the condition of perfect *Russell–Saunders-* or *L, S-coupling.*

The example of He shows that this condition is not sufficient; in spite of the extreme smallness of the triplet splitting in He compared with the triplet–singlet difference, the interval rule is not fulfilled. This is due to the existence of an interaction depending on the angle between **L** and **S** but not in the form of a scalar product. This is the magnetic interaction between the two spins which is linked with the orbital motion in a complicated way.

For Be and the iso-electronic ions, table III, 20 shows that in the lowest triplets $2\,^3P_{0,\,1,\,2}$ the interval rule is progressively better fulfilled with increasing Z. Table III, 21 shows that in Mg and Ca the P-terms obey the rule fairly well, but that in Sr and especially in Ba it holds less accurately. Similar conditions are found in the lower ^3D terms. In the atoms from Be to Ba, the interval ratio approaches the ideal value $\frac{3}{2}$ most closely for Ca and Sr (see table III, 18) and rises again for Ba.

TABLE 20

Triplet intervals in Be and isoelectronic ions (in cm^{-1})

	Be I	B II	C III	N IV	O V	F VI	interval rule
$2^3P_1 - 2^3P_2$	2·35	16·4	56·8	144·2	306·2	576	
$2^3P_0 - 2^3P_1$	0·68	6·4	23·0	63·2	136·7	260	
ratio	3·46	2·56	2·47	2·28	2·25	2·22	2·0

The deviations from the interval rule for the lightest atoms and for ions of lowest Z are obviously due to spin–spin interactions and relativity effects, leading in He even to a reversed order of the terms. Though Heisenberg's formula (77) only holds for helium-like atoms, the fact that spin–orbit interaction increases with a higher power of Z than spin–spin interaction can be expected to be more generally true; there is much evidence that the latter interaction is negligibly small for larger Z or Z_{eff}.[145, 146]

TABLE 21

Triplet intervals in two-electron spectra, lowest P-terms (in cm^{-1})

	He	Be	Mg	Ca	Sr	Ba	Ra
$^3P_1 - {}^3P_2$	−0·077	2·35	40·71	105·88	394·2	878·0	2689·2
$^3P_0 - {}^3P_1$	−0·988	0·68	20·06	52·16	186·83	370·5	920·9
ratio	0·078	3·46	2·03	2·03	2·11	2·37	2·92

The breakdown of the interval rule in heavier elements and also in higher terms is due to an entirely different cause. The magnetic interactions can, in these elements, not be regarded as very small compared with the electrostatic interactions. This is shown by the fact that the width of the triplet splitting is no longer small compared with the singlet–triplet difference (see table III, 18). In comparing the magnetic with the electrostatic interaction, we only refer to the direction-dependent part of the latter. This will be more fully explained in chapter V.A.

4. Intensity relations in triplets

The intensities, or strictly speaking the relative strengths of the lines of a triplet, are usually found to obey the *sum rule* of Ornstein, Burger and Dorgelo (see p. 170). If one term is single or has a splitting of negligible width, the sum rule is sufficient to give the relative strengths of the components. Thus the strengths of the transitions $^3P_{2,1,0} - {}^3S_1$ are found from this rule to be in the ratio 5: 3: 1.

If both terms are split, the sum rule is not sufficient. In a composite triplet $^3P - {}^3D$, e.g. (see fig. III, 27), the sum rule, when applied to both upper and lower term, provides four equations only for the five intensity ratios of the six lines.

As explained in chapter II.D.1, however, the relative intensities can be calculated with the sole assumption that L and S are good quantum numbers, i.e. that Russell–Saunders coupling is a good approximation for the terms involved. The following formulae which were first derived by correspondence methods[147–149] and later by means of Quantum Mechanics[150] apply not only to the triplet spectra considered in this chapter, but to any complex spectra obeying Russell–Saunders coupling:

For $L-1 \to L$:

$$I(J-1 \to J) = \frac{B(L+J+S+1)(L+J+S)(L+J-S)(L+J-S-1)}{J}$$

$$I(J \to J) = -\frac{B(L+J+S+1)(L+J-S)(L-J+S)(L-J-S-1)(2J+1)}{J(J+1)}$$

$$I(J+1 \to J) = \frac{B(L-J+S)(L-J+S-1)(L-J-S-1)(L-J-S-2)}{J+1}.$$

For $L \to L$: (88)

$$I(J-1 \to J) = -\frac{A(L+J+S+1)(L+J-S)(L-J+S+1)(L-J-S)}{J}$$

$$I(J \to J) = \frac{A[L(L+1)+J(J+1)-S(S+1)]^2(2J+1)}{J(J+1)}$$

$$I(J+1 \to J) = -\frac{A(L+J+S+2)(L+J-S+1)(L-J+S)(L-J-S-1)}{J+1}$$

Since these formulae are rather cumbersome to handle, it has become customary to use the very convenient tables of White and Eliason[151] (table app. 1) giving values which have been rounded off but are accurate enough for most purposes. For use with hyperfine structures and j, j-coupling, the values of L, given in bold type, include half-integral numbers.

The deviations from Russell–Saunders coupling in the heavier elements not only cause the interval rules to become inaccurate but also lead to the breakdown of the intensity rules. In contrast to the interval rule, however, no special assumptions on the nature of the magnetic interactions need be made in deriving the intensity rules and formulae, which can therefore be expected to hold accurately for the very light elements as helium and beryllium; this has in fact been found.

5. Intercombinations and metastable terms

The progressively decreasing validity of Russell–Saunders coupling in the sequence Be, Mg, Ca, Sr, Ba furnishes the explanation for yet another phenomenon, namely the increasing strength of intercombination lines, in the given order of these elements. The selection rule $\Delta S = 0$ gradually loses its validity and, as explained in B6, the strength of the intercombination lines can be quantitatively related to the coupling conditions.

Table III, 22 shows the ratio of the strength of the lowest singlet–singlet and triplet–singlet transitions. The degree of deviation from Russell–Saunders coupling was derived in two different ways,[152] namely from the interval ratio and from the absolute value of the difference between singlet term and centre of gravity of triplet term.

TABLE 22

Relative strengths of the lines $n\ ^1S_0 - n\ ^1P_1$ and $n\ ^1S_0 - n\ ^3P_1$

	Mg	Ca	Sr	Ba	Zn	Cd	Hg
$n =$	3	4	5	6	4	5	6
obs. ratio	—	33000	1660	146	7200	680	47
calc. ratio	4×10^5	30180	1580	169	6760	637	53

The values in the table are means between the two results which differ only moderately. The agreement with the experimental values is remarkably good. For Mg, the two methods give very different results; this supports the assumption that the deviation from the interval rule in the very light elements are due to spin–spin interaction.

The lowest $^3P_1$ level can hardly be regarded as metastable in the alkaline earth elements, with the exception of the lightest. The lowest $^3P_{0,\,2}$ levels, however, cannot combine with the ground term $^1S_0$ by dipole radiation at all, according to the selection rule for j; they are *metastable* in the strict sense.

6. The simple spectra of Zn, Cd and Hg

Most spectral lines of these elements in the visible and near ultraviolet range can be reduced to term diagrams of the same structure as those of the alkaline earth elements, and the same applies to their iso-electronic ions, with the usual shift towards shorter wavelengths, for increasing ionic charges. The lower terms are included in table III 18, and spectrum and term diagram of Hg are shown in fig. III, 28 and plate 7.

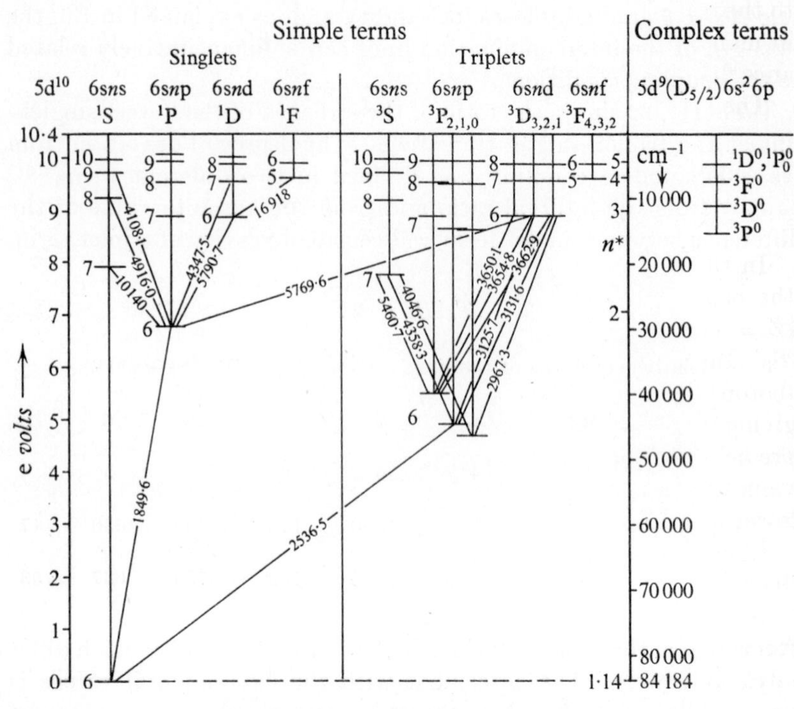

Fig. III, 28. Term diagram of Hg.

The structure of these spectra suggests that they have to be ascribed to a configuration of two electrons, one of which is always in its lowest state (4s, 5s and 6s for Zn, Cd and Hg). The remaining electrons form closed shells and do not contribute to the angular momenta. This view is in accordance with the chemical di-valency of these elements. The structure of the core will be discussed in IV.

The intercombination lines again increase in strength with increasing atomic weight, owing to increasing deviation from Russell–Saunders Coupling. The values in table III, 22 support this interpretation quantitatively in the way discussed above. For Hg, the values in table III, 18 show that the interval rule is not even approximately fulfilled.

The intercombination line 2537A ($6\,^1S_0 - 6\,^3P_1$), though fifty times weaker than the corresponding singlet–singlet line, appears as a prominent line in discharges and can easily be observed in resonance fluorescence. The mercury content of a normal laboratory atmosphere is usually enough for this line to appear in absorption. Owing to the convenience of handling mercury, the line 2537A has been used in many of the classical investigations of the phenomenon of resonance fluorescence and of the excitation of atoms by electron impacts.

The lowest $^3P_{0,2}$ levels of Zn, Cd and Hg are, of course, metastable.

E. THE SIMPLE SPECTRA OF TRIVALENT ELEMENTS (B, Al, AND Ga, In, Tl)

In the process of systematically building up the ground terms of the elements in order of their atomic number, we had reached Be ($Z = 4$) with the ground configuration $1s^2 2s^2$ and the ground term $^1S_0$. Increasing the charge to $Z = 5$ and adding a further electron (boron) we expect the lowest state available for the latter to be 2p giving the configuration $1s^2 2s^2 2p$ and a term ^2P as ground term. This prediction takes it for granted that the influence of l on the term value is smaller than that of n; otherwise the energy of 3s might be lower than that of 2p.

In fact most lines of the arc spectrum of B can be fitted into a term diagram (fig. III, 29) which resembles that of an alkali atom whose lowest ^2S term is missing, so that the lowest of the series of ^2P terms forms the ground term. The *sharp series* $n_0\,^2P - n\,^2S$ and the *diffuse series* $n_0\,^2P - n\,^2D$ appear in absorption and their limit has a shorter wave-length than that of the *principal series*. By comparison with the terms of the hydrogen spectrum, the value of the principal quantum number of the ground term is found to be 2, that of the lowest S-term 3. The term diagram can thus be explained qualitatively by the assumption that only the fifth electron is excited while the other four electrons remain in their closed shells $1s^2$ and $2s^2$, forming a spherically symmetrical core described by the term symbol $^1S_0$. This is often expressed by saying that the terms of B are *based* on the term $^1S_0$ of B^+.

Aluminium shows a very similar spectrum (fig. III, 30) which can be ascribed to the excitation of one electron whose lowest state is 3p. It is based on the ground term of Al^+ which, like that of the isoelectronic Mg, is $3s^2\,^1S_0$. In this term symbol, the remaining ten inner electrons have been omitted as irrelevant to the present qualitative discussion.

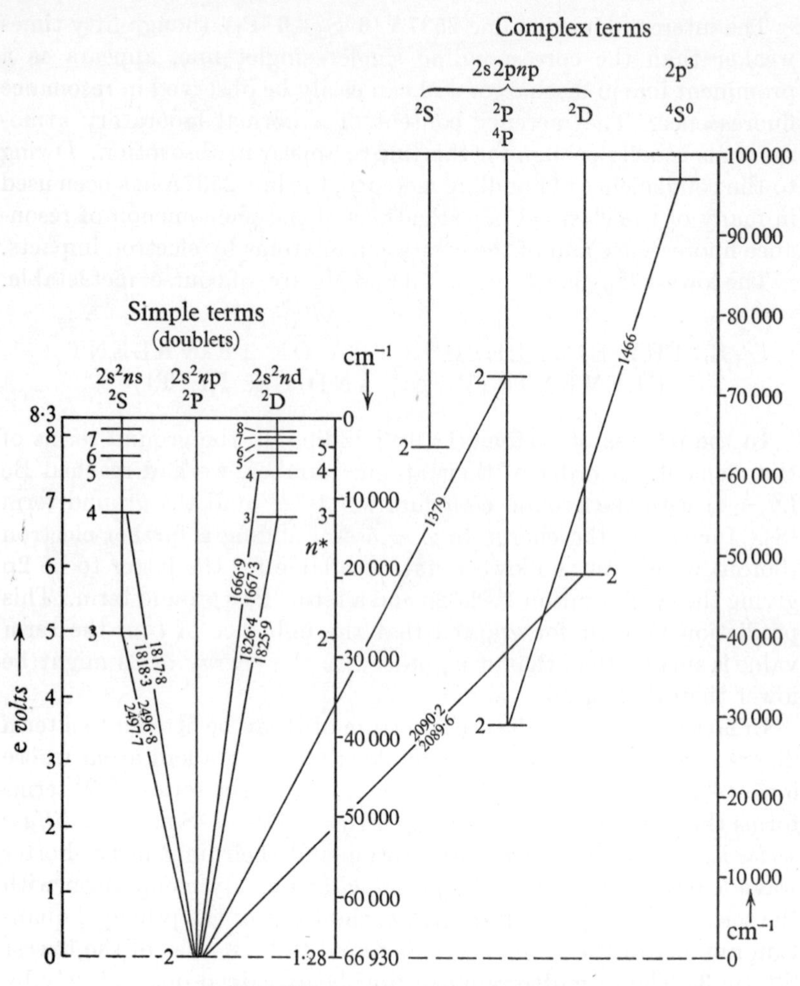

FIG. III, 29. Term diagram of B.

The elements Sc, Y and La belong chemically to the same column of tri-valent earth-elements, but their spectra are more complex, for reasons to be discussed in IV and V.

The elements in the subsidiary group, Ga, In and Tl, show the same, simple type of doublet spectrum as B and Al, with a ground term ^{2}P. The configurations of the ground term are $4s^24p$, $5s^25p$ and $6s^26p$ respectively. Table III, 23 gives the wavelengths of the resonance lines of the elements of the group, and fig. III, 31 shows the term diagram of Tl.

TABLE 23

Wavelengths of resonance lines (A.U.)

	n	$n\ ^2P_{1/2,3/2} - (n+1)\ ^2S_{1/2}$	$n\ ^2P_{1/2,3/2} - n'\ ^2D_{3/2,5/2}$
B	2	2496·8	1825·9
		2497·7	1826·4
Al	3	3944·0	3082·2
		3961·5	3092·8, 3092·7
Ga	4	4033·0	2874·2
		4172·1	2944·2, 2943·6
In	5	4101·8	3039·4
		4511·3	3258·6, 3256·1
Tl	6	3775·7	2767·9
		5350·4	3529·4, 3519·2

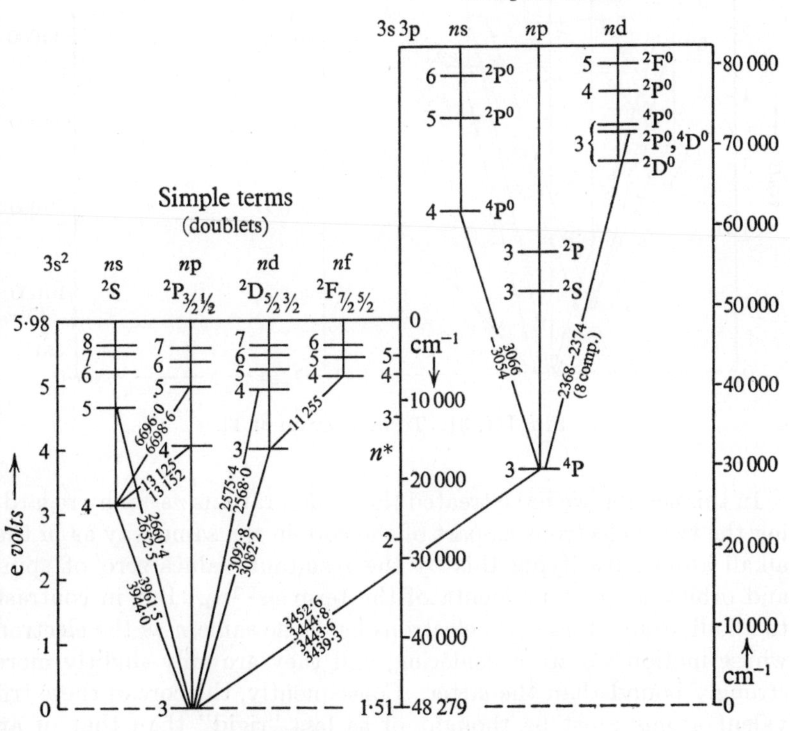

FIG. III, 30. Term diagram of Al.

The widths of doublet splitting and the intensity ratios of the components of the tri-valent atoms show the same regularities as those of the alkali spectra, and the same applies to the spectra of the iso-electronic ions. Table III, 24 gives some term values and n^* for these elements.

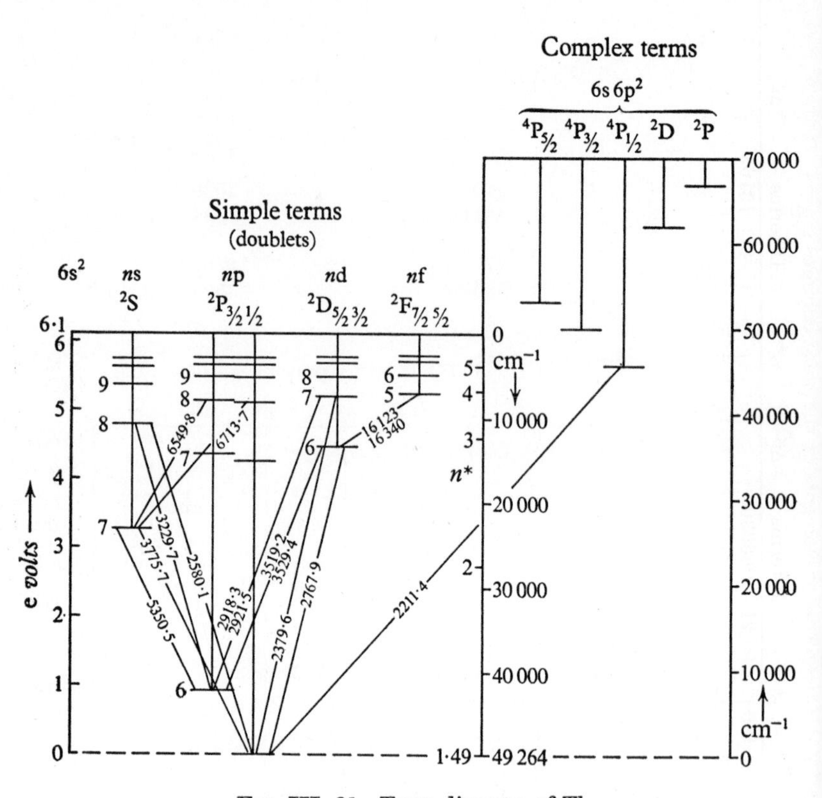

FIG. III, 31. Term diagram of Tl.

In this section we have treated the configurations ns^2np by regarding the two s-electrons as part of the core in the same way as in the alkali atoms, justifying this by the resultant values zero of spin- and orbital-angular momenta of the term $ns^2\,{}^1S_0$. But in contrast to alkali atoms, these two electrons have the same n as the electron whose motion we are considering, and they are only slightly more strongly bound than the latter. Consequently, the core of these tri-valent atoms must be thought of as less "rigid" than that of an alkali atom, though it has the same spherical symmetry.

Low terms of elements of the third column

	S terms	n^*	P terms	$\Delta\tilde{\nu}$	n^*	D terms	$\Delta\tilde{\nu}$	n^*
B	3S 26891	2·02	$2P_{1/2}$ 66930	15	1·280	3D 12163		3·00
	4S 11923	3·03	$2P_{3/2}$ 66914			4D 6932		3·98
Al	4S 22931	2·187	$3P_{1/2}$ 48279	112	1·508	$3D_{3/2}$ 15843	1·3	2·63
	5S 10589	3·219	$3P_{3/2}$ 48167		1·509	$3D_{5/2}$ 15842		
			$4P_{1/2}$ 15330	15·2	2·675			
			$4P_{3/2}$ 15314		2·677			
Ga	5S 23593	2·156	$4P_{1/2}$ 48380	826	1·51	$4D_{3/2}$ 13598	5·9	2·84
	6S 10797	3·187	$4P_{3/2}$ 47554			$4D_{5/2}$ 13592		
			$5P_{1/2}$ [15326]	[108]	2·68			
			$5P_{3/2}$ [15218]					
In	6S 22297	2·22	$5P_{1/2}$ 46670	2213	1·53	$5D_{3/2}$ 13777	23·5	2·82
	7S 10368	3·25	$5P_{3/2}$ 44457		1·57	$5D_{5/2}$ 13754		
Tl	7S 22787	2·195	$6P_{1/2}$ 49264	7793	1·493	$6D_{3/2}$ 13146	81·9	2·889
	8S 10518	3·230	$6P_{3/2}$ 41472		1·627	$6D_{5/2}$ 13064		2·898
			$7P_{1/2}$ 15105	1001	2·696			
			$7P_{3/2}$ 14103		2·790			

A direct evidence of this fact is the existence of anomalous terms which arise from the excitation of one of these s-electrons and lie not so very much higher than the regular terms of the simple spectra. They will be discussed in chapter V.

To some extent, even the normal terms are affected by the existence of the anomalous, low-lying terms which are the quantum-theoretical expression of the non-rigidity of the core: any quantitative treatment of width of doublet splitting, hyperfine structure or exact energy of a normal term requires the anomalous terms to be taken into account, e.g. in the form of perturbations which will be discussed in V.A.12.

The analysis of many of the spectra of tri-valent elements is not nearly as complete as it is often believed. To quote one example, it does not yet appear to be certain if the lowest D-term of Al I is to be ascribed to the configuration $3s^23d$ or to $3s3p^2$.

F. ZEEMAN EFFECTS IN DOUBLET SPECTRA*

1. Anomalous Zeeman effect and electron spin

The normal Zeeman effect showing the Lorentz triplet is found in hydrogen-like spectra provided the field is strong enough to allow fine structures to be neglected, and in all singlet lines of many-electron spectra; it is, in fact, observed whenever the influence of the electron spin is negligible. This is in accordance with quantum theory: for any kind of orbital motion of a point electron or even of several electrons, the ratio of the magnetic moment $\boldsymbol{\mu}$ to the angular momentum $\mathbf{L}$ has always the same value $-\gamma_0 = -e/2m_0c$. As a result of this, an external field causes all terms to split into energy levels with the same spacing $\mathscr{H}\mu_0 = \mathscr{H}eh/4\pi m_0c$, and causes the allowed transitions to form the *normal* triplet with the frequency differences $\nu_L = \mathscr{H}\mu_0/h = \mathscr{H}e/4\pi m_0c$.

The overwhelming majority of spectral lines do not fall under any of the categories mentioned, but form components of doublets, triplets or higher multiplets, or of complex groupings of lines of an even more general type. All these lines show the so-called *anomalous* Zeeman effect. Its structure can be very complex, but, in moderately weak fields, both the π and σ components are arranged symmetrically about the frequency of the undisplaced line; the π components have generally smaller displacements than the σ components (Plate 8). These facts suggest a formal description of the effect by means of the

* For Zeeman effects in triplet spectra, see V.A.11.

following assumption; the expression $m\mathscr{H}\mu_0$ for the magnetic energy is replaced by the more general expression

$$W_m = m\mathscr{H}\mu_0 g_J \tag{89}$$

where the factor g_J, known as "Landé g-factor", differs from level to level. This implies that the ratio of the magnetic moment to the angular momentum is no longer a universal constant, but is given by

$$\boldsymbol{\mu}/\mathbf{J} = -g_J e/2m_0 c = -g_J \gamma_0. \tag{90}$$

The assumption that, in multiplet terms, the total angular momentum $\mathbf{J}$ is due to the intrinsic spin of the electron as well as to orbital motion can account for this anomaly. It was largely the attempt to explain the anomalous Zeeman effect which led to the assumption of the double ratio of magnetic moment to angular momentum for the electron spin (see p. 30), namely

$$\boldsymbol{\mu}(S) = -2\gamma_0 \mathbf{S}, \quad \text{in contrast to} \quad \boldsymbol{\mu}(L) = -\gamma_0 \mathbf{L}. \tag{91}$$

According to the way in which $\mathbf{L}$ and $\mathbf{S}$ combine to give the resultant $\mathbf{J}$, the gyromagnetic ratio of the atom will generally differ from γ_0.

Figure III, 32 shows the term diagram of the Zeeman effect in the resonance lines of an alkali atom, $^2P_{\frac{1}{2},\frac{3}{2}} - {}^2S_{\frac{1}{2}}$. For the $^2S_{\frac{1}{2}}$ term the Landé g-factor has the value 2 since the angular momentum consists exclusively of the electron spin.

2. Elementary theory for weak fields

In the following sections of this chapter, we confine ourselves to alkali-like spectra in which L and S refer to a single electron. It will, however, be found that all essentials of the theory remain valid for the more general case of several electrons in Russell–Saunders coupling. By using capital letters and not confining S to the single value $\frac{1}{2}$, we provide for much of the contents of this chapter to apply also to complex spectra.

In an external field of strength H in the direction of the z axis, the following magnetic interaction energies have to be considered:

$$W_1 = \xi \mathbf{L}\cdot\mathbf{S}; \quad W_2 = -\mu_z^{(L)}\mathscr{H} = \gamma_0 L_z \mathscr{H}; \quad W_2' = -\mu_z^{(S)}\mathscr{H} = 2\gamma_0 S_z \mathscr{H} \tag{92}$$

where ξ has the same meaning as in (53). They depend on the angle between $\mathbf{L}$ and $\mathbf{S}$ and on the angles between either of these vectors and $\mathscr{H}$, and will generally vary during the motion in a complicated way.

We now assume that the external field is so weak that $W_1 \gg W_2$ and W_2'; to a first approximation, the motion can then be described (fig. III, 33) as a comparatively rapid precession of **L** and **S** about their resultant **J**. To a higher approximation, **J** itself precesses slowly about the z-axis. The angles (L, S) and $(\mathscr{H}, J)$ are constant,

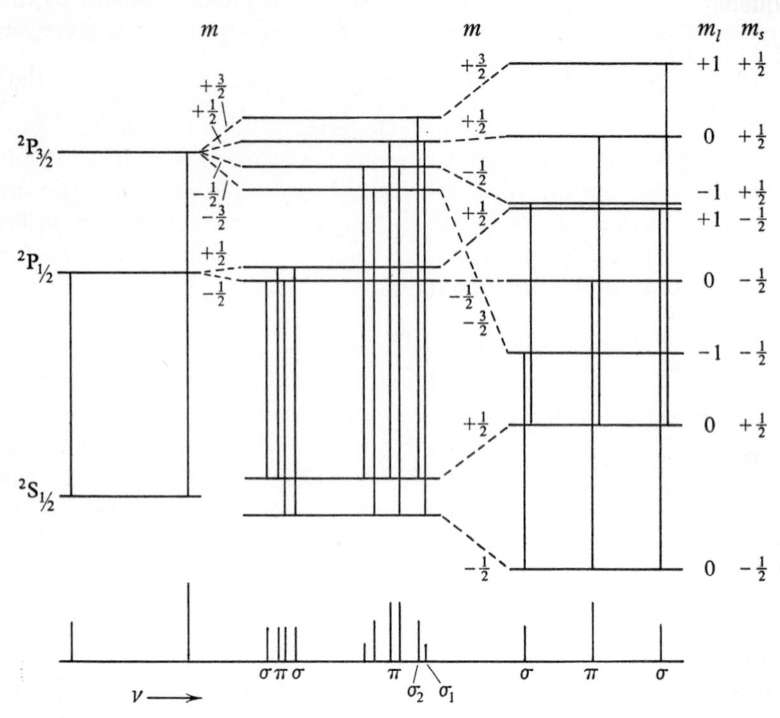

FIG. III, 32. Zeeman effect of simple doublet $^2S - {}^2P$ in weak and strong field. (H. G. Kuhn, *Atomspektren*, Akad. Verl. Ges. 1934.)

but not $(\mathscr{H}, S)$ and $(\mathscr{H}, L)$. The following angular momenta are constants of the classical motion and therefore quantised:

J_z rigorously,

$|\mathbf{J}|$ or J^2 as far as the effect of the field on the L, S precession can be neglected,

$|\mathbf{S}|$ or S^2 rigorously,

$|\mathbf{L}|$ or L^2 as far as the action of the field on the motion in the orbital plane can be neglected.

The corresponding quantum numbers m, J, S and L can be regarded as *good* quantum numbers.

The interaction energy W_1 is simply the energy causing the multiplet splitting of the term in the field-free atom. For the calculation of W_2 and W_2' the vector **L** can be replaced by the sum of its component AB (fig. III, 33) in the direction of **J**, and that at right angles to it. The latter, BD, performs a uniform rotation in a plane, with the frequency of the L, S precession. The procedure of the perturbation

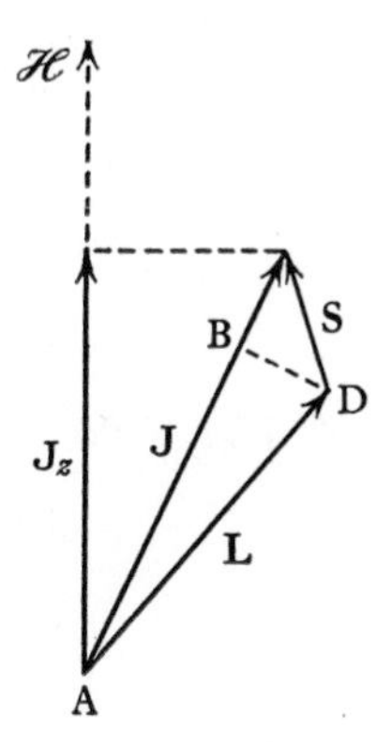

FIG. III, 33. Vector diagram of anomalous Zeeman effect in weak field.

theory is now as follows: for the calculation of the weak interaction W_2 we average over the motion caused by the stronger interaction (W_1). The average of BD then vanishes and only AB needs to be considered. Using the vectors $\boldsymbol{\mu}^{(L)}$ and $\boldsymbol{\mu}^{(S)}$ according to (91) and their resultant $\boldsymbol{\mu}$ and indicating their components in the z- and J-directions by suffices, we derive directly from the figure:

$$\mu_z = \frac{J_z}{|\mathbf{J}|}\mu_J = \frac{J_z}{|\mathbf{J}|}(\mu^{(L)} \cos \mathrm{L}, \mathrm{J} + \mu^{(S)} \cos \mathrm{S}, \mathrm{J}), \tag{93}$$

$$\mu_z = -J_z\gamma_0 \frac{\mathbf{L}\cdot\mathbf{J}+2\mathbf{S}\cdot\mathbf{J}}{\mathbf{J}^2} \tag{94}$$

and using $\mathbf{J} = \mathbf{L}+\mathbf{S}$ and $\mathbf{J}^2 = \mathbf{L}^2+\mathbf{S}^2+2\mathbf{L}\cdot\mathbf{S}$,

$$\mu_z = -J_z\gamma_0\left(1+\frac{\mathbf{J}^2+\mathbf{S}^2-\mathbf{L}^2}{2\mathbf{J}^2}\right).$$

If the squares of the angular momenta are expressed by quantum numbers according to the rules of the old quantum theory: $\mathbf{J}^2 = J^2\hbar^2$, $\mathbf{S}^2 = S^2\hbar^2$, $\mathbf{L}^2 = L^2\hbar^2$, the expression in the bracket does not correctly describe the observed facts. Landé[153] found empirically that the modified expression

$$g_J = 1 + \frac{J(J+1) + S(S+1) - L(L+1)}{2J(J+1)} \tag{95}$$

has to be used in the relation

$$\mu_z = -g_J m \mu_0 = -g_J \gamma_0 J_z \tag{96}$$

and in the expression for the energies of the states

$$W_m = \mathscr{H} m \mu_0 g_J = \mathscr{H} \gamma_0 \hbar m g_J, \tag{97}$$

in order to get agreement with the observed Zeeman patterns. Table III, 25 gives a list of g values for doublet levels.

TABLE 25

Landé g values of doublet levels

	L	$j = \frac{1}{2}$	$\frac{3}{2}$	$\frac{5}{2}$	$\frac{7}{2}$
^{2}S	0	2			
^{2}P	1	$\frac{2}{3}$	$\frac{4}{3}$		
^{2}D	2		$\frac{4}{5}$	[illegible]/5	
^{2}F	3			$\frac{6}{7}$	7

The selection and polarisation rules can be derived from the character of the classical motion. The presence of the external field merely adds a uniform precession about the z-axis to the classical motion. It has no effect on the z-coordinate of the electron and introduces a simple harmonic term into the coordinates x and y. In close analogy to the normal Zeeman effect, the rules are therefore:

$\Delta m = 0$ for the π components,

$\Delta m = \pm 1$ for the σ components.

Figure III, 34 shows the most common Zeeman patterns of doublet lines, with the normal Lorentz triplet for comparison. In the conventional way, π components are drawn above, σ components below the line. The displacements of the components are given on the right as fractions of the normal effect; the numbers in brackets refer to π components.

3. Elementary theory for strong fields

For spectral lines with very small multiplet splitting it is possible to use magnetic fields of such strength that another limiting case is closely approached: that the magnetic splittings are large compared with the multiplet splitting. The resulting pattern is found to be very similar to a normal Zeeman triplet.(154) The elementary theory of this *Paschen–Back effect* follows immediately from the assumption that the interactions between $\mathscr{H}$ and either **L** or **S** are now much larger than the interaction between **L** and **S**. To a first approximation the vectors **L** and **S** will precess rapidly and independently about

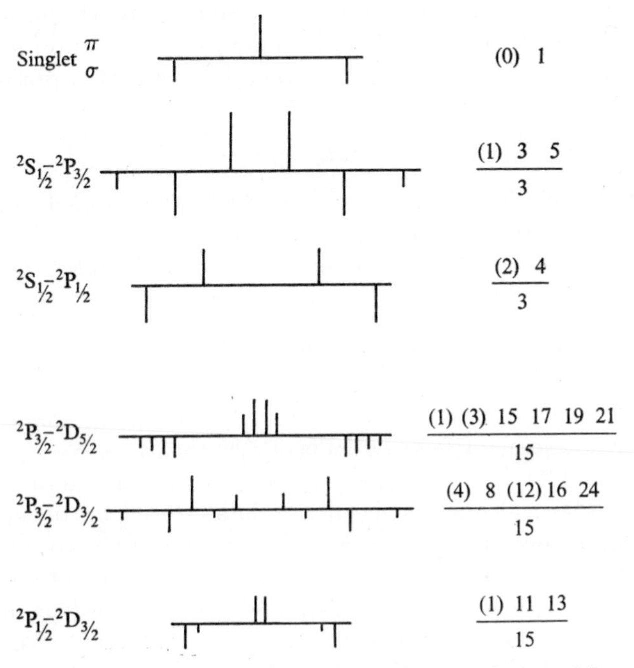

FIG. III, 34. Zeeman patterns for doublet lines, with normal Lorentz triplet for comparison.

$\mathscr{H}$, forming quantized z-components $m_L\hbar$ and $m_S\hbar$, where the quantum numbers m_L and m_S are restricted to the values L, $L-1$, $L-2$ .. $-L$ and S, $S-1$, .. $-S$ respectively. The energy of this interaction is

$$W(m_L, m_S) = m_L \mathscr{H}\mu_0 + m_S \mathscr{H} 2\mu_0. \tag{98}$$

In order to arrive at a description comparable in accuracy to that

of the weak field effect we should now have to add a term arising from the L, S interaction and causing a finer structure. The spectroscopic resolution in most experiments in strong fields is, however, so small that one can neglect this interaction. To this approximation, the energy states form a *normal* equidistant set with spacing $\mathscr{H}\mu_0$; the right side of fig. III, 32 shows the example of the alkali resonance lines.

The orbital motion is the same as in the normal effect, and the selection rules $\Delta m_L = 0$ and ± 1 for π and σ components follow immediately. The spin is no longer coupled to the orbital momentum **L** but to the external field and to the average field caused by the constant z-component of the orbital motion. Since emission and absorption of dipole radiation are due to orbital motion alone, the orientation of the spin will not change in a radiation process. The selection rule is therefore

$$\Delta m_S = 0,$$

so that the spin has no effect at all on the frequency of the radiation.

To this approximation, the pattern in very strong fields is the normal Lorentz triplet; this effect is known as *Paschen–Back effect*.

4. The quantum-mechanical theory of the anomalous Zeeman effect.[155, 156]

In contrast to the old quantum theory, quantum mechanics is able to describe Zeeman effects in fields of all strengths. We restrict ourselves to fields of not too great strength so that effects proportional to $\mathscr{H}^2$ can be neglected. This only excludes the effects of the field on the orbital motions apart from their orientation; these effects are responsible for diamagnetism, but in Zeeman patterns they become evident only in fields of extreme strength. We also restrict ourselves, in this chapter, to a single electron model, e.g. alkali atoms.

The Hamiltonian can then be written

$$H = -\frac{\hbar^2}{2m}\nabla^2 + V(r) + \xi(r)\mathbf{L}\cdot\mathbf{S} + \mathscr{H}\gamma_0(\mathrm{L}_z + 2\mathrm{S}_z). \tag{99}$$

In sufficiently weak fields the last term can be regarded as very small compared with the $\mathbf{L}\cdot\mathbf{S}$ term and can be treated as a perturbation acting separately on any one of the multiplet levels of given J. The perturbation energy can then be found as the expectation value of the last term in (99), calculated by means of eigenfunctions of the same J, in the representation in which J^2 and J_z

are diagonal, i.e. $|Jm\rangle$. If we write the perturbation term

$$\mathscr{H}\gamma_0(L_z+2S_z) = \mathscr{H}\gamma_0(J_z+S_z),$$

the component J_z is diagonal in the representation chosen and can be replaced by $m\hbar$, while for S_z the expectation value for the state $|J, m\rangle$ has to be substituted. The magnetic energy, referred to the multiplet level of given J in the absence of the field, is then

$$W_m = \mathscr{H}\gamma_0 m\hbar + \langle J,m|\mathscr{H}S_z|J,m\rangle\gamma_0.$$

Remembering that g_J is defined by $W_m = m\mathscr{H}\gamma_0\hbar g_J$, we see that the calculation of g_J has been reduced to the evaluation of $\langle J,m|S_z|J,m\rangle$. With the use of the eigenfunctions (47) one finds, e.g. the value g_J for $^2P_{\frac{3}{2}}$ most conveniently from the state $m = \frac{3}{2}$:

$$\mathscr{H}\gamma_0\langle\tfrac{3}{2}\tfrac{3}{2}|S_z|\tfrac{3}{2}\tfrac{3}{2}\rangle = \mathscr{H}\gamma_0(+)^*(+)\tfrac{1}{2}\hbar\int Y_{11}{}^*Y_{11}\sin\vartheta\,d\vartheta d\varphi = H\gamma_0\tfrac{1}{2}\hbar,$$

$$W_m = \mathscr{H}\gamma_0(\tfrac{3}{2}\hbar+\tfrac{1}{2}\hbar) = \tfrac{3}{2}\mathscr{H}\gamma_0\hbar g_J, \qquad g_J = \tfrac{4}{3}.$$

The reader can easily convince himself that any of the other functions $^2P_{\frac{3}{2}}$ of (47) give the same result, and that those of $^2P_{\frac{1}{2}}$ give $g_J = \frac{2}{3}$.

The close analogy with the classical procedure (p. 201) is more evident in the formal and more general method which may be briefly indicated: if $\mathbf{J}$ and $\boldsymbol{\mu}$ are two vector quantities obeying only very general commutation rules, e.g. any kind of angular momentum vectors, the following relation can be shown to hold for the matrix elements of these vectors and their components in a fixed (z-) direction

$$\langle J,m|\mu_z|J,m\rangle = \langle J,m|J_z|J,m\rangle\frac{\langle J||\mathbf{J}\cdot\boldsymbol{\mu}||J\rangle}{\hbar^2 J(J+1)}. \tag{100}$$

The scalar product $\mathbf{J}\cdot\boldsymbol{\mu}$ does not depend on the orientation and its matrix elements do not therefore depend on m, as indicated by the double vertical lines. For $\boldsymbol{\mu} = -\gamma_0\mathbf{L}-2\gamma_0\mathbf{S}$, (100) becomes

$$\mu_z{}^{(m)} = -m\hbar\frac{\langle J||\gamma_0\mathbf{L}\cdot\mathbf{J}||J\rangle+\langle J||2\gamma_0\mathbf{S}\cdot\mathbf{J}||J\rangle}{\hbar^2 J(J+1)} \tag{101}$$

in close analogy to (94). The eigenvalues of $\mathbf{L}\cdot\mathbf{J}$ and $\mathbf{S}\cdot\mathbf{J}$ can be found as in A.5.C. This gives

$$\mu_z^{(m)} = -m\hbar \frac{\gamma_0 \tfrac{1}{2}\hbar^2[J(J+1)+L(L+1)-S(S+1)] + \\ +2\gamma_0 \tfrac{1}{2}\hbar^2[J(J+1)+S(S+1)-L(L+1)]}{\hbar^2 J(J+1)}$$

$$= -m\gamma_0\hbar\left(1+\frac{J(J+1)+S(S+1)-L(L+1)}{2J(J+1)}\right), \tag{102}$$

which is identical with (95) combined with (96) and describes the Zeeman effect in weak fields.

In very *strong fields*, on the other hand, the $\mathbf{L}\cdot\mathbf{S}$ term in (99) can be neglected; it is then easy to calculate the perturbation by means of the eigenfunctions for which L_z and S_z are diagonal, with eigenvalues $m_L\hbar$ and $m_S\hbar$ where $m_L = L, L-1, \ldots -L$ and $m_S = \pm\tfrac{1}{2}$. The magnetic energies are found by substitution of these eigenvalues in the term $\gamma_0\mathscr{H}(L_z+2S_z)$. The eigenfunctions are simple products of space functions and spin functions, corresponding to the fact that spin and orbital motion are treated as entirely independent. For the example of a ^{2}P term ($l = 1$, $s = \tfrac{1}{2}$) one finds

$$\begin{aligned}
W_m &= 2\gamma_0\hbar\mathscr{H} && \text{for the function} && (+)Y_{1,1}\\
&= \gamma_0\hbar\mathscr{H} && \text{,, ,, ,,} && (+)Y_{1,0}\\
&= 0 && \text{,, ,, ,,} && (+)Y_{1,-1}\\
&= 0 && \text{,, ,, ,,} && (-)Y_{1,1}\\
&= -\gamma_0\hbar\mathscr{H} && \text{,, ,, ,,} && (-)Y_{1,0}\\
&= -2\gamma_0\hbar\mathscr{H} && \text{,, ,, ,,} && (-)Y_{1,-1}
\end{aligned}$$

In intermediate fields, or for the purpose of accurate calculation, the two last terms in (99) have to be treated together as a perturbation. The energy will not be diagonal in either of the representations J,m or m_L,m_S and the energy values have to be found as solutions of a secular equation.

The solution of the problem is greatly simplified by the fact, mentioned above, that the perturbation only connects states of the same m. This arises from the symmetry about the z-axis and the constancy of the sum of the z-components of all angular momenta in the classical model; the operator $J_z = L_z+S_z$ remains diagonal in fields of any strength. In the transition from weak to strong fields a state of quantum number m passes into a state $m_L+m_S = m$.

The argument now follows closely that of p. 53. For a fixed value of L, there are two levels $J = L \pm \frac{1}{2}$ in weak fields; if we write down all possible states, covering the range of m from $L+\frac{1}{2}$ to $-(L+\frac{1}{2})$ and from $L-\frac{1}{2}$ to $-(L-\frac{1}{2})$, there are two states for any value of m except for the highest and lowest which appear only once. Similarly, if we write down all possible strong field states defined by the combinations m_L, m_S, i.e. $m_L \pm \frac{1}{2}$, we find again two states for any given value of $m_L + m_S$ except for the highest and lowest which appear only once. In the perturbation matrix only pairs of states of equal $m = m_L + m_S$ have to be considered, leading to two-row matrices and to quadratic secular equations; this is due to the specific assumption of $S = \frac{1}{2}$ for doublet spectra. The states of highest and lowest m can be treated individually as non-degenerate systems; the expressions for their eigenfunctions and their energies are the same in strong and weak fields, and the latter are linear functions of $\mathscr{H}$.

It is convenient for the calculation to use the strong-field representation $|m_L, m_S\rangle$, so that the matrix elements of $\mathbf{L} \cdot \mathbf{S}$ in (99) are required in this representation. They can be derived quite generally by methods of matrix algebra, with the following result:

$$
\begin{array}{llllll}
m \rightarrow & & L+\frac{1}{2} & \overbrace{\qquad\qquad m \qquad\qquad} & & -(L+\frac{1}{2}) \\
\downarrow & (m_L, m_S) \rightarrow & (L, \frac{1}{2}) \;\; \cdots & (m+\frac{1}{2}, -\frac{1}{2}) & (m-\frac{1}{2}, +\frac{1}{2}) \;\; \cdots & (-L, -\frac{1}{2}) \\
 & \downarrow & & \cdot & \cdot & \cdot \\
L+\frac{1}{2} & (L, \frac{1}{2}) & \frac{1}{2}L\hbar^2 & \cdot & \cdot & \cdot \\
 & & & \cdot & \cdot & \cdot \\
m & \begin{cases} (m+\frac{1}{2}, -\frac{1}{2}) \;\cdots\cdots \\ (m-\frac{1}{2}, +\frac{1}{2}) \;\cdots\cdots \end{cases} & & \begin{matrix} A \\ C \end{matrix} & \begin{matrix} B \\ D \end{matrix} & \begin{matrix} \cdot \\ \cdot \end{matrix} \\
 & & & & & \cdot \\
-(L+\frac{1}{2}) & (-L, -\frac{1}{2}) & \cdots\cdots & \cdots\cdots & \cdots\cdots & \frac{1}{2}L\hbar^2
\end{array}
\qquad (103)
$$

with the abbreviations

$$
A = -\tfrac{1}{2}\hbar^2(m+\tfrac{1}{2}) \qquad B = \tfrac{1}{2}\hbar^2\sqrt{[(L+m+\tfrac{1}{2})(L-m+\tfrac{1}{2})]}
$$

$$
C = \tfrac{1}{2}\hbar^2\sqrt{[(L+m+\tfrac{1}{2})(L-m+\tfrac{1}{2})]} \qquad D = \tfrac{1}{2}\hbar^2(m-\tfrac{1}{2}).
$$

Elements having the same m are bracketed together, since only these are required for any one determinant.

The last term in (99) is diagonal and becomes $\gamma_0\mathscr{H}\hbar(m_L+2m_S)$ which assumes the values $\gamma_0\mathscr{H}\hbar(m-\frac{1}{2})$ and $\gamma_0\mathscr{H}\hbar(m+\frac{1}{2})$ for the two states $|m_L, m_S\rangle = |m+\frac{1}{2}, -\frac{1}{2}\rangle$ and $|m-\frac{1}{2}, +\frac{1}{2}\rangle$ respectively. The secular equation is then found to be

$$\begin{vmatrix} -\frac{1}{2}a(m+\frac{1}{2})+\hbar\mathscr{H}\gamma_0(m-\frac{1}{2})-W & \frac{1}{2}a\sqrt{[(L+m+\frac{1}{2})(L-m+\frac{1}{2})]} \\ \frac{1}{2}a\sqrt{[(L+m+\frac{1}{2})(L-m+\frac{1}{2})]} & \frac{1}{2}a(m-\frac{1}{2})+\hbar\mathscr{H}\gamma_0(m+\frac{1}{2})-W \end{vmatrix} = 0 \tag{104}$$

where

$$a = \hbar^2 \int_0^\infty |R(r)|^2\xi(r)r^2\,\mathrm{d}r \tag{105}$$

(see p. 167).

The two solutions are

$$W_m = \hbar\mathscr{H}\gamma_0 m - \tfrac{1}{4}a \pm$$
$$\pm\tfrac{1}{4}\sqrt{\{(4\hbar\mathscr{H}\gamma_0 m - a)^2 - 4[\hbar^2\mathscr{H}^2\gamma_0{}^2(4m^2-1) - 4\hbar a\mathscr{H}\gamma_0 m - a^2L(L+1)]\}}.$$

For the two extreme values $m = L+\frac{1}{2}$ and $m = -(L+\frac{1}{2})$, there is only one state in each case, and its energy is the same as that in the strong field approximation, except for an additional constant:

$$\begin{aligned} W_{L+\frac{1}{2}} &= \tfrac{1}{2}La + \hbar\mathscr{H}\gamma_0(L+1), \\ W_{-(L+\frac{1}{2})} &= \tfrac{1}{2}La - \hbar\mathscr{H}\gamma_0(L+1). \end{aligned} \tag{106}$$

It is not difficult to derive these relations directly without recourse to the general results of matrix algebra. This may be briefly indicated by the example of a ^{2}P term. The correlation of states in the two representations is shown in the following diagram:

Weak field J, m	$m = m_L + m_S$	strong field m_S, m_L
$\frac{3}{2}, \frac{3}{2}$	$\frac{3}{2}$	$\frac{1}{2}$, 1
$\frac{3}{2}, \frac{1}{2}$	$\frac{1}{2}$	$\frac{1}{2}$, 0
$\frac{3}{2}, -\frac{1}{2}$	$-\frac{1}{2}$	$\frac{1}{2}$, −1
$\frac{3}{2}, -\frac{3}{2}$		$-\frac{1}{2}$, 1
$\frac{1}{2}, \frac{1}{2}$		$-\frac{1}{2}$, 0
$\frac{1}{2}, -\frac{1}{2}$	$-\frac{3}{2}$	$-\frac{1}{2}$, −1.

If the perturbing terms in (99) are written

$$\xi(r)(L_xS_x+L_yS_y+L_zS_z)+\mathscr{H}\gamma_0(L_z+2S_z) \tag{107}$$

the two functions for the pair of states with $m = \frac{1}{2}$ are $(-)\ Y_{1,1}$

and $(+)\ Y_{1,0}$. Choosing these functions in the given order for numbering lines and columns, we find for the element in the second line and first column of the two-line, two-column matrix:

$$\langle(+)Y_{1,0}|\xi L_x S_x|(-)Y_{1,1}\rangle = \langle(+)|S_x|(-)\rangle\langle Y_{1,0}|\xi L_x|Y_{1\ 1}\rangle$$

because the spin operators do not act on space functions and vice versa. The first factor is the element in the first line and second row of the matrix S_x which has the value $\hbar/2$ (see table II, 3). The second factor combines with the corresponding factor of the y-component in (107) to give

$$\langle Y_{1,0}|\xi(L_x - iL_y)|Y_{1,1}\rangle.$$

In this matrix element ξ depends on r only, and the integration over this variable can be carried out independently, giving the factor $a/\hbar^2$ (see eq. 105). The remaining element is merely a linear combination of the elements in the second line and first column of the angular momentum matrices J_x and J_y for $J = 1$ and gives the result $\hbar^2/\sqrt{2}$. The term $L_z S_z$ vanishes since the off-diagonal elements of S_z are zero. The final result for the element $m_S = \frac{1}{2}$, $m_L = 0$; $m_S' = -\frac{1}{2}$, $m_L' = 1$ of the first perturbing term in (99) is thus found to be $a/\sqrt{2}$. The perturbation energy $2\gamma_0 S_z H$ vanishes for this element. The other elements are derived similarly and give the matrix

$$\begin{pmatrix} -\frac{1}{2}a & a/\sqrt{2} \\ a/\sqrt{2} & \gamma_0\hbar\mathscr{H} \end{pmatrix}.$$

The solutions of the secular equation are

$$W_{m=\frac{1}{2}} = \tfrac{1}{2}[\gamma_0\hbar\mathscr{H} - \tfrac{1}{2}a \pm (\gamma_0^2\hbar^2\mathscr{H}^2 + \gamma_0\hbar a\mathscr{H} + 9/4a^2)^{\frac{1}{2}}]. \qquad (108a)$$

The solutions for the pair of states $m = -\frac{1}{2}$ are

$$W_{m=-\frac{1}{2}} = \tfrac{1}{2}[-\gamma_0\hbar\mathscr{H} - \tfrac{1}{2}a \pm (\gamma_0^2\hbar^2\mathscr{H}^2 - \gamma_0\hbar a\mathscr{H} + 9/4a^2)^{\frac{1}{2}}]. \qquad (108b)$$

For the remaining two states one finds directly

$$W_{m=\frac{3}{2}} = \tfrac{1}{2}a + 2\gamma_0\hbar\mathscr{H} \quad \text{and} \quad W_{m=-\frac{3}{2}} = \tfrac{1}{2}a - 2\gamma_0\hbar\mathscr{H}. \qquad (108c)$$

Figure III, 35 shows the energies of these six states of a ^{2}P-term as functions of $\gamma_0\hbar\mathscr{H}/a$. It gives an example of a rule which can be shown to hold quite generally in any magnetic transformation: *states of the same m do not cross.*

The validity of this rule is based on the property of m as a *good* quantum number in fields of any strength, arising from the constancy of $J_z = L_z + S_z$ in the classical model. This behaviour can be understood in terms of the general perturbation theory (p. 57). States having the same m will perturb one another, and only those; this leads to a *repulsion* of the energy levels which increases with decreasing energy difference. The plot of the levels as function of the perturbation parameter, e.g. of H, thus shows a curvature tending to bend each level away from the other.

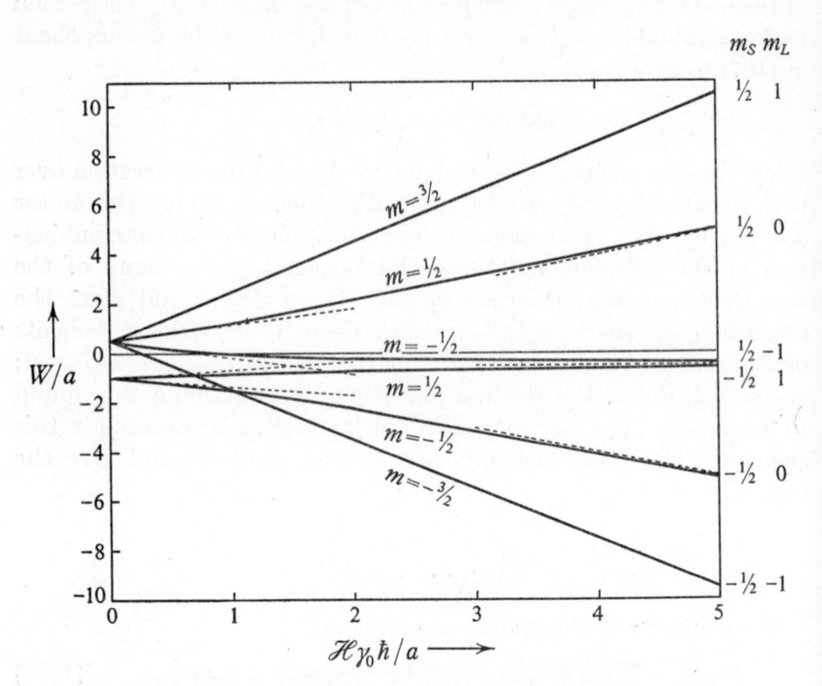

FIG. III, 35. Zeeman levels of ^2P term. conversion from weak to strong fields (E. U. Condon & G. H. Shortley, *The Theory of Atomic Spectra*, Cambridge Univ. Press 1935).

5. Intensities of Zeeman components

The relative intensities of Zeeman components in weak fields can be derived directly from the correspondence principle. In the presence of the field the states are no longer degenerate and are all equally populated provided the magnetic energies are small compared with the thermal energy kT. All statistical weight factors are then unity and the concepts of strength and intensity can be used indiscriminately.

We define

$$\sum_{m,m'} I_\pi$$

as the sum of the intensities of all π-components of a given line in *transverse* observation, and

$$\sum_{m,m'} I_\sigma$$

as the sum of all σ-components in *transverse* observation. Since the light must become unpolarised in the limit of vanishing field strength we must expect:

$$\sum_{m,\, m'} I_\sigma = \sum_{m,\, m'} I_\pi. \tag{109}$$

We further assume that the Zeeman pattern is symmetrical, in accordance with its nature as a phenomenon of modulation of the field-free motion by the Larmor precessions. Any Zeeman component need then be numbered only on one side of the pattern.

A definite value of m of the atom in the excited state corresponds to a definite orientation of the total angular momentum in space, for weak fields. Since the orientation cannot affect the rate at which the atom radiates we must assume that the total energy of radiation emitted in all Zeeman components arising from one excited state of given m is the same for any state. The rate of emission for any π-component, averaged over all directions is found as the sum of the intensities observed in the two transverse directions (x or y) and the longitudinal direction (z):

$$\overline{I_\pi} = 2I_\pi + 0.$$

Correspondingly we find, by considering that a σ-component of given frequency corresponds to one oscillator when viewed from any transverse direction, and to two oscillators, e.g. along the x and y axes, when viewed longitudinally:

$$\overline{I_\sigma} = 4I_\sigma.$$

In comparing total emission rates of energy, we thus have to give twice the weight to σ-components as to π-components, both in transverse observation. This leads to the sum rule for the total emission from two levels m_1 and m_2[157]:

$$\sum_{m'} [I_\pi(m_1) + 2I_\sigma(m_1)] = \sum_{m'} [I_\pi(m_2) + 2I_\sigma(m_2)] \tag{110}$$

where the summation is to be carried out over all values m' of the lower level.

The method may be illustrated by the examples of the transitions $^2S_{\frac{1}{2}}-{}^2P_{\frac{1}{2}}$ and $^2S_{\frac{1}{2}}-{}^2P_{\frac{3}{2}}$ (see fig. III, 32). In the first of these, eq. (109), together with the symmetry of the pattern, shows at once the equality of all four components in transverse observation. For the second doublet line we have to apply not only (109) but also (110), the latter, e.g. for the two levels $m = \frac{1}{2}$ and $\frac{3}{2}$:

$$I_{\sigma 1}+I_{\sigma 2} = I_{\pi}$$
$$2I_{\sigma_1}+I_{\pi} = 2I_{\sigma_2},$$

giving the result

$$I_{\pi} : I_{\sigma_2} : I_{\sigma_1} = 4:3:1.$$

The same arguments can be applied to the general case of a transition $J \rightarrow J$ and $J \rightarrow J+1$ and lead, after elementary but somewhat involved calculation, to the formulae

$$\begin{aligned} J \rightarrow J, \quad & m \rightarrow m \pm 1, \quad & (\sigma) &= A(J \pm m+1)(J \mp m) \\ & m \rightarrow m, \quad & (\pi) &= 4Am^2 \\ J \rightarrow J+1, \quad & m \rightarrow m \pm 1, \quad & (\sigma) &= B(J \pm m+1)(J \pm m+2) \\ & m \rightarrow m, \quad & (\pi) &= 4B(J+m+1)(J-m+1). \end{aligned} \tag{111}$$

Since final and initial states can be interchanged, transitions $J \rightarrow J-1$ are included in these formulae.

The underlying postulates are so general that the relations (111) hold not only for Russell–Saunders multiplets but for transitions in any kind of coupling scheme, provided the field is small enough to preserve the symmetry of the pattern.

Quantum-mechanical calculation of the matrices of the electric dipole moment leads to the same results; the assumptions made above in terms of the correspondence principle are introduced in the form of certain commutation relations. The procedure is similar to that outlined in the discussion of the selection rules in II.D.1.

In weak fields, the eigenfunctions $|J, m\rangle$, such as (47), have to be used for calculation of the matrix elements of the dipole moment, in strong fields the functions $|m_L, m_S\rangle$. In intermediate fields, linear combinations of the latter have to be formed whose coefficients depend on the field strength. Only the states with $|m| = J$ have the same eigenfunctions in fields of arbitrary strength.

Most of the Zeeman components will therefore change their intensities with change of field strength, and some transitions which are allowed in weak fields become forbidden in strong fields and vice versa. Figure III, 36[158] shows the conversion of a doublet $^2S - {}^2P$ from weak to strong fields, with the intensities indicated by the width of the curves.

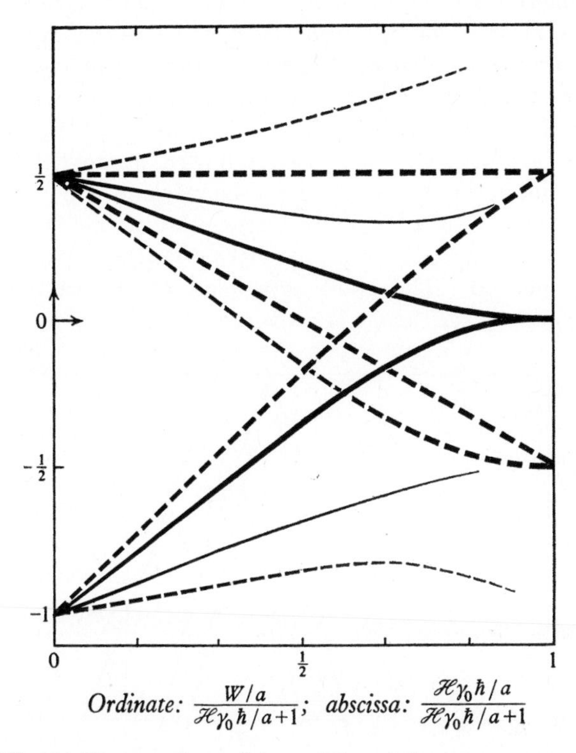

FIG. III, 36. Frequencies and intensities of Zeeman components in conversion from weak to strong fields (K. Darwin, ref. 158).

The fundamental sum rules (109) and (110) are valid at arbitrary field strength, but the pattern can no longer be assumed to be symmetrical, and the relations (111) lose their validity.

6. Observation of Zeeman effects

The theory discussed above applies to all doublet spectra, and the effects have been observed in many spectra of the atoms in the first and third column of the periodic table and their iso-electronic ions. In order to produce splittings considerably larger than the Doppler width in ordinary discharge sources and to allow good resolution with grating spectrographs, one normally uses fields of over 10,000 gauss; with the latter value, the Larmor splitting is, according to (22) about 0·47 cm^{-1}. Plate 8 shows the anomalous Zeeman effect of the resonance lines of Na in a weak field.

The transition to the Paschen–Back effect can only be observed if the magnetic splitting is larger than the doublet splitting. The practical limitation of magnetic fields restricts such measurements to the very smallest doublet splittings such as occur in the lightest elements. In the resonance lines of Li, $2\ ^2S_{\frac{1}{2}} - 2\ ^2P_{\frac{1}{2},\frac{3}{2}}$, λ6708, the magnetic transformation has been followed up to fields of 44,000 gauss,[159] and in the corresponding resonance line of Be^+, λ3131, up to fields of 300,000 gauss.[160] The extremely high fields in this latter work were achieved only intermittently, by means of a special short circuiting method. The good agreement of the results with the quantum mechanical theory is shown in fig. III, 37.

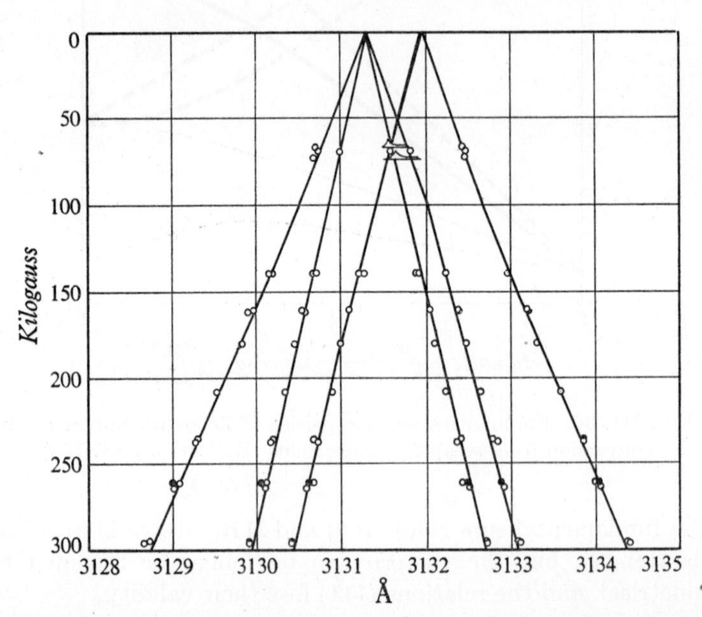

Fig. III, 37. Conversion of Zeeman pattern from weak to strong field in Be^+, σ-components only. (P. Kapitza, P. G. Strelkow & E. Laurman, ref. 160.)

Indications of the finer structure, due to the *l*, *s*-interaction, have been found in the strong-field pattern of the resonance lines of Li.[161]

Also in the forbidden alkali lines $^2S_{\frac{1}{2}} - {}^2D_{\frac{3}{2},\frac{5}{2}}$ Zeeman effects have been observed[138] (see p. 174) and have allowed the character of the radiation to be established.

For electric quadrupole radiation the selection rule for m is

$$\Delta m = 0, \pm 1, \pm 2,$$

while for magnetic dipole radiation the rule

$$\Delta m = 0, \pm 1$$

holds, but in the polarisation of the latter the magnetic vector takes the place of the electric vector, so that the polarisations of the transitions 0 and ± 1 are reversed as compared with the electric dipole radiation.

7. The quadratic effect

In the normal or the anomalous Zeeman effect the energies and frequencies were found to be linear functions of $\mathscr{H}$ if the region of transformation to Paschen–Back effect is excluded. In the elementary treatment the magnetic energy appears either as kinetic energy of the Larmor precession or as potential energy of the fixed magnetic dipole of the orbit in the external field. In the quantum-mechanical treatment the degeneracy of the states of different m causes, in the first-order perturbation, magnetic energies proportional to $\mathscr{H}$. The eigenfunctions, or the shapes of the orbits are, to this approximation, unaffected by the field.

In the second-order approximation the change of the orbits has to be included. In a classical orbit, a magnetic field causes a change of the magnetic moment proportional to $\mathscr{H}$ and to the area A of the orbit. This is due to the induction effect and gives rise to the diamagnetic susceptibility. The energy is increased in proportion to $\mathscr{H}^2A$. Quantum mechanics leads to similar relations which can be expected to apply rigorously to atomic systems; the effect is proportional to $\mathscr{H}^2\overline{r^2}$ or approximately to $\mathscr{H}^2n^{*4}$, where n^* is the effective quantum number. It was therefore expected to be observable in large fields and for highly excited states.

The quadratic Zeeman effect was observed in the high members of the absorption series of alkali metals[162,163] and was found to be in good agreement with the theory.[164]

8. The anomalous g value of the electron spin

In the Zeeman effect, the magnetic energy forms only an extremely small fraction of the total energy of the light quantum.

The Larmor frequency thus appears superimposed, as a modulation, on the much higher optical frequency. Doppler width and radiation width, both of which increase with frequency, are therefore appreciable and set a limit to the accuracy to which magnetic splittings can be measured optically.

They can be measured much more accurately by means of transitions between the magnetic states of the ground level. These are allowed as *magnetic dipole* radiation, their frequencies are of the order of radio-frequencies, and they can be observed by means of the atomic beam resonance method.[165]

While the main importance of such work lies in the study of hyperfine structures which will be described later (VI) one very important result of a general nature has been found by Kusch and Foley[166]: the value of the Landé factor g of the electron is not exactly 2 as Dirac's theory requires but has the value 2·0024. This fact has now been well established by measurement of the magnetic splitting of the ground term of several alkali atoms and other doublet ground states. It can be explained by quantum electrodynamics[167] and is closely connected with the "Lamb shift" (p. 120). The anomaly has also been established, though with much lower accuracy, by optical methods.[168] The theoretical formula

$$g_s = 2(1+\alpha/2\pi - 0{\cdot}328\alpha^2/\pi^2 = 1{\cdot}00116 \qquad (112)$$

agrees well with the observed values, especially if various corrections are taken into account.[169]

G. THE QUADRATIC STARK EFFECT[170,171]

With the exception of hydrogen-like spectra, all atomic spectra show displacements—and sometimes splittings—in an electric field which are proportional to the *square* of the field strength F. Even hydrogen-like spectra show quadratic effects if the field is so weak that its effects are small compared with the fine structure.

These facts and the most important features of the quadratic Stark effect can be derived immediately from the treatment given in III.A.4. The perturbing term in the Hamiltonian, zeF, is an odd function, commuting with L_z. This causes all matrix elements H' to vanish except those which connect states of opposite parity and equal m. Linear perturbation, due to degeneracy, can only occur if states of different parity, and thus of different l for a single-electron atom, have the same energy; this applies to hydrogen-like atoms in fields of such strength that the energy differences of the fine structure can be neglected compared with the perturbation energy. In other atoms, however, there is no such degeneracy and the perturbation of lowest order is given by the term

$$\sum_n{}' |H_{mn}|^2/(E_m - E_n)$$

in (II.86). Owing to the influence of the denominator it is often enough to consider, in this sum, only the closest of the perturbing terms whose effect consists in a "repulsion" of the perturbed term proportional to F^2.

The displacement depends on m^2 so that the level will generally split into several states, one for each $|m|$. The remaining degeneracy for states $\pm m$ is due to the fact that an electric field cannot define the sense of a rotation or precession.

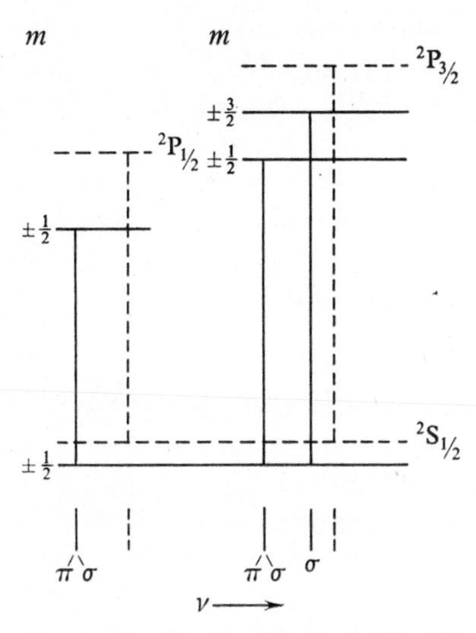

Fig. III, 38. Quadratic Stark effect of $^2S - {}^2P$ transition.

Taking as an example a line of the principal series $n\ {}^2S_{\frac{1}{2}} - n'\ {}^2P_{\frac{1}{2},\frac{3}{2}}$ in an alkali spectrum, we find that there is no other term of different parity near the ground term $n\ {}^2S_{\frac{1}{2}}$ so that its Stark displacement will be very small. For the excited term $n'\ {}^2P$, the nearest term will be $n'\ {}^2D$ which is slightly higher and will depress the states of the term $n'\ {}^2P$. The result is a red-shift and splitting of the lines considered. The resulting pattern is shown in fig. III, 38. In transverse observation the polarisation of the components is the same as in the Zeeman effect and is indicated by π and σ, but in longitudina observation the light is entirely unpolarised owing to the $\pm$ degeneracy.

Since the arrangement of levels is qualitatively similar in other terms and in all alkali-like spectra, the red shift is most commonly found in these.

In the principal series of the alkali metals the Stark effect can be conveniently and accurately measured in absorption. In weak fields the patterns have to be observed by interferometric methods and show the influence of hyperfine structure.[172,173] The Stark effect constant increases very rapidly with the principal quantum number and in the high members of the absorption series of alkali metals very small fields have been found to cause marked effects.[139,140,174] While the effects on the energy levels can be observed as line shifts, the effect on the eigenfunctions, the mixing of functions of different l, accounts for the observed changes in the strength of lines; the selection rule for l is no longer valid and lines of the type $n\ {}^2\mathrm{S}-n'\ {}^2\mathrm{D}$, $n\ {}^2\mathrm{S}-n'\ {}^2S$ .. are observed. The photometer tracing in Plate 9 shows the absorption series of Rb in a field of 300 volts per cm.; the forbidden lines are seen to become more pronounced with increasing n.

Stark effects in other spectra can be understood in a similar way. In the heavier atoms the effects are generally much smaller; this can be ascribed to the increased penetration effects causing greater separation of terms of equal n but different l . For highly excited terms, on the other hand, the effects increase rapidly as the series limit is approached, owing to the closing up of the terms.

In the lighter atoms the term shifts due to the field can even be made to exceed the differences between terms of the same n and different l. These terms can then be regarded as degenerate and, in accordance with the theory, the Stark effect is found to pass from the quadratic to a linear effect. This transformation has been most thoroughly studied in the spectrum of helium by Foster[176] who was the first to disclose the full patterns of Stark effects and to apply quantum-mechanical theory to them.

Plate 10 shows the effect in the line $2\ {}^1\mathrm{P}-6\ {}^1\mathrm{D}$ (4144 A). The field was caused by the cathode fall in a discharge and increases downward in the figure to a maximum value of 85000 volts per cm. In weak fields, near the top, a shift towards decreasing wave numbers is observed, representing the quadratic effect of the field on the term $6\ {}^1\mathrm{D}$. With increasing field, the breakdown of the l-selection rule causes at first the appearance of the "forbidden" transitions from the closely adjacent terms $6\ {}^1\mathrm{P}$, $6\ {}^1\mathrm{F}$, $6\ {}^1\mathrm{G}$, $6\ {}^1\mathrm{H}$ ($l = 1, 3, 4, 5$) the first of which is particularly conspicuous. In still larger fields, the pattern tends towards the symmetrical arrangement of components of the linear effect in which all the terms mentioned have lost their identity; the field is then acting on the level $n = 6$ which is to be regarded as degenerate in l, except for the value $l = 0$ which still forms the separate term $6\,^1S$. The latter suffers essentially a quadratic displacement effect.

The apparently anomalous position of the terms $6\,^1P$ and $6\,^1D$ relative to one another has been discussed on p. 144.

Useful formulae for the Stark effect constants of the terms of simple spectra have been derived by the method of the old quantum theory[177] and by wave mechanics.[178,179] The accuracy of the latter depends on the wave functions available.

The perturbation operator eFz of the Stark effect is the same as the operator of the z-component of the dipole moment which forms the amplitude in the emission or absorption of light (see p. 69). Measurements of Stark effects therefore offer a possibility of studying transition probabilities. The method only gives the sum of several dipole amplitudes, but in spite of this restriction it has been used.[180]

The constant of the Stark effect of a state is closely connected with the polarizability α of an atom in this state, as defined by

$$P = \alpha F$$

where P is the induced dipole moment. The energy change of the atom is then

$$\Delta W = \int F\, \mathrm{d}P = \tfrac{1}{2}\alpha F^2.$$

In wave-mechanical terms, the quadratic effect is due to a unilateral deformation of the wave function and thus of the charge distribution. In Bohr–Sommerfeld's theory the precession of the orbital ellipse is slowed down on the down-field side causing the centre of gravity of the negative charge in the time-average to be displaced. The effect is the greater the slower the precession, i.e. the smaller the penetration effect.

Observations of the quadratic effect and its conversion into the the linear effect in hydrogen-like spectra require very high resolving power, and the patterns cannot usually be resolved completely.[181,182] The theory has to take into account the removal of degeneracy which is due to the Lamb shift.[183]

CHAPTER

IV. Periodic Table and X-ray Spectra

A. THE PERIODIC PROPERTIES OF THE ELEMENTS*

1. Pauli principle and shell structure

It has long been known that many properties of the elements vary in a markedly periodic manner as functions of Z. Examples are the chemical properties, the ionisation potentials and the characteristic features of the optical spectra. These facts pointed to some kind of shell structure in the arrangement of the electrons. The inert gases, e.g., with an apparently very stable arrangement of electrons, are followed by alkali metals in which one electron, the *valence electron*, is much more loosely bound and could be imagined as outside the stable configuration or *shell* structure formed by the other electrons. This valency electron is mainly responsible for the chemical behaviour of the element and for its optical spectrum. The knowledge of the spectroscopic term structure of the elements allowed a more quantitative approach; comparison of the term values of alkali atoms with those of hydrogen showed, e.g. that the valence electron in the ground states of Li, Na, K, Rb and Cs has the principal quantum number 2, 3, 4, 5 and 6 respectively.

On the basis of Bohr's quantum theory and of Pauli's exclusion principle it became finally possible to deduce, at least in a qualitative manner, the complexity of these periodic phenomena from a few very general and simple postulates. This mainly qualitative deduction which is the subject of this section forms one of the most impressive successes of quantum theory. The most important contributions to it are due to Kossel,[1] Ladenburg,[2] Bohr[3] and Stoner.[4] Modern quantum mechanics has made it possible to tackle the almost unlimited task of deducing the chemical and physical properties of the elements quantitatively from first principles.

The theoretical basis of the method has been described in the treatment of the Li atom (p. 150): the third electron cannot be in the same quantum state as the two 1s electrons without violating the

* See app. table 2, (from C. E. Moore, *Atomic Energy Levels*, vol. III, table 36).

Pauli principle. The state of lowest possible energy for the third electron is then 2s so that it is much more loosely bound than the other two; it acts chemically as a *valence electron* and is responsible for the entire spectrum of Li in the *optical* (infra-red, visible and ultra-violet) range. With He, $Z = 2$, the shell of electrons with $n = 1$ or K-shell, is said to be completed.

Quite generally the maximum number of electrons having the same n is described as a *complete shell*, those of the same n and l form a *sub-shell*. Electrons having the same values of n and l are called *equivalent*.

The higher the value n of an electron the higher will generally be its energy, and the greater the average value of its distance from the nucleus. In this sense one speaks of lower and higher or inner and outer shells; but these attributes must not be taken too literally, because energy and average distance depend on l as well as on n. The attraction due to the nuclear charge would cause the energy to depend on n alone, but the electrostatic effect of the other electrons cause the energy to be the lower the smaller the value of l, for any given n; this is essentially due to the penetration effect (see p. 150). On the whole, especially in the lighter atoms, the influence of n on the energy predominates over that of l, but the competitive influence of the two quantum numbers leads to some complications and irregularities in the structure of the periodic table.

An electron has four degrees of freedom, one of which is due to the spin, and its state is described by four quantum numbers. The validity of this statement does not depend on any assumptions on the coupling forces. According to the adiabatic hypothesis we can always imagine the forces to be changed slowly into any imaginary, simple coupling condition, e.g. weak or strong external fields to be applied and interactions between the electrons to be made weaker or stronger. The number of states will always remain the same in this process, and even some general properties connected with symmetry, such as the parity.

In merely counting the states we can thus choose any convenient set of quantum numbers, e.g. n, l, m_l, m_s for an individual electron, though sometimes the set n, l, j, m is preferred. Pauli's principle then states that any one set of these quantum numbers identifies a quantum state and this can only be occupied by one electron. The values of n and l for all individual electrons, described as a *configuration*, determine the energy of the atom to a first approximation. It includes the electrostatic attraction between each electron and the nucleus and also the major part of the mutual repulsion between the electrons, namely that part which can be described by a central force field, e.g. in the form of an *effective* nuclear charge or a screening constant. In this section we shall be primarily concerned with *configurations* and their energies. The energy differences within each sub-shell, i.e. the terms and levels for each configuration, require a more detailed treatment of the electrostatic repulsion and also the inclusion of magnetic forces. This treatment which also involves further quantum numbers will be deferred to chapter V.

TABLE 1

Numbers of electrons in the shells

	s $l = 0$	p $l = 1$	d $l = 2$	f $l = 3$	g $l = 4$	total
K-shell $n = 1$	2					2
L-shell $n = 2$	2	6				8
M-shell $n = 3$	2	6	10			18
N-shell $n = 4$	2	6	10	14		32
O-shell $n = 5$	2	6	10	14	18	50

The permissible number of equivalent electrons of quantum numbers n, l is given by the number of states: there are $2l+1$ possible values of m_l and for each of these two values $m_s = \pm\frac{1}{2}$, giving $2(2l+1)$ states. The same result is obtained with the use of the set n, l, j, m where j has the two values $l\pm\frac{1}{2}$; the number of states is then $2(l+\frac{1}{2})+1+2(l-\frac{1}{2})+1 = 2(2l+1)$. The maximum numbers of electrons in the shells and sub-shells, resulting from this relation, are shown in table IV, 1. The expressions K-shell, L-shell, etc. have their origin in X-ray spectroscopy but are often used in other contexts.

The occurrence of equal numbers of positive and negative values of m_l and m_s in any closed sub-shell shows that its resultant angular momentum is always zero; this applies, of course, also to a complete shell.

If the energy of an electron always depended predominantly on n and only very little on l, as would occur if Z far exceeded the number of electrons, table IV, 1 would give immediately the electron configuration of the ground state of any element; one could simply proceed from line to line until the number of electrons characteristic of the element was reached. The more complicated structure of the periodic table is due to the double dependence of the energy on n and l.

2. The first three periods

The ambiguity starts at the element $Z = 5$, boron (see p. 194), but any estimate of the crudest kind, either from terms of other spectra or from screening considerations, makes it obvious that the influence of n must be expected to predominate so that 2p has a much lower energy than 3s. This is, in fact, found and the term 3s ^{2}S is as much as 40000 cm^{-1} above the ground term.

In the further progress of *building up* successive atoms, each step adds a full unit charge to the nucleus, and one electron which will only partly screen the latter. We must then expect that the arrangement of energy states, owing to the increasing preponderance of the nuclear attraction, becomes even more hydrogen-like so that all the 2p electrons would be filled in, which is a fact found to occur (see app. table 2). The spectra of these elements from C to Ne are much more complex than those treated in III; this is due to the fact that more electrons than one have finite values of l.

While postponing the discussion of the term structure we merely consider the value of the ionisation potential, W^I, as a rough indication of the state of attachment of the outer electrons. App. table 3 and fig. IV, 1 show that W^I rises from B to Ne, with a slight irregularity at N and O, confirming the view that the increase in electron number only partly compensates for the increase in Z. In the next element, $Z = 11$, the lowest energy state of the last electron must be 3s, as the state of lowest n and l. This explains the property of Na as an alkali metal and the large drop of W^I from Ne to Na. The lowest excited terms are 3P and 3D, in that order, as expected. The properties of Mg follow in the same way as due to the configuration 3s^2: chemical di-valence, slightly higher W^I, ground term $^1S_0$, lowest excited terms 3P and 3D.

For Al the same applies as for B: decrease of l by one unit lowers the energy less than decrease of n by one unit, so that 3s^{2}3p forms the ground term ^{2}P while 3s^{2}4s ^{2}S is an excited term, about 4 eV above the ground term. In the elements from Si to A the properties change in the same way as in the preceding period. The charge of W^I (fig. IV, 1) in the configurations 3p to 3p^6 is remarkably similar to that in the second period.

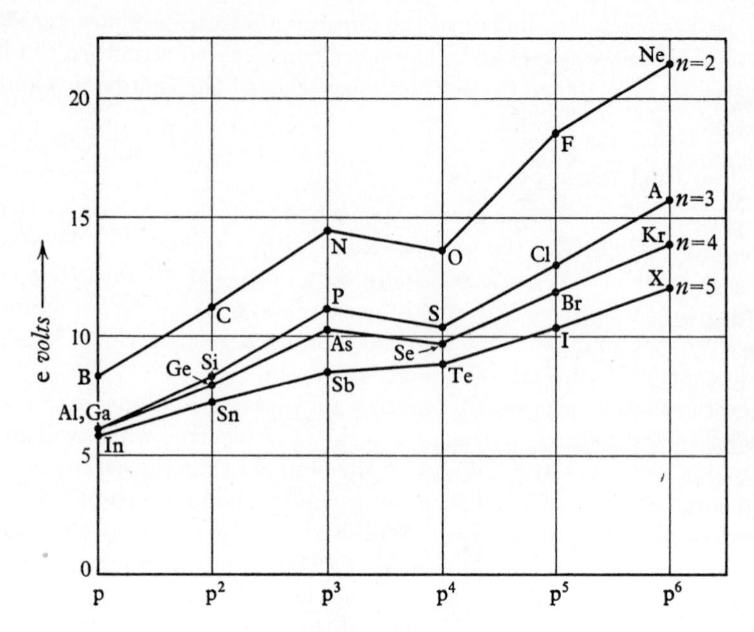

FIG. IV, 1. Ionisation potentials of elements with ground configurations p^n.

3. The 3d sub-group and the fourth period

Up to this point in the periodic table the ground term was always found to have the configuration of lowest possible n for each electron. In K, $Z = 19$, a new situation arises: compared with the state of lowest possible n, which is now 3d, the state 4s of the last electron has a value of l lower by two units. It is therefore not surprising that the latter forms the ground state 4s $^2S_{\frac{1}{2}}$ so that K is an alkali metal. The influence of l, in comparison with that of n, has now become so strong that even 4p ^{2}P and 5s ^{2}S are lower than 3d ^{2}D.

It is interesting to compare the order of terms in the iso-electronic sequence K, Ca$^+$, Sc^{++}. In writing the electron states in order of their energy we find that the increase in nuclear charge, at constant number of electrons, restores the hydrogen-like order:

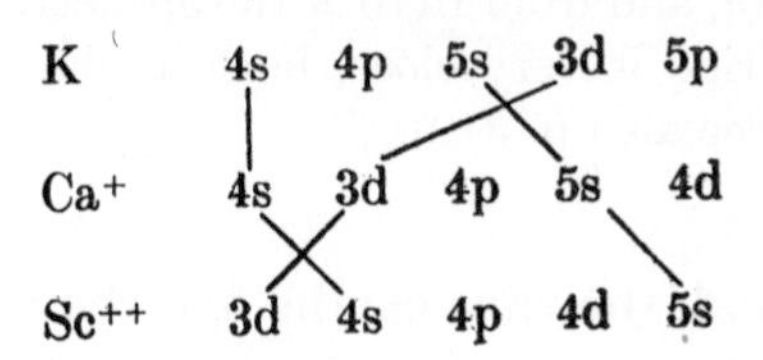

The following element, Ca, has a ground state $4s^2\,{}^1S_0$ and the chemical properties of an alkaline earth, and of the two lowest excited terms, 4P is a little lower than 3D.

In proceeding further we must expect that, again owing to the imperfect screening of the increased nuclear charge, a tendency to favouring the states of lower n will develop. In fact, the ground term of scandium, $Z = 21$, has the configuration $4s^2 3d$. Again it is interesting to note that the ground configurations of the iso-electronic ions Ti^+ and V^{++} are $4s3d^2$ and $3d^3$.

In the elements from Sc to Ni the 3d states continue to be the most stable, as expected, and the ground configurations of these elements contain one or two 4s electrons and an increasing number of 3d electrons. The spectra and chemical properties of the elements of this sub-group, sometimes called *iron-group*, do not correspond to any of the elements in the preceding groups. The relative stability of the states 3d and 4s varies in an irregular fashion and will be discussed later (see p. 318). In the last two elements, Cu and Zn, the sub-shell $4d^{10}$ is complete, and the ground states are $3d^{10}4s\,{}^2S_{\frac{1}{2}}$ and $3d^{10}4s^2\,{}^1S_0$. The chemical properties and the lower terms of these elements are somewhat similar to those of alkali and alkaline earth elements; but owing to the comparative ease of exciting one of the 3d electrons, they have also complex spectral terms, and copper can be di-valent as well as mono-valent.

The elements Ga to Kr in which the 4p electrons are built in resemble the elements Al to A in their properties and spectra. They complete the sub-shells $4s^2 4p^6$ the development of which had been interrupted by the completion of the 3d sub-shell.

The plot of W^I shows again the same trend as in the second and third period, also the same drop after the completion of the half subgroup at p^3 (fig. IV, 1).

4. The 4d sub-group and the fifth period

While the N-shell still lacks the electrons 4d and 4f, the fifth period is started by the alkali metal Rb, $Z = 39$. The 5s state thus proves to have lower energy than 4d. The same is true for Sr. In the following groups of elements Y to Pd, known as *palladium group*, the 4d electrons are filled in, and from In to X the 5p electrons. The analogy to the previous period is very close, both in chemical and spectroscopic properties (see also p. 319).

5. The sixth period, the rare earths and the 5d sub-group

At $Z = 55$, the sixth period starts with the elements Cs and Ba, while the N-shell still lacks its 4f electrons and the O-shell its 5d, 5f and 5g electrons. This shows to what extent the influence of l has gained on that of n in this higher part of the periodic table.

In La the ground configuration is $6s^2$ 5d, but after this element the filling in of 5d electrons is interrupted. Instead, 4f electrons enter successively with increasing Z, though not in an entirely regular fashion, since the energies of the single-electron states 4f, 5d and 6s are very similar. The elements from La to Yb are known as *rare earth* elements. They are di- and tri-valent, have also generally similar chemical properties and are therefore not easy to separate from one another. This similarity is most pronounced in their salts and is due to the fact that all their ions have the ground configuration $4f^x$. The properties of the elements in other respects, e.g. their vapour pressure, are by no means as similar as is often thought.

Though considerations of energy would lead to the 4f electrons being considered as outer electrons, they appear to play the part of inner electrons in the chemical behaviour of the elements. This can be ascribed to the fact that the charge distribution of a 4f state in any central force field is confined to much smaller values of r than that of any of the states 6s, 5d, 5p or 5s; in Bohr–Sommerfeld's theory a 4f orbit is circular. An electron in this state thus acts essentially as an inner electron, and its addition in the building-up process screens the additional nuclear charge almost completely.

From Hf to Rn, the 5d and 6p-electrons are built in. The relative stability of the states 6s and 5d fluctuates somewhat as seen from app. table 3. Radon has the properties of an inert gas.

6. The seventh period

The last period is very incomplete, owing to the increasing instability of the nuclei. All its elements are radioactive, and some are very short-lived and have not been found in nature at all. Their spectra are very complex and have not been analysed very fully.

7. Ionisation potentials and magnetic properties

In the comparison of the elements of a vertical column, the fact that the number of electrons outside closed shells is the same accounts for the similarity of the elements in a purely formal way; but it is by no means obvious why, e.g. the alkali atoms, with the valence electron in the quantum state $n = 2, 3, 4, 5, 6$ respectively should have ionisation energies lying within such narrow limits (5·37 to 3·87 eV). For an inverse square law of the attracting force, a ratio of $6^2:2^2 = 9:1$ would be expected for the W^I values of Li and Cs. It so happens that the effect of the increased quantum number n is very nearly compensated by the penetration effect, and this applies to the outer electrons of most elements: in the whole of the periodic table few ionisation potentials lie outside the limits 5 and 15 eV.

The elements in which the sub-shells 3d, 4d, 5d and 4f are partly filled are known as *transition elements.* Apart from their complex chemical behaviour they are characterised by the magnetic and optical properties of their ions in solutions and crystals. Many of them are strongly *paramagnetic*, in contrast to other elements whose ions consist of closed shells. This is due to the resulting spin and orbital momenta of the incomplete shells (see V.B.4).

The solutions and ionic crystals containing transition elements have often a *characteristic colour.* The electron configuration of an incomplete shell can form a number of fairly widely spaced terms; though transition between these are forbidden by dipole radiation in the free ion, the density of ions in compact matter is so high that dipole radiation enforced by the crystal field may occur and the absorption bands due to quadrupole radiation is sufficient to give the appearance of colour. Also transitions between "competing" configurations may cause such absorption bands.

B. THE CALCULATION OF ELECTRON STRUCTURES[5, 6, 7]

1. Coulomb effects and exchange effects

It was shown in the preceding section that the application of Pauli's principle, together with crude estimates of screening and penetration effects, lead to a qualitative interpretation of the periodic properties of the elements. This section deals with the quantum-mechanical methods whose aim it is to calculate the relative energies of different *configurations* and also to provide a quantitative description of the distribution of electronic charge in the core of the atom. The details of the methods are so involved that no more than an outline of the fundamental ideas and some typical results can be given, apart from references for further study.

We restrict ourselves, in this section, to electrostatic effects. Quantum numbers are ascribed to the single electrons and Pauli's principle restricts the number of electrons having any given set of quantum numbers. To a first approximation we can imagine each individual electron to have a definite wave function and therefore a definite charge distribution, and we can calculate the repulsion between the charge clouds of the different electrons by classical electrostatics. The resulting effect is usually called Coulomb repulsion. Owing to the fact, however, that electrons are indistinguishable, the existence of several electrons in different quantum states causes *exchange degeneracy*. This was shown, for the example of helium, to give rise to electrostatic effects causing energy difference between different terms such as the singlet- and triplet-terms which belong to any given configuration. Even when the exchange effects do not give rise to new terms they have an influence on the value of the energy. In a Na atom e.g. the exchange degeneracy between the 3s electron and the 2s and 2p-electrons affect the energy of the ground level of Na though it does not cause any splitting of terms. The effect is generally the larger the more the two wave functions overlap. The exchange effect between the two 1s-electrons and the 3s electron in Na would thus be extremely small.

2. Self-consistent field methods

The most important methods of calculating electronic structures of atoms are based on the principle of the *self-consistent field* first introduced by Hartree(8). If we choose the helium atom in its ground state $1s^2\ ^1S_0$ as the simplest possible example we can describe the influence of one of the two electrons on the other by means of a central force field, for which we first assume a suitably chosen potential function $V_0(r)$ as a zero approximation. The eigenfunction of lowest energy, ψ_0, for an electron moving in this field is then found, e.g. by methods of numerical integration. The function $\rho_0(r) = 2e\psi_0{}^*\psi_0$ gives the charge density which would exist if the electrons were correctly described by ψ_0. The electrostatic potential $V_1(r)$ caused by $\frac{1}{2}\rho_0(r)$, together with the central positive charge, is then calculated and compared with $V_0(r)$. The closer the agreement between the two the better had been the choice of $V_0(r)$. A better potential is then chosen and by a process of successive approximation a field can finally be found which will produce such a function that its charge density will reproduce the field. A field having this property is called *self-consistent*.

The state of motion described by a self-consistent field is clearly a kind of equilibrium, and it is not very surprising that the condition of self-consistency has been shown to be identical with the extremum condition of the energy function which is used in the variation method (see p. 58). In the calculation of electron structures the extremum condition is now most frequently used for judging how "good" an eigenfunction is. For special purposes, however, where high accuracy of the eigenfunction in certain parts of space, e.g. near the nucleus, is required, the energy is often not the most suitable criterion.

Different methods of calculating electronic structures are distinguished not so much by application of either the variation principle or the original self-consistency criteria but rather by (i) the use of *analytical* eigenfunctions whose parameters are determined by the variation principle, (ii) the calculation of eigenfunctions by methods of *numerical* integration without restriction to analytical functions.

Methods of the first type are simpler if functions with few parameters are chosen, and it is capable of high accuracy for the lighter elements. The second type is generally laborious but capable of great accuracy even for heavier atoms.

The ground state of helium was chosen as an exceptionally simple example owing to the spherical symmetry of the charge distribution of an s-electron. As a more realistic example, showing up the restrictions of the methods, we consider the core of Na, i.e. the configuration $1s^2 2s^2 2p^6$ with $Z = 11$. If we assume hydrogen-like wave-functions for each of the 2p electrons, the charge distribution of the group $2p^6$ is spherically symmetrical (see p. 96). If the interactions between the electrons are taken into account, it can be shown that this symmetry is still retained rigorously, even if the magnetic interactions are included. It is connected with the fact that the resultant angular momentum of a complete sub-shell vanishes, owing to $\Sigma\, m_l = 0$ and $\Sigma\, m_s = 0$. Each 2p electron, however, moves in a field produced by the non-closed configuration of the other five 2p electrons; the assumption that this field has spherical symmetry is one of the approximations which has to be made in all calculations of atomic structure involving electrons with $l > 0$.

The procedure in its simplest form is this: trial potentials $V(r)$ are chosen for the motion of the three kinds of electrons, 1s, 2s and 2p, and wave functions are computed numerically for each; they depend on the angles in the form of the appropriate spherical harmonics.

The total wave function is then taken as the simple product of the one-electron functions and the energy integral is made as small as possible by successive adaptation of the initial potential functions.

The simple product function $u_a(1)u_b(2)u_c(3)$ . . where each of the indices a, b . . stands for a quadruple of quantum numbers n, l, m_l, m_s for a single electron and the number refers to the electron, attributes a definite quantum state to a definite electron, in violation of the principle of the identity of electrons. If exchange effects are to be included the simple product has to be replaced by the determinant

$$\begin{vmatrix} u_a(1) & u_b(1) & u_c(1) & — & — \\ u_a(2) & u_b(2) & u_c(2) & — & — \\ — & — & — & & \end{vmatrix}. \tag{1}$$

As a linear combination of the product functions it is a solution of the wave equation for central fields. It is a property of a determinant to change its sign on interchange of any two elements; this causes the function to be antisymmetric, in accordance with Pauli's principle. The method of using determinants is due to Slater[9] and is a generalisation of the functions (II,44) used in helium. When more than one term arises from one configuration the wave function can consist of a sum of determinants. The method of the self-consistent field has been adapted for determinant wave functions by Fock.[10] Such calculations "with exchange" are much more laborious and have been mostly confined to the lighter atoms.

The results of self-consistent field calculations are usually given in tabular form, either as radial charge density, defined as charge per radial increment dr, or as effective charge number Z_p, in both cases as a function of r/a_0. Z_p is defined by equating the potential to eZ_p/r. Extensive tables of Z_p have been computed by Freeman.[11]

Figure IV, 2 shows the radial charge densities, calculated without exchange, of the core of Rb[8] and of its individual sub-shells. For Ca^{++} the self-consistent field has been calculated with exchange, and some wave functions of the valence electron, i.e. of the alkali-like states of Ca^+, have been computed.[12] The influence of the exchange effects on the 3d wave functions was found to be remarkably large. Another example of a calculation with exchange, that of Be, was mentioned in III.D.1.

Slater[13] has recently proposed a simplification of the self-consistent field method. It includes the exchange effect in a slightly less rigorous but simpler way.

3. Other methods

Variation methods with analytical functions have been used by numerous authors,[14, 15] especially for lighter elements. The important calculations on He, by Hylleraas and others, which led to results of extreme accuracy were referred to in III.B.3. Of more recent work, only that by Morse, Young and Haurwitz[16] and its correction and extension by others,[17, 18] and finally that by Löwdin[19] may be mentioned.

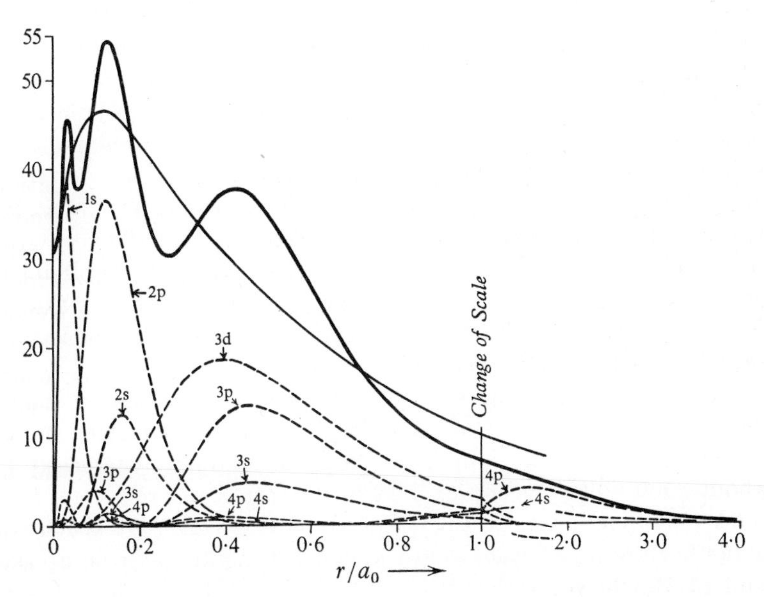

FIG. IV, 2. Radial charge density of Rb^+ (D. R. Hartree, ref. 8).

An entirely different approach is due to Thomas[20], Fermi[21] and Dirac.[22] These authors apply the methods of quantum statistics for calculating the equilibrium of a cloud of electron in the vicinity of a nucleus. This leads to a differential equation for the potential $V(r)$ which can be solved numerically. The results cannot be expected to represent the potential in as detailed a manner as calculations of the Hartree-Fock type, but it is so much simpler that a large number of atomic states can be calculated without undue labour. A recent, extensive, comparison of such calculations with experimental term values[23] has shown a remarkable degree of agreement. The method describes correctly the relative stability of competing electron states in atoms and gives a good estimate of the absolute term values.

C. X-RAY SPECTRA

1. The origin of X-ray spectra

When the anticathode of an X-ray tube is bombarded with electrons it emits two kinds of radiation. One of them has a continuous spectrum whose intensity distribution depends little on the material of the anticathode; it is known as Bremsstrahlung. The other type of radiation can be shown, by an X-ray spectrometer, to consist of distinct spectral lines whose frequencies are characteristic of the element forming the target. Only this *characteristic radiation* is of direct interest for the study of atomic structure.

The order of magnitude of the frequencies, which are about 1000 times higher than optical frequencies, indicates that the very firmly bound electrons of the inner, complete shells are responsible for the emission of X-ray spectra. This view agrees with the fact that the X-ray spectra of a free element and its chemical compounds are the same, apart from some very fine details.

The way in which the characteristic spectra are excited may be described by means of the example of a target of tungsten used as anticathode while the applied voltage is gradually increased. At about 2500 volts a group of lines in the soft X-ray region, of about 6·5A (6500X.U.) appears; it is known as *M-series*. At 12000 volts a further group, of wavelength about 1·3A., the *L-series*, appears, and at 70,000 volts a very hard radiation, the *K-series*, of about 0·2A. Each series has thus a clearly defined *excitation potential*. With the use of suitable equipment, an even softer series can be detected at very low voltages, but no radiation harder than the K-series.

The most striking feature of the characteristic X-ray spectra is their dependence on the atomic number of the emitting element. In contrast to optical spectra, the frequencies of the lines, and also their excitation potentials, increase steadily, not periodically, with increasing Z. The square roots of the frequencies are an approximately linear function of Z. This important relation, discovered by Moseley,[(24)] is shown in fig. IV, 3.

A further contrast to optical spectra is found in the relation of X-ray emission- to absorption-spectra. The latter consist essentially of continuous regions of absorption, bounded on the long wave-length-side by sharp edges. There is a K-edge, an L-edge, etc., at frequencies slightly higher than those of the K.., L.. emission lines. In the spectrograms plate 11 the K-absorption edges of a sequence of elements can be seen to move towards shorter wavelengths, towards the undeflected beam in the centre of the spectrograms, as Z increases. The K-edge due to the silver in the photographic emulsion can be seen in all spectrograms, but inverted: Increase of absorption in the emulsion itself causes an increase in photographic density. The K-emission lines of the tungsten anti-cathode are also visible.

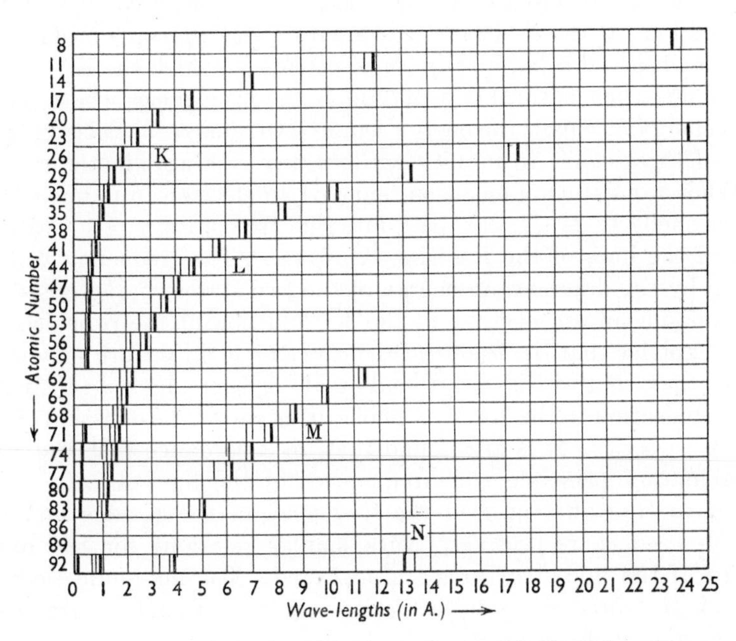

FIG. IV, 3. Moseley's law for X-ray lines. (M. Siegbahn, *Spectroscopy of X-rays*, Oxford Univ. Press, 1925.)

Emission of X-ray spectra can also be caused as the result of irradiation with X-rays, as fluorescence. X-rays absorbed in the M-band cause emission of the M-series, irradiation in the L-band the L- and M-series and irradiation in the K-band excites all the series in emission.

The interpretation of these fundamental features of the X-ray spectra was first given by Kossel[1,25] and contributed greatly to the development of our views of the electronic structure of atoms as sketched in the preceding sections. Accepting these views in advance we can interpret the facts quite readily. An electron in one of the inner shells of an atom can only be excited to a quantum state which is not occupied by other electrons, i.e. to one of the optical levels on the "surface" of the atom or to the continuous states of positive energy. On the scale of X-ray energies, the entire energy range of optical levels is so small that it can be neglected unless finer details are being studied. The process of excitation thus consists of the ejection of an electron from one of the closed shells. The result of this internal ionisation is a structure in which one electron is missing from one of the closed shells. It forms a highly excited state of the atomic ion. If the excitation is caused by electromagnetic waves it can be observed as an absorption spectrum in which the K . . , L . . edges find their natural explanation as the frequencies whose quantum energy $h\nu$ equals the energy required to remove an electron from one of these shells to infinity without any residual kinetic energy.

When a K-electron has been removed an electron from one of the outer shells L, M, N, can pass into the free state $n = 1$ emitting one of the lines of the K-series, K_α, K_β . .. The frequency will be given by the difference in energy of the two states characterised by the missing of an electron in the K- and one of the other shells. This explains that the frequencies of the emission lines are somewhat lower than that of the corresponding absorption edge. In a similar way, ionisation in the L-, M- . . shells gives rise to the emission of the L-, M- . . series. It is also found that, in accordance with the Ritz combination principle, the frequencies of the lines K_α, K_β . . are equal to the differences of the frequency of the K-edge and the L- M- . . edges respectively, with similar relations for the other series, if the fine structures are neglected. The minimum electron energies required to excite the various series are found to agree with the $h\nu$ values of the edges.

That the emission of the K-series is always accompanied by the emission of the other series is obviously due to the vacancies created by the emission of the lines K_α, K_β, K_γ in the L-, M-, N-shells, and similarly the appearance of the M- and N-series are linked with the emission of the L-lines. The M- and N-series only appear in atoms of medium and large Z in which the corresponding shells are filled.

The fact that the X-ray frequencies increase with increasing Z without any marked periodicity can be easily understood in a qualitative way. It is the outer electrons whose properties change periodically with their number, but they have only a very small screening effect on the inner electrons whose energy will thus steadily increase with Z.

2. Screening effects and Moseley's law

In X-ray spectroscopy the normal state of the atom is chosen as reference level for the energies and term values. They are usually quoted in units of the Rydberg constant R, i.e. in units of 109737 cm^{-1}, or as energies in keV (1000 electron volts). In accurate work the slight difference between an A.U. and 1000 X.U. is to be remembered (see p. 9). The following relation holds:

$$E(\text{keV}) = 12{\cdot}3964/\lambda(\text{A.U.}) = 12{\cdot}3964/1{\cdot}00202\lambda(\text{k.X.U.})$$

A term arising from the removal of an electron is described by a symbol referring to the state of this electron, and the term- or energy-value is the higher the more firmly bound the electron. We thus write for the wave number of the K_α line

$$\tilde{\nu} = \text{K} - \text{L}.$$

In later sections, the use of a one-electron symbol for a"hole" in a complete shell will be justified more fully.

In order to estimate the energy of the state of one of the inner electrons and thus, by reversal of the sign, the X-ray level due to its removal, we have to extend the concept of screening and include the effect of charges outside the electron orbit or wave function we are considering. If a nuclear charge Ze is surrounded by a spherically symmetrical distribution of charge density $\rho(r)$, the potential at distance r_0 from the nucleaus is

$$V(r_0) = \frac{Ze}{r_0} + \frac{q_{r<r_0}}{r_0} + \int_{r_0}^{\infty} \rho(r) 4\pi r \, dr \tag{2}$$

where

$$q_{r<r_0} = \int_0^{r_0} \rho(r) 4\pi r^2 \, dr$$

is the total space charge inside the sphere of radius r_0. The values of e and ρ are of opposite sign, so that the second and third term in (2) are numerically opposite in sign to the first.

Though the electric force inside a spherical distribution of charges vanishes, the potential has the finite value given by the last term in

(2). We can define an average radius $\bar{r}_{ext.}$ for the charge distribution for $r > r_0$ by

$$\int_{r_0}^{\infty} dq/\bar{r}_{ext.} = \int_{r_0}^{\infty} dq(r)/r \tag{3}$$

and can then write (2) as

$$V(r_0) = \frac{Ze}{r_0} + \frac{q_{r<r_0}}{r_0} + \frac{q_{r>r_0}}{\bar{r}_{ext.}}. \tag{4}$$

While the inside screening effect, given by the second term, does not depend on the details of the charge distribution, the external screening effect does depend on it and is the smaller the greater the average radius of the charge distribution.

The example of one of the two electrons in the K-shell of an element of medium or large Z may be chosen to show how the order of magnitude of screening effects can be estimated. If r_0 is the average radius of the K-electron whose motion is being considered the total screening effect of the other K-electron will be rather smaller than e/r_0, the value it would have if its charge were entirely inside that of the first electron; it will then contribute an amount a little below 1 to the value of the screening constant σ. Since the major axis of hydrogen-like orbits and the average radius of the charge density of the corresponding wave functions of s-electrons is proportional to n^2 the eight electrons in the L-shell will have an average radius of about $4r_0$; it will in fact be rather less for the p-electrons, but increased by the partial screening of the nuclear charge. The contribution of these electrons to σ is almost entirely due to external screening effects and will be about $8/4 = 2$. With similar assumptions the effect of the complete M-shell is $18/9 = 2$.

These crude estimates show that the external screening effect is considerably smaller than the internal screening by the same number of electrons, though not negligible for purposes of more accurate calculations of term values.

On the basis of the concept of screening, the energy state of an inner electron can then be approximately described by the formula

$$W = -Rhc(Z-\sigma)^2/n^2 \tag{5}$$

and the X-ray term value by

$$T = R(Z-\sigma)^2/n^2 \tag{6}$$

where σ is of the order of the number of electrons in the same and in

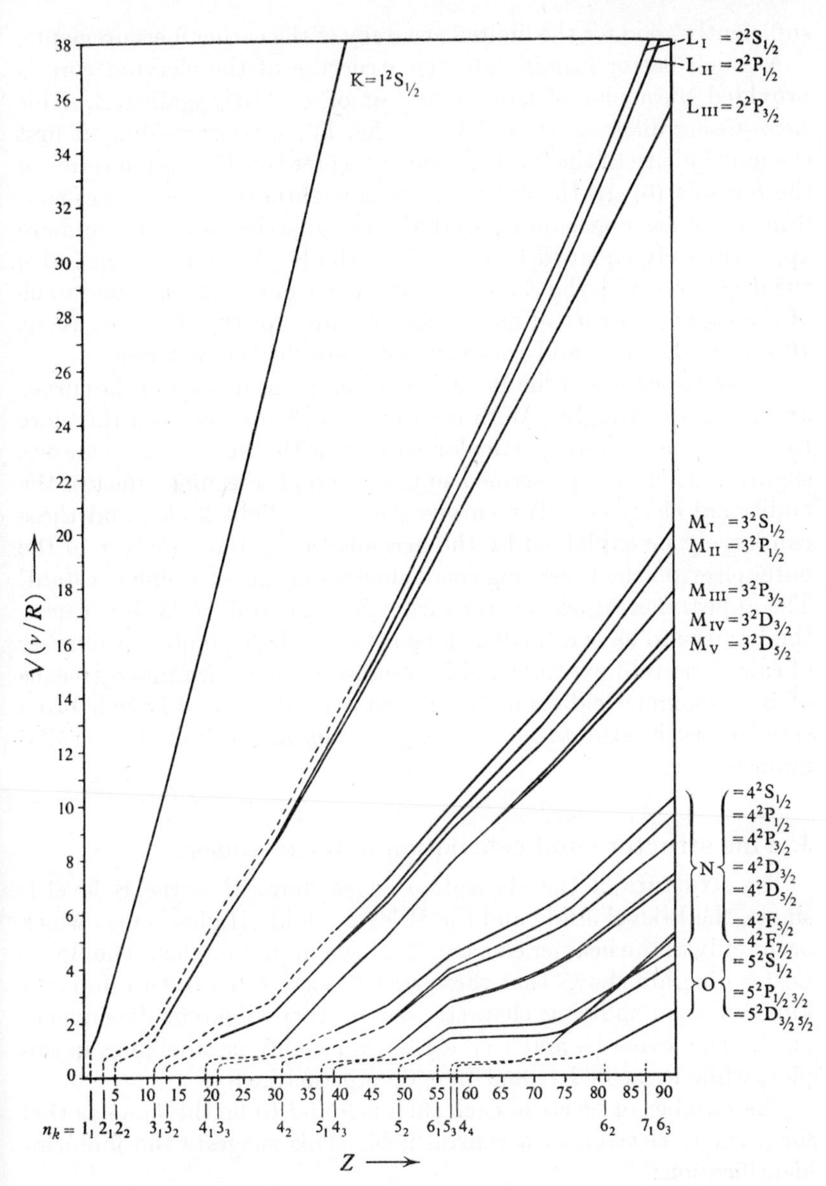

FIG. IV, 4. Bohr–Coster diagram. (N. Bohr & D. Coster, ref. 26.)

lower shells. Moseley's law of the linear variation of the square root of the frequencies of the lines with Z follows with the simplifying assumption that σ is the same for both states. This assumption was sufficiently good for the limited accuracy of the earlier measurements.

A much better insight into the structure of the electron core is provided by a plot of *term values*, or of $\sqrt{(T/R)}$, against Z. This *Bohr–Coster* diagram[26] is shown in fig. IV, 4. Disregarding at first the multiplicity of the levels we can interpret the diagram in terms of the formula (6): (i) the approximate straightness of the curves confirms the dependence on Z, (ii) the slopes give the values $1/n$ and are approximately equal to 1, $\frac{1}{2}$, and $\frac{1}{3}$ for the K-, L- and M-terms. (iii) the intercepts with the Z-axis give the value of σ which is found to be of the expected order of magnitude: about 3 for the K-terms, about 10 for the L-terms and considerably more for the M-terms.

Closer inspection of the Bohr–Coster diagram shows that the curves are not quite straight. With increasing Z their slope, and therefore the intercept of any particular section with the Z-axis, increases slightly. This can be ascribed to the external screening effect of the additional electrons. The curves also show slight kinks, and these can largely be explained by the periodicities in the structure of the outer electron shells causing small changes in the screening constant. The appearance of 3d-electrons from $Z = 21$ to $Z = 28$ thus causes the L-curve to have a smaller slope between these points. A number of calculations of external and internal screening constants by means of Bohr–Sommerfeld orbits have been carried out and have led to a satisfactory interpretation of many details in the Bohr–Coster diagram.[27]

3. Fine structure and calculation of term values

The structure of the absorption edges shows that the K-level is single, the L-level 3-fold and the M-level 5-fold. Higher levels which occur only in the heavier elements are even more complex. The Bohr–Coster diagram shows that these structures are due to two different kinds of term splitting characterised by their different dependence on Z. The levels L_I and L_{II}, e.g. form roughly parallel lines in this plot, while those of L_{II} and L_{III} diverge markedly.

The number of levels in each shell is found to be the same as that for a single electron in a central field. This suggests the following identification:

$n = 1$		2				3		
K	L_I	L_{II}	L_{III}	M_I	M_{II}	M_{III}	M_{IV}	M_V
$1\,^2S_{\frac{1}{2}}$	$2\,^2S_{\frac{1}{2}}$	$2\,^2P_{\frac{1}{2}}$	$2\,^2P_{\frac{3}{2}}$	$3\,^2S_{\frac{1}{2}}$	$3\,^2P_{\frac{1}{2}}$	$3\,^2P_{\frac{3}{2}}$	$3\,^2D_{\frac{3}{2}}$	$3\,^2D_{\frac{5}{2}}$

This is confirmed by the number and relative strength of the lines of the emission spectrum which can be fully explained by the given identification, together with the selection rules $\Delta l = \pm 1$, $\Delta j = 0, \pm 1$. The K-series, e.g. consists of doublets (K_{α_1}, K_{α_2}), (K_{β_1}, K_{β_2}) . . . ., with intensity ratios 2:1. The other series are more complex, but by means of the combination principle connecting frequencies of lines with differences of frequencies of absorption edges, and by reference to the intensities, the analysis leads quite easily and unambiguously to the identification of terms shown in fig. IV, 5. The conventional nomenclature of X-ray lines is somewhat unsystematic, and it is preferable to specify lines by giving the term symbols.

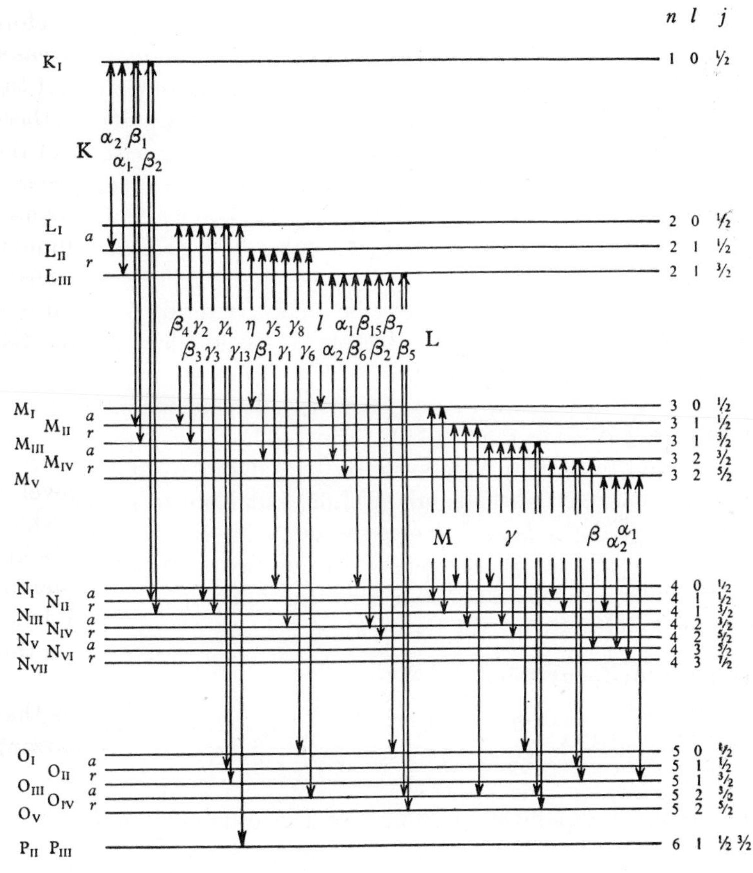

FIG. IV, 5. X-ray terms and lines; the level differences are not shown in the true ratios. (M. Siegbahn, *Spectroscopy of X-rays*, Oxford Univ. Press 1925.)

The X-ray terms thus correspond closely to alkali terms with their order reversed, and with no restriction for the lowest possible value of n. One is thus led to the conclusion that a vacancy in a closed shell produces the same levels as a single electron, but in reverse order. Since X-ray states are defined by the missing of an electron these results mean that the energy levels of an electron in a closed shell are in the same order as those of a single alkali electron: the removal of a 2s electron, e.g., as compared with a 2p electron, leads to a higher X-ray term because the 2s electron is more firmly bound and has a lower energy in the complete shell. Figure IV, 6 illustrates the arrangement of terms by the example of W. The diagram is similar to an inverted term diagram of an alkali atom.

The very far-reaching analogy between a "hole" and a single electron can be fully understood by quantum-mechanical methods and will be discussed in chapter V. At this point, attention may only be drawn to the plausible fact that the removal of a particle from a complex of zero angular momentum leaves the latter with the same values of spin- and orbital-angular momentum as the single particle; the momenta of particle and hole are equal and opposite in direction before removal, and this applies to spins and orbital momenta separately.

We have seen that one can describe an X-ray term quantitatively by calculating the energy of an electron in a field which is, to a first approximation, a Coulomb field ($V \sim 1/r$). In closer approximation, the screening effects will make the energy dependent on σ, though not as strongly as in most alkali terms. The high value of Z or $Z-\sigma$ causes relativity effects to be considerable. We can then apply the hydrogen-like formula (III.65) but have to replace Z by $Z-\sigma$, where σ depends on l. Since the screening constant represents an average over a range of effective Z-values we shall have to expect slightly different screening constants for expressions of different power of Z and write for the term value

$$T_{n\,l,\,j} = R(Z-\sigma)^2/n^2 - \\ -R\alpha^2(Z-\sigma')^4/n^3\left[\frac{3}{4n} - \frac{1}{l+\frac{1}{2}} + \frac{j(j+1)-l(l+1)-s(s+1)}{l(2l+1)(l+1)}\right]. \quad (7)$$

The doublet separation is found as the difference of the levels $j = l \pm \frac{1}{2}$,

$$\Delta T = R\alpha^2(Z-\sigma')^4/[n^3 l(l+1)] \quad (8)$$

in obvious analogy to (III.62). The dependence on the fourth power

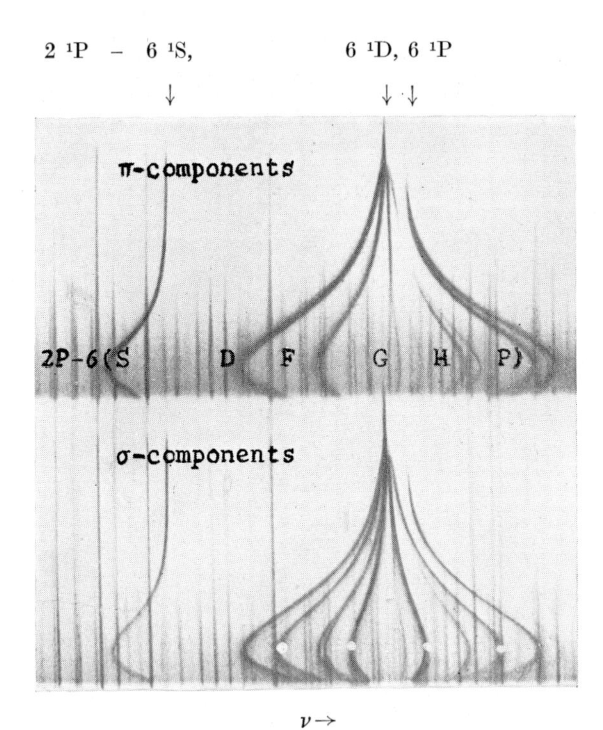

PLATE 10. Stark effect in He line 2 ¹P – 6 ¹D (4144 A) in fields up to 85000 volts/cm. (J. S. Foster, *Can. J. Phys.*, **37**, 1202, 1959).

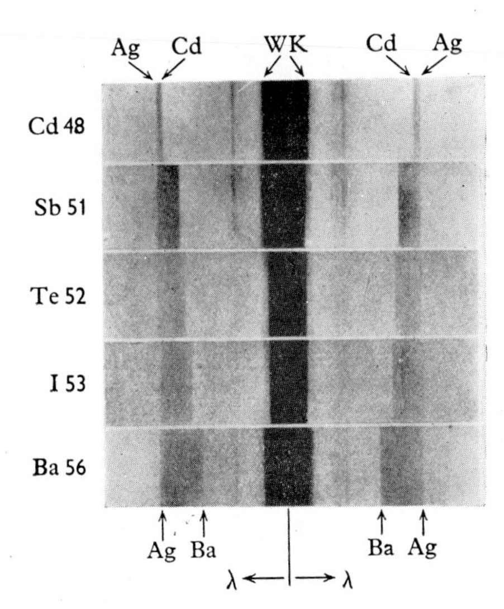

PLATE 11. X-ray absorption spectra. (M. Siegbahn, *Spectroscopy of X-rays*, Oxford Univ. Press, 1925).

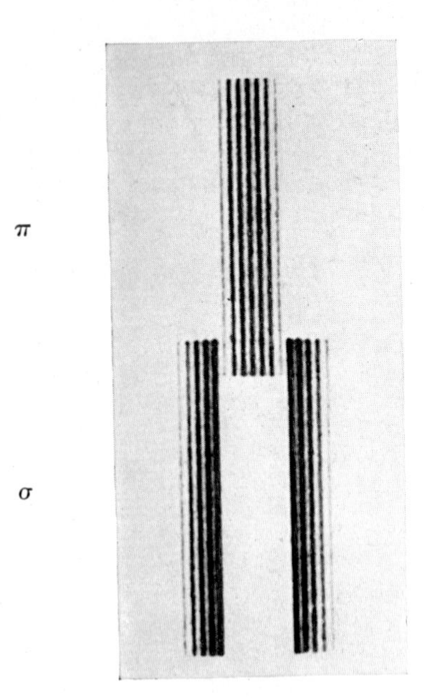

PLATE 12. Zeeman effect of septett line $^7S_3 - {}^7P_4$ in Cr (4254 A). (E. Back & A. Landé, *Zeemaneffekt u. Multiplettstruktur*, Springer, 1925).

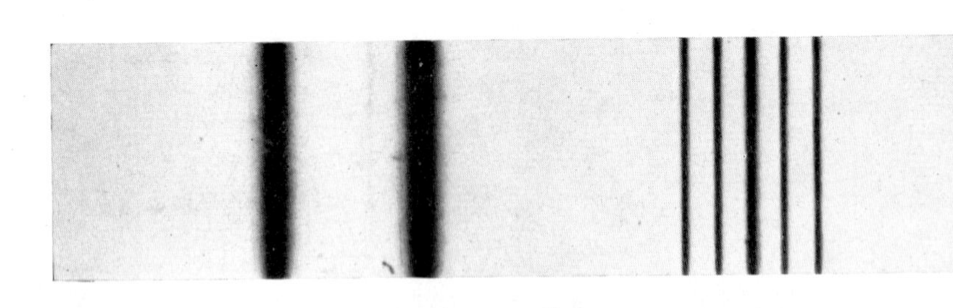

PLATE 13. PP′-triplet 3s3p $^3P^0 - 3p^2\ ^3P$ in MgI and resonance doublet 3s ^2S – 3p ^2P in MgII. (A. Gatterer & J. Junkes, *Atlas d. Reststrahlen*, Vol. I, Specola Vaticana, 1947).

of $Z-\sigma'$ explains the divergence of the lines representing these so-called *spin-doublets* in the Bohr–Coster diagram; it is known as the law of *regular* (or spin) doublets.

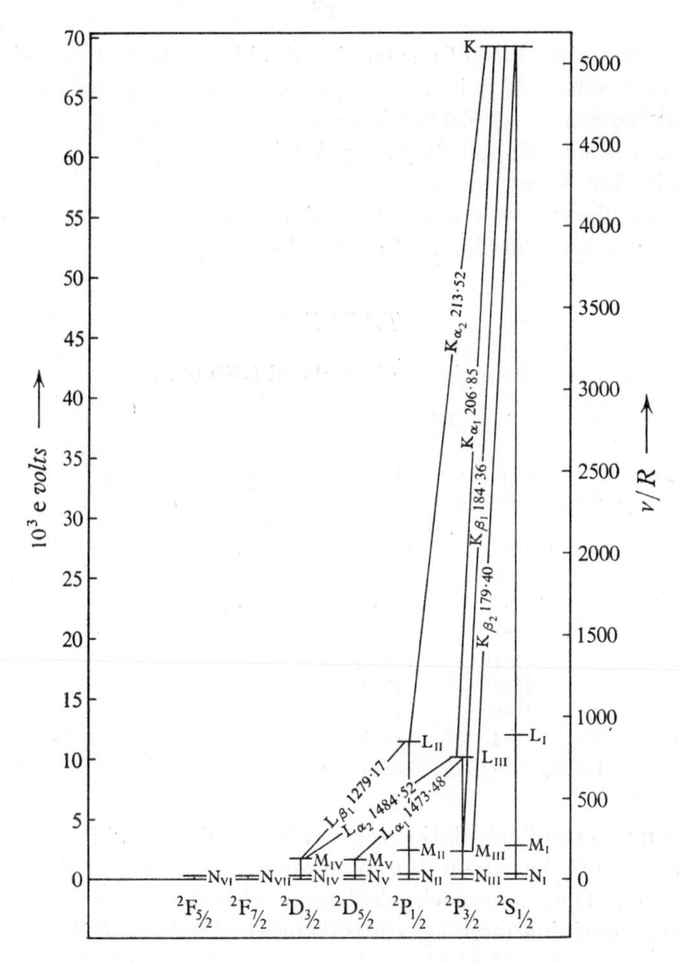

FIG. IV, 6. X-ray term diagram of W. (W. Grotrian, *Graph. Darst. d. Spektren* II, J. Springer 1928.)

For two levels of equal n and j—and therefore different l—the term value depends on l only through σ and σ'. The first of the two terms in (7) by far predominates, and the difference in σ explains the almost parallel appearance of the curves representing such *screening doublets* in the Bohr–Coster diagram. By differentiation of (7) one obtains the relation for screening doublets

$$\Delta T' = \frac{2R(Z-\sigma)}{n^2}\Delta\sigma \qquad (9)$$

The term difference of a screening doublet is, in fact, found to be a linear function of Z in X-ray spectra, in a similar way as in the optical spectra of iso-electronic sequences (see III.C.2). Such pairs of levels are sometimes called *irregular doublets*. The similarity of the relations for X-ray spectra and sequences of iso-electronic ions, which applies to spin doublets and to screening doublets, shows the comparative smallness of external screening effects.

TABLE 2

X-ray levels of Rb[7] and K[26] ($\tilde{\nu}/R$)

	Rb		K		
	$\epsilon_{n,l}$	obs.	$\epsilon_{n,l}$	rigor. calc.	obs.
K	1116	1119	267·3	265·0	265·0
L_I	149·8	152·0			
L_{II}	137·5	137·2	21·97	16·8	21·7
L_{III}	133·3	132·8		16·6	21·5
M_I	22·19	(24·1)			
M_{II}	17·4	18·2			
M_{III}	16·8	17·4			
$M_{IV,V}$	8·4	(8·4)			

The methods of calculating term values as described in the preceding pages are based on the semi-empirical concept of screening constants. Quantum-mechanical approximation methods offer the means of deriving more accurate theoretical values of X-ray terms.

In most of the calculations of atomic structures, such as those of the self-consistent field and the variation method, the total energy of the atom is found as a sum of energies $\epsilon_{n,l}$ of the single electrons in the field produced by the presence of the other electrons. To a first approximation these values can be expected to be equal to the negative X-ray term energies. Table IV, 2 shows the comparison for Rb.[8] The closeness of the agreement found in this and other elements was quite unexpected, because the values $\epsilon_{n,l}$ give the energy required to remove an electron from the specified state while the other electrons are left in their original energy states. This does not take account of the re-arrangement of the other electrons which will accompany the removal of the inner electron in an X-ray absorption. The energy difference involved in this process can only be found accurately by means of a calculation of the self-consistent field of the new configuration with a vacancy in one shell. This rather involved calculation has been carried out for potassium,[28] with the results shown in table IV, 2. While the agreement is extremely good for the K-level it is less satisfactory for the L-levels. Again, the simpler, inaccurate use of the values $\epsilon_{n,l}$ gives unexpectedly good agreement.

The eigenfunctions of the X-ray states of potassium have also been used for calculating the absorption coefficients in the continua of the K- and L-edges.[29]

4. Structures due to outer electrons

It has long been known that absorption edges are not quite sharp and that the absorption does not always decrease uniformly with decreasing wavelength but often shows banded structures; they depend on the physical and chemical state of the element used as absorber. This is not unexpected: between the completely filled electron shells and the free states of electrons there must be a range of energies where the electron can exist in an "optical", previously unoccupied quantum state. X-ray absorption and emission spectra are usually observed in the solid state, very often in metals. In these conditions the permitted energy levels form broad bands, partly separated by energy bands which are forbidden. For electrons of sufficiently high energy these bands can be regarded as due to interference of de Broglie waves scattered by the crystal lattice, and some of the observed structures and their dependence on temperature were explained in this way.[30,31,32]

For lower energies the quantum states of the electrons have to be calculated by means of more detailed assumptions on periodically varying potentials. Some results of this kind have been found to be in very good agreement with observations.[33]

The electronic levels in metals can also be studied from structures which have been observed in the emission spectra of the lighter elements in the soft X-ray range.[34,35] Some studies have also been made on heavier elements.[36]

Conditions are much simpler when the element whose X-ray spectrum is observed is in the form of free atoms. Figure IV, 7 shows the structure of the K-absorption edge of gaseous argon, observed with very high spectroscopic resolution.[37] The states produced as the result of a transition of an electron from the K-shell to one of the free, optical orbits can be expected to be very nearly the same as those of the potassium atom; it can be imagined to arise from the latter by the transfer of one negative charge from the K-shell to the nucleus, with obviously very little effect on the potential near the periphery of the atom. Owing to the l-transition rule, the electron would have to pass into a p-state. In fact, the distance between the

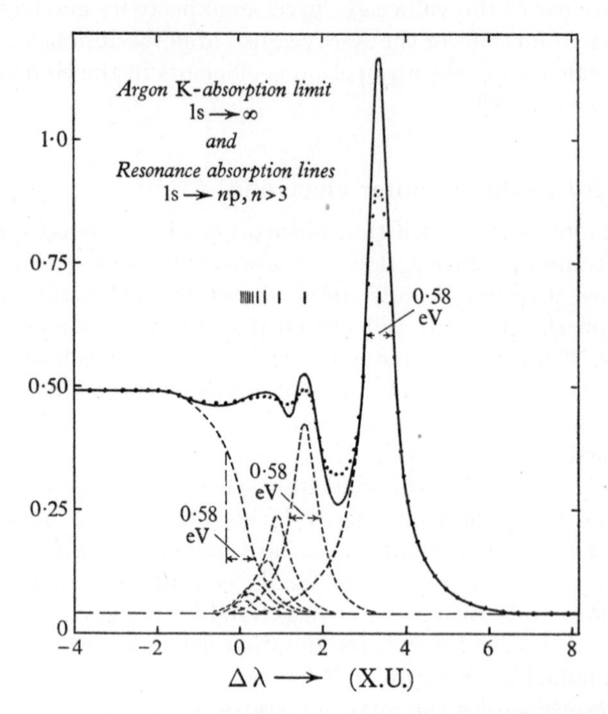

FIG. IV, 7. X-ray absorption edge in argon. (L. G. Parratt, ref. 37.)

first two maxima in fig. IV, 7 corresponds exactly to the difference between the terms 4P and 5P of K. The figure also shows how the entire absorption curve can be constructed as the sum of transitions to the various states nP and to the states of positive energy. The structure covers a range of about 4 X.U., in an edge of wavelength 3866 X.U., so that the energy range of the optical levels is only

1/1000 of the energy of a K-electron. Similar experiments with krypton have shown less distinct structures.[38]

5. Satellite lines and Auger effect

The lines in X-ray spectra are often found to be accompanied by faint satellites, also known as *non-diagram lines*, on their short wavelength side. They are usually several orders of magnitude weaker than the main lines and therefore somewhat difficult to study. A few of these satellites could be explained as quadrupole transitions;[39–42] the intensity of this type of radiation relative to dipole radiation is generally greater in the X-ray range than in the optical range, because the wavelength decreases more rapidly than the linear dimensions of electron orbits, in passing from optical to X-ray spectra (see p. 76).

The majority of the very large number of satellite lines, however, have a different origin[43,44,45] as is shown by their excitation potential which is found to be somewhat higher than that of the corresponding main line: they are due to double internal ionisation. If, e.g. the bombarding electron has sufficiently high energy it can remove an electron from the L-shell as well as one from the K-shell, producing a state of ionisation which can be described as a KL state. A transition of an electron from the L-shell to the K-shell will then cause the emission of a line of wave-number

$$\tilde{\nu} = \mathrm{KL} - \mathrm{LL}.$$

Compared with the normal line K–L, this transition takes place while one electron is missing from the L-shell; whereby the screening is reduced and all term values are increased. If we consider that the absolute value of the K-term is far greater than that of the L-term, it is plausible that the effect on the former predominates and can cause a satellite of higher wavenumber. Similar satellites are observed in the other series, and the fine structure of the levels accounts for an often quite complex fine structure of the satellite lines.

While the excitation of the satellites in the K-series is exclusively due to double ionisation in a single electron impact, those in the other series are partly excited by a process which is known as *Auger effect* or *auto-ionisation* and is also encountered in optical spectra (see V.A.12). In observing the effect of hard X-rays on argon in a cloud chamber, Auger[46] found not only tracks arising from ejection of photo-electrons from the K-shell, but also shorter tracks originating at the same point in space. He ascribed them to a radiationless transition in which an electron passes from the L-shell into the vacancy in the K-shell and the excess energy is used for ejecting an electron from one of the higher shells, e.g., the M-shell:

$$K \rightarrow LM + \text{kin. energy of free electron.}$$

In this way the Auger effect produces a doubly ionised state which can then give rise to the emission of satellites, in this particular example of the L-series.

CHAPTER

V. Complex Spectra

A. THE GENERAL STRUCTURE OF COMPLEX SPECTRA

1. Normal multiplets

The formation of series is the most conspicuous feature of simple spectra and has provided the basis for their analysis and interpretation. The multiplet- —i.e. doublet- and triplet-structure—appears of secondary importance.

In the complex spectra the appearance of multiplets as groups of lines with characteristic spacings and strengths is the most striking feature and generally forms the starting point of the term analysis. Singlet terms are therefore most difficult to identify and are often missing in otherwise fairly complete analyses of spectra. Series of lines and of terms also occur in complex spectra, but their higher members are generally difficult to identify. This is partly due to the complexity of the term structure, partly to the fact that the multiplet structure sometimes changes profoundly in the higher members of a term series.

Since the extrapolation of series limits is less certain, ionisation energies are less accurately known than in simple spectra. This has led to the custom of counting term values from the ground level as zero-point in tables of all except the simplest spectra.

Our present knowledge of atomic structure has been built up in successive steps of empirical discoveries of regularities, and application of theories to atomic models. Instead of following strictly the interesting, but often confusing historical development, we shall find it easier to introduce theoretical interpretations at an earlier stage.

As a typical example of the term analysis of a multiplet, a group of lines in the spectrum of Mn, in the wavelength region of 3800 A, may be chosen. The grouping of the lines (bottom of fig. V, 1) suggests a relation between them, and in fact some of the differences of their wave-numbers agree with one another within the accuracy of the measurement. The Rydberg–Ritz combination principle then leads to the term diagram shown in fig. V, 1, and to the two-dimensional diagram of table V, 1, where combinations with a common upper or

TABLE 1

Two-dimensional array of wave-numbers in normal sextet $^6D - ^6F$ with inverted term order

J		9/2		7/2		5/2		3/2		1/2
		$i+4$		$i+3$		$i+2$		$i+1$		i
11/2	$i+5$	26260·9								
		(115·3)								
9/2	$i+4$	26376·2	(229·6)	26146·6						
		(95·6)		(95·5)						
7/2	$i+3$	26471·8	(229·7)	26242·1	(169·5)	26072·6				
				(71·4)		(71·3)				
5/2	$i+2$			26313·5	(169·6)	26143·9	(116·9)	26027·0		
						(49·0)		(48·9)		
3/2	$i+1$					26192·9	(117·0)	26075·9	(68·6)	26007·3
								(28·6)		(28·5)
1/2	i							26104·5	(68·7)	26035·8

lower level appear in one line or column respectively; the differences are shown in brackets. If we assign, in formal analogy to doublet- and triplet-spectra, a quantum number J to each level, with the selection rule $\Delta J = \pm 1$, the values of J are determined except for an additional constant i.

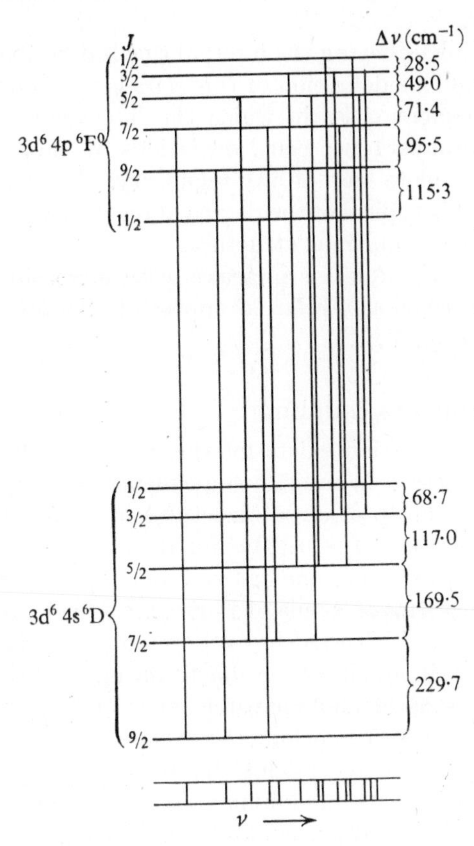

FIG. V, 1. Term diagram of sextet in spectrum of Mn.

The marked regularity in the spacings of the levels invites an attempt to apply the interval rule which was found to hold for triplet levels. The sets of values of J which agree best with the interval rule are those found by putting $i = \frac{1}{2}$, which gives the "theoretical" ratios 3 : 5 : 7 : 9 : 11 and 3 : 5 : 7 : 9 for upper and lower term respectively. The agreement is by no means within the limits of the errors of measurement and would hardly be convincing as the sole basis of the assignment of J-values. Investigation of intensity ratios can give additional evidence, but only the study of Zeeman patterns allows the J-values to be established beyond any doubt. One ambiguity remains even then: if the term diagram of fig. V, 1 is inverted as a whole, it gives the same wave-numbers, intensity ratios and Zeeman patterns. Only combinations with other terms can remove this ambiguity.

A term multiplet obeying the interval rule is described as a *normal multiplet*, though the definition of this class is by no means distinct. From the infrequent cases in which the interval rule holds to an accuracy of perhaps 1 per cent, deviations of varying degree are found up to complete lack of any regularity. The quoted example of the Mn sextet is typical of the degree of regularity often found and described as *normal* multiplet structure.

If the energy of the levels increases with increasing J, the term multiplet is described as *regular*, as opposed to the *inverted* multiplet terms shown in fig. V, 1.

2. Russell–Saunders coupling

No definite convention exists for the distinction between simple and complex spectra. We define conveniently a complex spectrum as one which is due to electron structures in which more than one electron has a value of $l > 0$. By this definition all simple spectra are characterised by either one or two non-combining term systems, each having one series of S- one of P- etc. terms as in an alkali atom. They can be treated, as was shown in chapter III, as one-electron spectra with certain modifications due to the effect of the spin. Some of the higher terms of alkaline earth- and earth-spectra, known as *dashed* or *anomalous* terms, belong to the class of complex spectra and will accordingly be treated in this chapter.

Normal multiplets are predominant in the lower terms of most elements of low and medium values of Z. Their structure is such as would be produced by a fictitious electron with an orbital quantum number L ($= 1, 2, ..$) and a spin quantum number S which can assume various integral values for an even number of electrons and half-integral values for an odd number of electrons. In an atom with several electrons outside closed shells this could be explained by the assumption, made entirely *ad hoc*, that the orbital momenta of the different electrons are coupled to form a resultant orbital momentum, according to the vector relation

$$\mathbf{L} = \mathbf{L}_1 + \mathbf{L}_2 + \ .. \qquad (1)$$

with the quantum numbers

$$L = l_1+l_2+ \,..; \quad l_1+l_2+ \,..-1; \quad ...$$

Similarly the spins would have to be coupled to form a spin resultant, with the quantum numbers

$$S = s_1+s_2+ \,.., \quad s_1+s_2+ \,..-1, \quad s_1+s_2+ \,..-2, \,..$$

down to 0 or $\frac{1}{2}$ for an even or odd number of electrons respectively. A much weaker coupling would have to be assumed between spins and orbital momenta, leading in second approximation to the vector relation

$$\mathbf{J} = \mathbf{L}+\mathbf{S} \tag{2}$$

with the possible values of the quantum number $J = L+S$, $L+S-1, \,..|L-S|$. The latter interaction would have to be magnetic in order to explain the validity of the interval rule.

These relations are formed in obvious generalisation of those used to explain the doublet- and triplet-structure in simple spectra. The value $2S+1$ is the *multiplicity*, i.e. the number of levels for a sufficiently large value of L. This purely formal theory explains the variation of multiplicities in the periodic table as mentioned in I.5.

This scheme of vector addition expressing the strong coupling of the orbital momenta among themselves and of the spins among themselves, compared with the weak spin–orbit coupling, is known as *Russell–Saunders coupling.*[(1)] The old quantum theory is not able to account for the nature of these coupling forces; especially the magnetic forces between two spin moments are far too small to explain the apparently strong coupling leading to the quantum number S. This can, in fact, only be understood as a consequence of the Pauli principle, while the strong coupling between the L vectors can be ascribed directly to electrostatic interaction. Both effects are intimately connected and can be treated satisfactorily by quantum-mechanical methods. It will be seen that Russell–Saunders coupling arises from the predominance of electrostatic over magnetic interactions.

3. The basic assumptions of the theory

The motions of the electrons in an atom are governed by the following forces: (1) the *electrostatic* forces of attraction between nucleus and electrons and of repulsion between pairs of individual electrons, (2) *magnetic* forces due to the orbital motions and the spins of the electrons.

The basis of the theoretical treatment is the many-particle Schrödinger equation with a Hamiltonian containing various terms due to all these forces. These terms are operators derived by analogy with classical expressions. If the equation could be solved rigorously, the eigenfunctions, of the Pauli type, would describe the possible stationary states of motion, and the eigenvalues would give the energies of these states to a high degree of accuracy. The procedure would not be entirely rigorous because the treatment of the electron spin by the Pauli method as the effect of a magnetic dipole is an approximation, though an exceedingly good one for most purposes.

In this form, however, the problem is mathematically unmanageable, and some drastic assumptions have to be made for even approximate solutions to become feasible. With the exception of very few calculations on simple atoms, one generally assumes the only magnetic interaction between electrons to be that between the spin of each electron and the field caused by its *own* orbital motion. "*Spin–other orbit*" and *spin–spin* interactions are generally—often tacitly—neglected. Further, some purely mathematical approximations have to be made, and perturbation methods are used.

Considering at first the electrostatic repulsion and describing the distance of two electrons from one another by r_{ij}, we find that the potential energy of an electron,

$$\sum_{j} e^2/r_{ij},$$

is of the same order of magnitude as the potential due to the nuclear charge, $-Ze^2/r_i$, and cannot directly be treated as perturbation. The major part of the repulsion, however, is clearly a function of r_i only, owing to the approximately spherical symmetry of the entire electron cloud, and can be described as a central force field. It can be considered as a screening effect, but with a screening constant depending on r_i.

In the zero-order approximation, each electron is then regarded as a single particle moving in its respective central field. This motion is mathematically expressed by a wave function which is a product, or a linear combination of products, of single-particle wave functions. To this approximation, the energy depends only on the values of the set of two quantum numbers n, l for each electron, so that a high degree of degeneracy remains for any specified *configuration.*

Degeneracy arises, however, not only from the fact that the energy does not depend on the remaining two quantum numbers, for each electron, in the central field approximation, but also from the identity of the electrons. The latter causes the *exchange degeneracy* the simplest form of which was discussed for two-electron spectra (III.B.4). In order to ensure fulfilment of the Pauli principle one has to replace the simple product of functions by a determinant (see IV.B.1) which makes the total wave function antisymmetrical. In actual fact, a sub-determinant including only the electrons outside closed shells can usually be used, which means that the electrons in closed shells are only treated in the zero-approximation of the central field.

In proceeding to the next approximation one has to introduce (1) the remaining part of the repulsion potential which depends on the angles between the vectors $\mathbf{r}_i$ and $\mathbf{r}_j$ and is therefore time-dependent; (2) magnetic interactions.

4. Electrostatic interaction between two electrons

The quantum-mechanical theory of electrostatic interaction between electrons, even in its simplest form of first-order perturbation, is rather abstract and involved. It is possible, however, to understand some of the results qualitatively in a simple, pictorial way, somewhat analogous to the method by which the Landé formula for the anomalous Zeeman effect can be explained on a vector model. This may be illustrated as a preliminary to the quantum-mechanical theory, for the configuration of two non-equivalent p-electrons in Russell–Saunders coupling.

Ignoring at first the existence of the spin, we try to form a picture of the charge distributions of two p-electrons whose **L**-vectors form resultants 0, $\hbar$ and $2\hbar$ giving rise to S-, P- and D-terms. The charge distribution of a p-electron whose axis is fixed in space is described by the function $\rho = e\psi\psi^*$ of the state $l = 1$, $m = \pm 1$ (see p. 40) in which the charge is concentrated about the orbital plane normal to **L**. For the terms D and S in which the **L** vectors are parallel and antiparallel, the orbital planes coincide and we get the greatest possible overlap of charge and the highest energy due to repulsion. In the P-term the planes form an angle of about 60° and the energy will be lower. The dependence of the energy on the angle ω between the **L** vectors has a period of π and will be a function of $\cos^2\omega$, as compared with the proportionality to $\cos\omega$ in magnetic interactions. As in the latter case, any formulae based on the vector model will apply asymptotically for large quantum numbers and hold only approximately for the small quantum numbers in which we are mainly interested.

In qualitative agreement with this picture, it is found that the P-terms of a configuration $np\ n'p$ are always lower than the S- and D-terms, though the latter do not coincide.

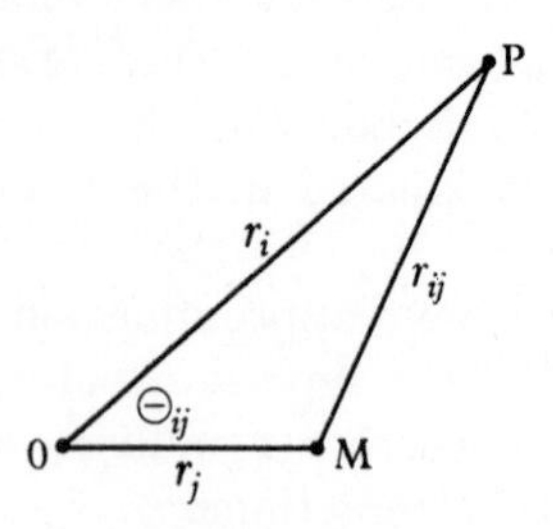

FIG. V, 2. Electrostatic repulsion between two electrons placed at M and P.

The quantitative treatment makes use of the symmetry properties of interactions in a central field by expanding them in terms of spherical harmonics. We first consider the classical repulsion between two point electrons. In fig. V, 2 where 0 is the nucleus, and the origin of the polar coordinate system, the two electrons are first assumed to be point charges placed at M and P. The potential energy of the electron at P due to the repulsion by that at M is

$$V_{ij} = e^2/r_{ij} = e^2/(r_i^2+r_j^2-2r_ir_j\mu)^{\frac{1}{2}} \tag{3}$$

where $\mu = \cos\theta_{ij}$. If $r_j < r_i$ this can be expanded in terms of r_j/r_i:

$$V_{ij} = \frac{e^2}{r_i}\left(1+\frac{r_j}{r_i}P_1+\frac{r_j^2}{r_i^2}P_2+\ ..\right) \tag{4}$$

where the coefficients

$$P_1 = \mu, \qquad P_2 = \tfrac{1}{2}(3\mu^2-1), \qquad P_3 = \tfrac{1}{2}(5\mu^3-3\mu)\ ..$$

are the Legendre polynomials mentioned above (II.C). If the point charge at M is replaced by a charge "cloud" having definite symmetry properties about the axis OP, the integration over the whole volume of the charge cloud causes some of the terms in the expansion to vanish. If the distribution has rotational symmetry about OP and is symmetrical with regard to the equatorial plane ($\perp$ OP), positive and negative values of μ occur equally often in the integration and all odd polynomials vanish. This applies, e.g. to P_1 which describes the dipole interaction.

If both charges are replaced by rigid charge clouds of densities ρ_i and ρ_j of well-defined axial symmetries, the electrostatic potential

$$V_{ij} = \iint \rho_i \rho_j \frac{1}{r_{ij}} d\tau_i d\tau_j \tag{5}$$

can be similarly expanded and expressed as a function of the angle θ_{ij} between the two axes of symmetry.

If this classical calculation is applied to the electrostatic interaction of charge clouds whose densities are given by the square of the modulus of Schrödinger functions, the energy appears as a function of the angles θ_{ij} between the angular momentum vectors $\mathbf{L}_i$ and $\mathbf{L}_j$. For two p-electrons this function is of the type of the interaction between two quadrupole moments ($\sim \cos^2\theta_{ij}$); for other electrons it may include interactions of higher, even powers of $\cos\theta_{ij}$, i.e. the functions P_4, P_6 ... The different θ_{ij} correspond to different values of the resultant $\mathbf{L} = \mathbf{L}_i + \mathbf{L}_j$.

This quasi-classical treatment does not account for the occurrence of two terms for each value of L, one with parallel the other with antiparallel spins. This effect is due to the non-static nature of the electron clouds in which the electrons can oscillate in phase or in antiphase, in a manner which cannot be satisfactorily described in classical language. The greater or smaller overlap of the wave functions of the two electrons in the two states of motion causes the difference in their energies of electrostatic repulsion. According to Pauli's Principle, these two types of motion are linked with parallel or antiparallel spins, giving rise to a triplet- or singlet-term respectively.

In quantum mechanics, the electrostatic interaction between two electrons is described by a matrix whose elements have the general form

$$\langle ab|q|cd\rangle = e^2 \int \psi_1{}^*(a)\psi_2{}^*(b)\frac{e^2}{r_{12}}\psi_1(c)\psi_2(d)\, d\tau_1 d\tau_2 \tag{6}$$

where each of the letters a, b, c, d stands for a quadruple of single electron quantum numbers n, l, m_l, m_s of electrons 1 and 2.

In the zero approximation of a central force field the energy depends on n and l, but not on m_l and m_s. The first-order perturbation energy will then contain only matrix elements linking degenerate states, i.e. states of the same n and l. The electrostatic interaction between two electrons does not involve the spin directly and it does not affect the sum $L_{i_z} + L_{j_z}$. Correspondingly, there are only matrix elements connecting states for which m_s has the same value for any one electron and for which the sum $m_l{}^{(1)} + m_l{}^{(2)}$ is the same. As a result of these special properties of the interaction, all elements (6) are of the two types

$$J = \langle \mathrm{ab}|q|\mathrm{ab} \rangle \text{ and } K = \langle \mathrm{ab}|q|\mathrm{ba} \rangle.$$

An example of this was found to be the electrostatic interaction of two s-electrons in He (p. 142) where J and K took the simple form of integrals involving the radial wave functions only. The lack of dependence on the angles was due to the spherical symmetry of the s-wave functions. J was identical with the classical electrostatic interaction of two rigid charge clouds whose densities are given by the values of $e\psi^*\psi$, while K had no simple classical analogue.

For electrons with $l > 0$ the dependence on the relative orientation of the two l vectors causes the matrix elements J and K to consist of integrals containing, apart from r, the angles θ and φ in the form of spherical harmonics. The integration over θ and φ can be carried out and leads to coefficients a^{k} and b^{k} which are somewhat involved expressions of the sets of values l, m_l and l', m_l' of the two electrons; they have been tabulated.*

Further constants D_k depend only on l and l'.

The "direct" integrals J and the "exchange" integrals K can be written as sums:

$$J(\mathrm{a}, \mathrm{b}) = \langle \mathrm{ab}|q|\mathrm{ab} \rangle = \sum_{\mathrm{k}} a^{\mathrm{k}}(l^{\mathrm{a}}m_l^{\mathrm{a}}, l^{\mathrm{b}}m_l^{\mathrm{b}}) D_k F_k(n^{\mathrm{a}}l^{\mathrm{a}}, n^{\mathrm{b}}l^{\mathrm{b}}), \quad (7)$$

$$K(\mathrm{a}, \mathrm{b}) = \langle \mathrm{ab}|q|\mathrm{ba} \rangle = \delta(m_s{}^{\mathrm{a}} m_s{}^{\mathrm{b}}) \sum_{\mathrm{k}} b^{\mathrm{k}}(l^{\mathrm{a}}m^{\mathrm{a}}, l^{\mathrm{b}}m^{\mathrm{b}}) D_{\mathrm{k}} G_{\mathrm{k}}(n^{\mathrm{a}}l^{\mathrm{a}}, n^{\mathrm{b}}l^{\mathrm{b}}). \quad (8)$$

F_{k} and G_{k} are radial integrals which are independent of m_l and m_s:

$$F_{\mathrm{k}} = \frac{e^2}{D_{\mathrm{k}}} \int_0^\infty \int_0^\infty \frac{r^{\mathrm{k}}}{r^{\mathrm{k}+1}} R_1{}^2(n^{\mathrm{a}}l^{\mathrm{a}}) R_2{}^2(n^{\mathrm{b}}l^{\mathrm{b}})\, \mathrm{d}r_1 \mathrm{d}r_2, \quad (9)$$

$$G_{\mathrm{k}} = \frac{e^2}{D_{\mathrm{k}}} \int_0^\infty \int_0^\infty \frac{r^{\mathrm{k}}}{r^{\mathrm{k}+1}} R_1(n^{\mathrm{a}}l^{\mathrm{a}}) R_1(n^{\mathrm{b}}l^{\mathrm{b}}) R_2(n^{\mathrm{a}}l^{\mathrm{a}}) R_2(n^{\mathrm{b}}l^{\mathrm{b}})\, \mathrm{d}r_1 \mathrm{d}r_2. \quad (10)$$

The F_{k} and G_{k} can only be evaluated if the central field acting on each electron is known; they are often regarded as unknown parameters as will be seen below. *Both F_{k} and G_{k} are always positive, also $G_{\mathrm{k}} = F_{\mathrm{k}}$* for equivalent electrons.

* See E. U. Condon and G. H. Shortley, *Theory of Atomic Spectra.*

The coefficients a^k and b^k have non-vanishing values only for a few values of k so that the sums J and K reduce to a small number of terms. Table V, 2 gives a few examples of values of $D_k a^k$ and $D_k b^k$.

TABLE 2

($D_0 = 1$ for ss, $D_1 = 3$ for sp, $D_2 = 5$ and 25 for sd and pp resp.)

			$k = 0$		$k = 1$		$k = 2$	
	m_l	m_l'	$D_k a^k$	$D_k b^k$	$D_k a^k$	$D_k b^k$	$D_k a^k$	$D_k b^k$
ss	0	0	1	1	0	0	0	0
sp	0	±1	1	0	0	1	0	0
	0	0	1	0	0	1	0	0
sd	0	±2	1	0	0	0	0	1
	0	±1	1	0	0	0	0	1
	0	0	1	0	0	0	0	1
pp	±1	±1	1	1	0	0	1	1
	±1	0	1	0	0	0	−2	3
	0	0	1	1	0	0	4	4
	±1	∓1	1	0	0	0	1	6

Remembering that each term of J, of a given k, corresponds to one term in the expansion, in terms of spherical harmonics, of the classical interaction between two electron clouds we find the following result: the expression J describing the electrostatic interaction between a p-electron and a p-, d-, f- . . electron contains only terms with k = 0 and 2. The latter, i.e. the term with F_2, corresponds to the classical *quadrupole–quadrupole* interaction between charge clouds. The interaction between a d-electron and a d-, f-, . . electron contains only terms with k = 0, 2 and 4; the last of these, i.e. F_4, corresponds to a quadrupole–16 pole interaction.

The exchange integrals K have no simple corresponding term in the classical interaction. The factor $\delta(m_s{}^a, m_s{}^b)$ in (8) indicates that all terms of the sum vanish except those for which $m_s{}^a = m_s{}^b$. It means that there is *no exchange effect between electrons in states with different* m_s.

Both J and K contain the first-order perturbation only; the perturbing potential is, in accordance with the rules given in II.C.3, calculated with the use of the degenerate eigenfunctions of the unperturbed one-electron states in central force fields. Any deformations of the charge distributions resulting from their mutual interactions, such as polarisation effects, are disregarded. Their treatment requires the inclusion of matrix elements linking states of different configurations.

5. Identification of terms[2,3,4,5]

The matrix components $\langle ab|q|cd \rangle$ describe the electrostatic interaction between two electrons in the representation defined by the quantum numbers $n, l, m_l, m_s, n', l', m_l'$ and m_s' of the two individual electrons, each of the letters a, b, c, d standing for a particular set n, l, m_l, m_s. If we indicate the electrostatic energy of all the electrons by

$$Q = \sum_{i,j} e^2/r_{ij} \tag{11}$$

its matrix element can be written as $\langle A|Q|B \rangle$, where each of the symbols A and B stands for a complete set of quantum numbers $a, b, \ldots$ assigned to all the electrons in such a way that Pauli's principle is fulfilled. For most purposes, only electrons outside closed shells need be included.

Of main importance are the diagonal elements of Q which are found to be

$$\langle A|Q|A \rangle = \sum_{ij} J(i, j) - K(i, j) \tag{12}$$

where the sum is to be taken over all pairs of electrons concerned, and J and K can, according to (7) and (8) be expressed in terms of the radial integrals F_k and G_k and the coefficients a^k and b^k.

In confining ourselves to the diagonal elements of Q, we assume the values n_i and l_i of the individual electrons to be fixed. This is justified if the energy differences between different configurations are large compared with the energy differences between terms within one configuration. Otherwise the effect of non-diagonal matrix elements, known as *configuration interaction*, has to be considered (see p. 269).

In the zero-order approximation the energy is independent of all the m_l and m_s so that the system is highly degenerate. If now electrostatic and magnetic interactions are to be calculated as first-order perturbations, the usual diagonalisation methods have to be applied to the perturbation matrix.

Assuming at first that the magnetic interactions can be entirely neglected—the case of *perfect Russell–Saunders coupling*—we have to find a representation in which Q only is diagonal. The methods of doing this are based on the fact that, with exclusively electrostatic interaction, the resultant orbital angular momentum $\mathbf{L} = \Sigma \mathbf{L}_i$

and the resultant spin momentum $\mathbf{S} = \Sigma \mathbf{S}_i$ are constant in classical mechanics, as well as their vector sum $\mathbf{J} = \mathbf{L}+\mathbf{S}$. To this corresponds the quantum-mechanical relation that the operators of their absolute squares and of all their components commute with the Hamiltonian. The quantum numbers L, S, M_L, M_S, M can then be defined by

$$\mathbf{S}^2 = S(S+1)\hbar^2 \qquad \mathbf{L}^2 = L(L+1)\hbar^2 \qquad \mathbf{J}^2 = J(J+1)\hbar^2$$

$$\mathrm{S}_z = M_S\hbar \qquad \mathrm{L}_z = M_L\hbar \qquad \mathrm{J}_z = M\hbar$$

where the range of permissible values is limited in the usual way: $|M_L| \leqslant L$, etc. It is obvious from the classical model that only four of these quantum numbers are independent. The following two *complete sets* are used: L, S, M_L, M_S and L, S, J, M.

From the fact that $\mathbf{L}$ and $\mathbf{S}$ commute with the Hamiltonian it can be shown by general matrix methods that the energy is independent of M_L and M_S, in accordance with the classical model. Since the states of given L and S in the second scheme are linear combinations of those of the same values L and S in the first scheme, it follows that, in the second scheme, the energy is also independent of J and M. Also the number of states, for given values L and S, must be the same in both schemes.

Since the electrostatic interaction has no matrix elements connecting two states differing in either M_L or M_S these latter are always good quantum numbers and equal to $\Sigma\, m_l$ and $\Sigma\, m_s$ respectively.

In order to show how the number and characteristics of terms, defined by L and S, can be derived from a given configuration, we choose the configuration of two *non-equivalent* p-electrons: npn'p. We first identify all the possible states by writing down all possible combinations of individual particle quantum numbers nlm_lm_s (table V, 3, first columns). The next columns show the values $M_L = m_l+m_l'$ and $M_S = m_s+m_s'$ and, in brackets, the number of states for each pair M_LM_S. The bottom part of the list, with $M = -1, -2$ and -3 is obviously analogous to the part with positive M and has been omitted.

The right-hand part of the table contains the description of the terms in the usual nomenclature where S, P, D . . indicate the values of $L = 0, 1, 2$ . . and the superscript stands for $2S+1$. The common value of $M = M_L+M_S$ for each section is shown in the last column. The values of J are left unspecified, but we make use of the known number of states: J can have all values between $L+S$ and $L-S$ and there are $2J+1$ states for any given J. If this is

TABLE 3

Terms of the configurations npn'p and np^2
(the latter without brackets)

$(m_l m_s)(m_l' m_s')$		M_L	M_S								M
$(1\ \frac{1}{2})(1\ \frac{1}{2})$*		2	1	(1)	(^{3}D)						3
$(1\ \frac{1}{2})(1\ -\frac{1}{2})$	$(1\ -\frac{1}{2})(1\ \frac{1}{2})$*	2	0	(2)	(^{3}D)	^{1}D					2
$(1\ \frac{1}{2})(0\ \frac{1}{2})$	$(0\ \frac{1}{2})(1\ \frac{1}{2})$*	1	1	(2)	(^{3}D)		^{3}P				
$(1\ \frac{1}{2})(0\ -\frac{1}{2})$ $(1\ -\frac{1}{2})(0\ \frac{1}{2})$	$(0\ -\frac{1}{2})(1\ \frac{1}{2})$* $(0\ \frac{1}{2})(1\ -\frac{1}{2})$*	1	0	(4)	(^{3}D)	^{1}D	^{3}P	(^{1}P)			1
$(1\ \frac{1}{2})(-1\ \frac{1}{2})$ $(0\ \frac{1}{2})(0\ \frac{1}{2})$*	$(-1\ \frac{1}{2})(1\ \frac{1}{2})$*	0	1	(3)	(^{3}D)		^{3}P		(^{3}S)		
$(1\ -\frac{1}{2})(1\ -\frac{1}{2})$*		2	-1	(1)	(^{3}D)						
$(1\ -\frac{1}{2})(0\ -\frac{1}{2})$	$(0\ -\frac{1}{2})(1\ -\frac{1}{2})$*	1	-1	(2)	(^{3}D)		^{3}P				0
$(1\ \frac{1}{2})(-1\ -\frac{1}{2})$ $(1\ -\frac{1}{2})(-1\ \frac{1}{2})$ $(0\ \frac{1}{2})(0\ -\frac{1}{2})$	$(-1\ -\frac{1}{2})(1\ \frac{1}{2})$* $(-1\ \frac{1}{2})(1\ -\frac{1}{2})$* $(0\ -\frac{1}{2})(0\ \frac{1}{2})$*	0	0	(6)	(^{3}D)	^{1}D	^{3}P	(^{1}P)	(^{3}S)	^{1}S	
$(0\ \frac{1}{2})(-1\ \frac{1}{2})$	$(-1\ \frac{1}{2})(0\ \frac{1}{2})$*	-1	1	(2)	(^{3}D)		^{3}P				
— — —	— — —	—	—	—	—		—		—		-1

taken into account, the number of states on the left is found to be equal to that on the right; only states appearing in the same line (or bracketed together) are connected by electrostatic interaction.

The terms are found as follows. The highest values of M_L and M_S are 2 and 1 respectively and appear together in one state only. This shows that the highest values of L and S must be 2 and 1, arising from a term ^{3}D, with possible values $J = 3, 2, 1$. This accounts for $7+5+3 = 15$ states, one in each line of the table. The second line of the table now leaves one state $M_L = 2, M_S = 0$ to be accounted for. This can only be due to a term with $L = 2$. If we assume it to be a ^{1}D term (i.e. $S = 0$) it will just account for any one of the states with $M_S = 0$ and $M_L = 1, 0, -1, -2$. This leaves one state in the line $M_L = 1$, $M_S = 1$ which must be due to a ^{3}P term, and the other states due to the latter can be entered.

In this way all the states can be uniquely accounted for by the terms ^{1}S, ^{3}S, ^{1}P, ^{3}P, ^{1}D, ^{3}D. The two non-equivalent p-electrons are thus found to give rise to exactly those terms which the simple vector diagram would predict. This applies quite generally to any number of non-equivalent electrons; all possible terms are found by combining any one of the possible values $L = \Sigma\, l_i, \Sigma\, l_i - 1, \ldots$ to $L_{\text{min.}}$ with any one of the possible values of $S = \Sigma\, S_i, \Sigma\, S_i - 1, \ldots$ to $S = 0$ or $\frac{1}{2}$, for an even or odd number of electrons respectively.

If we now consider two *equivalent* p-electrons, so that $n = n'$, all the entries marked by an asterisk in table V, 3 are forbidden by the *Pauli principle*, either because both m_l and m_s are the same for both electrons or because, with $n = n'$, the interchange of order in the specification $(m_l, m_s)(m_l', m_s')$ no longer represents two different states. When the terms are now identified in the way described, all those given in brackets have to be omitted. We find that the configuration p^2 gives rise to the terms ^{1}D, ^{3}P and ^{1}S, a result which cannot be derived by ordinary vector addition.

It is obvious that the method can be extended to other configurations of equivalent electrons. Table V, 4 and V, 5 show this in a slightly simplified way for the configurations p^3 and p^4. After half of the shell of 6 p-electrons has been filled in, the number of allowed combinations becomes less again, and p^4 and p^5 have the same terms as p^2 and p respectively. Quite generally, if N is the number of electrons in a complete sub-shell, N-n electrons produce the same terms as n electrons. Tables V, 6 and V, 7 show all possible terms for equivalent p- and d-electrons. It will be noted that a given configuration d^n can have several terms of the same kind, so that L and S do not always specify a term completely.

TABLE 4

Terms of the configuration np^3

$m_l^{1,2,3}$	$m_s^{1,2,3}$	M_L	M_S	
1 1 0	$\frac{1}{2}$ $-\frac{1}{2}$ $\pm\frac{1}{2}$	2	$\pm\frac{1}{2}$	^2D, ^2P, ^4S
1 1 −1	$\frac{1}{2}$ $-\frac{1}{2}$ $\pm\frac{1}{2}$	1	$\pm\frac{1}{2}$	
1 0 0	$\pm\frac{1}{2}$ $\frac{1}{2}$ $-\frac{1}{2}$	1	$\pm\frac{1}{2}$	
1 0 −1	$\pm\frac{1}{2}$ $\pm\frac{1}{2}$ $\pm\frac{1}{2}$	0	$\pm\frac{1}{2}$ $\pm\frac{1}{2}$ $\pm\frac{1}{2}$ $\pm\frac{3}{2}$	
1 −1 −1	$\pm\frac{1}{2}$ $\frac{1}{2}$ $-\frac{1}{2}$	−1	$\pm\frac{1}{2}$	
0 0 −1	$\frac{1}{2}$ $-\frac{1}{2}$ $\pm\frac{1}{2}$	−1	$\pm\frac{1}{2}$	
0 −1 −1	$\pm\frac{1}{2}$ $\frac{1}{2}$ $-\frac{1}{2}$	−2	$\pm\frac{1}{2}$	

TABLE 5

Terms of the configuration np^4

$m_l^{1,2,3,4}$	$m_s^{1,2,3,4}$	M_L	M_S	
1 1 0 0	$\frac{1}{2}$ $-\frac{1}{2}$ $\frac{1}{2}$ $-\frac{1}{2}$	2	0	^1D, ^3P, ^1S
1 1 0 −1	$\frac{1}{2}$ $-\frac{1}{2}$ $\pm\frac{1}{2}$ $\pm\frac{1}{2}$	1	1, 0, 0, −1	
1 1 −1 −1	$\frac{1}{2}$ $-\frac{1}{2}$ $\frac{1}{2}$ $-\frac{1}{2}$	0	0	
1 0 0 −1	$\pm\frac{1}{2}$ $\frac{1}{2}$ $-\frac{1}{2}$ $\pm\frac{1}{2}$	0	1, 0, 0, −1	
1 0 −1 −1	$\pm\frac{1}{2}$ $\pm\frac{1}{2}$ $\frac{1}{2}$ $-\frac{1}{2}$	−1	1, 0, 0, −1	
0 0 −1 −1	$\frac{1}{2}$ $-\frac{1}{2}$ $\frac{1}{2}$ $-\frac{1}{2}$	−2	0	

TABLE 6

Terms of the configuration p^k

p		^2P			
p^2	^1S		^1D ^3P		
p^3		^2P		^2D	^4S
p^4	^1S		^1D ^3P		
p^5		^2P			
p^6	^1S				

TABLE 7

Terms of the configuration d^k

d	^{2}D
d^2	^{1}S ^{1}D ^{1}G ^{3}P ^{3}F
d^3	^{2}D ^{2}P ^{2}D ^{2}F ^{2}G ^{2}H ^{4}P ^{4}F
d^4	^{1}S ^{1}D ^{1}G ^{3}P ^{3}F ^{1}S ^{1}D ^{1}F ^{1}G ^{1}J ^{3}P ^{3}D ^{3}F ^{3}G ^{3}H ^{5}D
d^5	^{2}D ^{2}P ^{2}D ^{2}F ^{2}G ^{2}H ^{4}P ^{4}F ^{2}S ^{2}D ^{2}F ^{2}G ^{2}J ^{4}D ^{4}G ^{6}S
d^6	^{1}S ^{1}D ^{1}G ^{3}P ^{3}F ^{1}S ^{1}D ^{1}F ^{1}G ^{1}J ^{3}P ^{3}D ^{3}F ^{3}G ^{3}H ^{5}D
d^7	^{2}D ^{2}P ^{2}D ^{2}F ^{2}G ^{2}H ^{4}P ^{4}F
d^8	^{1}S ^{1}D ^{1}G ^{3}P ^{3}F
d^9	^{2}D
d^{10}	^{1}S

If a configuration consists of a group of equivalent electrons and some further electrons which are neither equivalent to one another nor to those in the group, one has to establish at first the terms due to the equivalent electrons. To each of these, the values l and s of the non-equivalent electrons are then added according to the rules of the vector diagram. The terms of the configuration $n\mathrm{p}^2n'\mathrm{p}$ are, e.g. found as follows:

$n\mathrm{p}^2$	$^1\mathrm{D}$	$^3\mathrm{P}$	$^1\mathrm{S}$
$n'\mathrm{p}$	$^2\mathrm{P}$ $^2\mathrm{D}$ $^2\mathrm{F}$	$^2\mathrm{S}$ $^4\mathrm{S}$ $^2\mathrm{P}$ $^4\mathrm{P}$ $^2\mathrm{D}$ $^4\mathrm{D}$	$^2\mathrm{P}$

In this example, the three different terms $^2\mathrm{P}$ are distinguished by their *genealogy*. The *parent* terms are $^1\mathrm{D}$, $^3\mathrm{P}$ and $^1\mathrm{S}$, for the three terms, and this fact can often be used in estimates of term values. The group of terms arising from one parent term by the addition of one electron is sometimes described as a *polyad*; addition of an s-, p-, d-electron leads to a *monad, triad, pentad* respectively. Transitions between two polyads arising from the same parent term form a *super-multiplet*, of which the following is an example:

3d4p	$^3\mathrm{P}$ $^3\mathrm{D}$ $^3\mathrm{F}$	$^1\mathrm{P}$ $^1\mathrm{D}$ $^1\mathrm{F}$
	a b c	a′ b′ c′
	↘ ↓ ↙	↘ ↓ ↙
3d4s	$^3\mathrm{D}$	$^1\mathrm{D}$

6. The term differences

In calculating the energies of the terms one can make use of the fact that the electrostatic interaction matrix only connects states with the same values of M_L and M_S. The secular equation then only needs to be written down separately for all the states of a given pair of values M_L, M_S, i.e. for any one line or bracket of table V, 3.

For the simpler configurations the eigenvalues of the energy can be found quite easily by Slater's method of *diagonal sums.*[(5)] It is based on the general theorem stating that the sum of the diagonal elements of a matrix is equal to the sum of its eigenvalues.

The configuration $n\mathrm{p}n'\mathrm{p}$ may again be used as an example. Since there are only two electrons, the relation (12) simplifies to

$$\langle \mathrm{A}|Q|\mathrm{A}\rangle = \langle \mathrm{ab}|q|\mathrm{ab}\rangle - \langle \mathrm{ab}|q|\mathrm{ba}\rangle = J(\mathrm{ab}) - K(\mathrm{ab}). \tag{13}$$

The diagonal elements $\langle m_l m_s, m_l' m_s'|\mathrm{Q}|m_l m_s, m_l' m_s'\rangle$ will be written briefly as $(m_l m_s;\, m_l', m_s')$. The first line of table V, 3 contains only one state, so that the diagonal sum reduces to a single element and the first-order perturbation energy of the term ^{3}D is found directly:

$$^3\mathrm{D} = (1\ \tfrac{1}{2},\ 1\ \tfrac{1}{2}).$$

The second line gives by the sum rule

$$^3\mathrm{D}+{}^1\mathrm{D} = (1\ \tfrac{1}{2},\ 1\ -\tfrac{1}{2})+(1\ -\tfrac{1}{2},\ 1\ \tfrac{1}{2}).$$

If these equations are written down for successive lines of the table, it is found that each new equation introduces not more than one new term, so that by simple elimination each of the terms can be expressed by the diagonal elements of Q in the m_l, m_s representation. Each of them has the form $J-K$ and can be derived from table V, 2 and equations (7) and (8).

For the configuration np^2 the procedure is the same, but the starred and bracketed states have to be omitted, so that we find:

$$^1\mathrm{D} = (1\ \tfrac{1}{2},\ 1\ -\tfrac{1}{2})$$
$$^3\mathrm{P} = (1\ \tfrac{1}{2},\ 0\ \tfrac{1}{2})$$
$$^1\mathrm{S} = (1\ \tfrac{1}{2},\ -1\ -\tfrac{1}{2})+(1\ -\tfrac{1}{2},\ -1\ \tfrac{1}{2})+(0\ \tfrac{1}{2},\ 0\ -\tfrac{1}{2})-(1\ \tfrac{1}{2},\ 0\ \tfrac{1}{2})- -(1\ \tfrac{1}{2},\ 1\ -\tfrac{1}{2}).$$

Table V, 2 gives for $(1\ \tfrac{1}{2},\ 1\ -\tfrac{1}{2})$, i.e. for $m_l = 1$, $m_l' = 1$,

$$J = F_0+F_2$$

but, since $m_s = -m_s'$, $K = 0$, according to (8), so that

$$^1\mathrm{D} = F_0+F_2.$$

For $(1\ \tfrac{1}{2},\ 0\ \tfrac{1}{2})$ K does not vanish, but $G_k = F_k$ for equivalent electrons:

$$J = F_0-2F_2, \qquad K = 3F_2, \qquad \text{giving } ^3\mathrm{P} = F_0-5F_2.$$

Similarly we find $^1\mathrm{S} = F_0+10F_2$. This example shows how the results of the theory can be compared with experimental data, without any knowledge of the radial integrals: for the configuration np^2 the relation should hold:

$$(^1\mathrm{S}-{}^1\mathrm{D})/(^1\mathrm{D}-{}^3\mathrm{P}) = 1{\cdot}5. \tag{14}$$

This result goes, of course, well beyond the qualitative information gained from the semi-classical picture. Experimentally, the ratio is found to lie between 1·5 and 1·13 in typical cases (see V.B).

For more complex configurations the diagonal sum method becomes tedious, and whenever one configuration leads to several terms of the same kind the method gives only the sum of their energies. It leads, however, directly to some useful general results connected with the analogy between the configurations l^{N-n} and l^n where l stands for one of the symbols s, p, d, . . and N is the number of electrons in the complete sub-shell. It is found that the electrostatic interaction energy is the same for the terms of these two configurations of n electrons and n "holes".

More powerful methods have been developed by Racah.[6,7] He transforms the angle-dependent part of the interaction matrix, i.e. $P_k(\cos \omega)$, from the $n,l,m_l, m_s, n', l'm_l' m_s'$ scheme into the $nn'll'LS$ scheme by means of matrix methods based on tensor operators. This leads to closed expressions for the coefficients of the F_{k} and G_{k} in simpler cases and has been used for the calculation of these coefficients for more complicated configurations. By this method the values of different terms of the same L and S can be obtained separately, though often by means of rather laborious solutions of determinant equations.

Results for some configurations are given in table V, 8, but for a more comprehensive list the reader must be referred to the literature.

Since there are generally more equations than unknown parameters for every configuration, some comparison with experimental data is possible. Ratios of differences of terms can be compared as in the example discussed above (p. 265) or the best values of the parameters can be determined by the least squares method, and all term values compared together.[8] The values of F_{k} and G_{k} have a marked tendency to decrease with increasing k. Some detailed use of the theory will be made in section V.2, but a few points of a general nature may be mentioned here.

It has been found empirically that the lowest term of a configuration of equivalent electrons is that which has the highest value of S and, if there are several of these, that with the highest value of L. This is known as *Hund's rule.*[2] The lowest terms in table V, 8 are underlined.

Table V, 8 shows that in the configuration pd, the singlet term is higher than the triplet term in P and F, but the order is reversed in D. This is an example of a general rule that the order singlet–triplet alternates with increasing L within one configuration.

It is interesting to compare the term orders given by the theory, which are in general agreement with experience, with the semi-classical picture of interacting charge clouds (p. 253), though we cannot expect exchange effects, i.e. the terms in G_{k}, to be explained by the latter. In the two-electron configurations p^2, pp, pd, pf . . there are no F_{k} terms for $\mathrm{k} > 2$, so that the interaction is entirely of the quadrupole type, and the considerations of p. 253 can be expected to apply. In fact it is found that the lowest and highest values of L form the highest terms while the lowest term has an intermediate value of L. In configurations such as d^2, dd, df and ff the expression in F_2 still appears to predominate and the terms are found to be similarly arranged. Nearly always, however, F_2 is found to be algebraically smallest for a value of L rather larger than $\frac{1}{2}(L_{max}+L_{min})$.

TABLE 8

Relative term values

ps, ds, ...

$$^1P, {}^1D, \ldots = F_0+G_0$$
$$^3\underline{P}, {}^3\underline{D}, \ldots = F_0-G_0$$

$npn'p$

$$^1S, {}^3S = F_0+10F_2 \pm (G_0+10G_2)$$
$$^1\underline{P}, {}^3P = F_0-5F_2 \mp (G_0-5G_2)$$
$$^1D, {}^3D = F_0+F_2 \pm (G_0+G_2)$$

pd

$$^1P, {}^3P = F_0+7F_2 \pm (G_1+63G_3)$$
$$^1\underline{D}, {}^3D = F_0-7F_2 \mp (3G_1-21G_3)$$
$$^1F, {}^3F = F_0+2F_2 \pm (6G_1+3G_3)$$

p^2

$$^1S = F_0+10F_2$$
$$^1D = F_0+F_2$$
$$^3\underline{P} = F_0-5F_2$$

p^3

$$^4\underline{S} = 3F_0-15F_2$$
$$^2P = 3F_0$$
$$^2D = 3F_0-6F_2$$

d^2

$$^1S = F_0+14F_2+126F_4$$
$$^3P = F_0+7F_2-84F_4$$
$$^1D = F_0-3F_2+36F_4$$
$$^3\underline{F} = F_0-8F_2-9F_4$$
$$^1G = F_0+4F_2+F_4$$

p^2s

$$^2S = F_0+10F_2-G_1$$
$$^2P = F_0-5F_2+G_1$$
$$^4\underline{P} = F_0-5F_2-2G_1$$
$$^2D = F_0+F_2-G_1$$

where $F_0 = F_0(p^2)$, $F_2 = F_2(p^2)$, $G_1 = G_1(ps)$

This quantitative departure from the semi-classical prediction is not unexpected since we are dealing with small quantum numbers.

It is interesting to note how the coefficient of one particular F_k changes with L within one configuration. This is shown in fig. V, 3 where the coefficients of F_2, F_4 and F_6 are plotted for the configuration ff from tables given by Condon and Shortley and joined by smooth curves. In the range from L_{min} to L_{max} the angle ω between the two **L** changes classically from 0 to π, and the three curves show the existence of one, two and three minima respectively according to quadrupole, 16-pole and 64-pole interaction ($\cos^2 \omega$, $\cos^4 \omega$ and $\cos^6 \omega$).

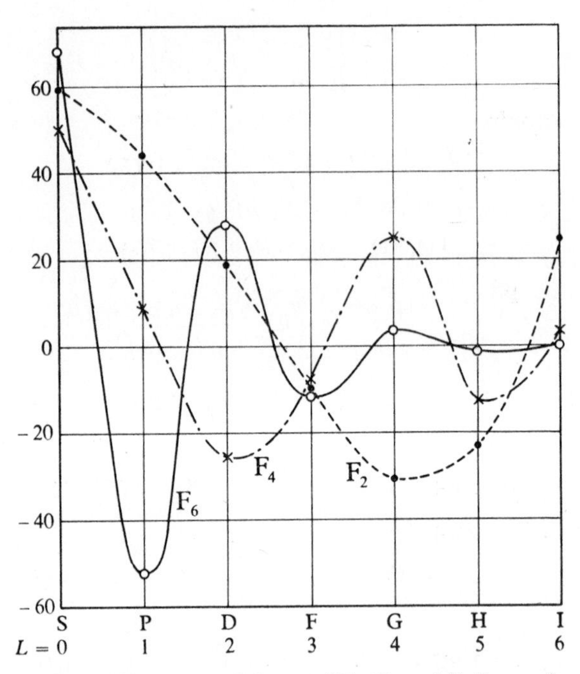

Fig. V, 3. Plot of Slater coefficients of F_2, F_4 and F_6 for configuration ff. The coefficients of F_4 and F_6 have been scaled down by factors 4 and 25 respectively.

It was found empirically[9] that in the d-electron spectra the agreement between experiment and calculation can be greatly improved if a term proportional to $L(L+1)$ is added to the electrostatic interaction terms. According to Racah[10,11] this can be explained by a distortion of the orbital wave functions of the individual electrons under the influence of mutual interaction.

The description of such effects in terms of central field wave functions requires second-order terms in the perturbation and thus involves other quantum states of the electrons, i.e. other configurations. The proper treatment of this effect of *configuration-interaction* in complex spectra is difficult, and semi-empirical methods are generally employed.[(12)] More extensive studies of configuration interaction have been made for two electron systems (ref. III, 68).

Failures of the theoretical formulae due to insufficient validity of the assumption of Russell–Saunders coupling will be treated in V.B.

7. The absolute term values

In the preceding sections, the radial integrals F_k and G_k have been regarded as empirical, adjustable parameters. If the actual term differences, as opposed to their ratios, are to be calculated, the radial functions $R(r)$ have to be known for the electrons outside closed shells. The absolute term values themselves also involve the potential energy due to the core. For term values as defined relative to the state in which all *outer* electrons are removed to infinity, only the screening effect of the core on these electrons is required—though strictly speaking of the core as affected by the outer electrons. The absolute value of the term as referred to the state in which all particles are at infinity involves the interaction between *all* the core electrons.

In any of these cases, each eigenfunction depends on the charge distribution due to the other electrons and the latter depends in its turn on the eigenfunctions to be calculated. Approximation methods for carrying out such calculations, especially the methods of the *self-consistent field* type, have been described in IV,B. Expressions for some Slater integrals have recently been derived by these methods.[(13)]

In view of the difficulty and limited accuracy of such purely theoretical calculations, semi-empirical procedures are often used in spectroscopy. The method due to Bacher and Goudsmit[(14)] has proved especially useful, partly owing to its simplicity. It aims at deriving term values of an atom from known term values of the first and higher stages of ionisation.

$W(\mathrm{a})$ and $W(\mathrm{b})$ may be defined as the energy of an ion with a single outer electron in the quantum state a or b respectively, and $W(\mathrm{a}, \mathrm{b})$ as the energy of the neutral atom with its two electrons in these two quantum states; the state with both electrons at infinity is defined as zero energy. In order to explain the principle of the method we neglect, at first, degeneracies and assume that a definite state of the ion with one electron in the state a passes into a definite two-particle state a, b on addition of a second electron in the state b. The energy of the atom $W(\mathrm{a, b})$ can then be expressed as the sum of the single-electron states and an additional term $w(\mathrm{a, b})$ expressing the energy of interaction between the two electrons:

$$W(\mathrm{a, b}) = W(\mathrm{a}) + W(\mathrm{b}) + w(\mathrm{a, b}). \tag{15}$$

An example would be the ground state of Mg^+, with one electron in the 3s state, and the ground state of Mg, with two 3s electrons, in which case $W(\mathrm{a}) = W(\mathrm{b})$.

For three electrons we can write similarly, using (15),

$$\begin{aligned} W(\mathrm{a, b, c}) &= W(\mathrm{a}) + W(\mathrm{b}) + W(\mathrm{c}) + w(\mathrm{a, b}) + w(\mathrm{a, c}) + w(\mathrm{b, c}) \\ &= W(\mathrm{a, b}) + W(\mathrm{a, c}) + W(\mathrm{b, c}) - W(\mathrm{a}) - W(\mathrm{b}) - W(\mathrm{c}). \end{aligned} \tag{16}$$

The latter relation expresses the energy of an atomic state with three electrons in terms of energies of states of the singly and doubly ionised atom. While (15) neglects only the effect of the electrons in question on any core electron which may also be present, (16) further neglects the effect which the presence of a third electron has on the interaction between the other two. It is obvious, however, that the method can be extended by the addition of a further term—the *triple energy*—allowing for this effect.

If, e.g. a 4-electron state is to be expressed in terms of the known states of the one- and two-electron system, only the pair energies $w(\mathrm{a, b})$ . . can be used. If, however, also the states of the 3-electron system are known, these data can be utilised by the introduction of triple-energy terms. The method can quite generally be made the more accurate the more information on the ionic states is available.

The high degree of degeneracy of atomic terms makes the actual calculation slightly more complicated than it might appear from the preceding explanation. In correlating a term of an atom, with given values L, S, to a term of the ion, we have to consider the states individually. In the example of two equivalent p-electrons in the atomic state a, b, the pair of ionic states a and b can be taken, in the $n^{\mathrm{i}}l^{\mathrm{i}}m_l^{\mathrm{i}}m_s^{\mathrm{i}}$ scheme, as a pair of brackets in the first column of table V, 3, with the omission of the starred entries. Each of these states with M_L, M_S values 2, 0; 1, 1; 0, 1; 1, −1; −1, 1 belongs to only one term—the term symbols in brackets having to be omitted—so that the equations (15) can be written down directly for each state. For the M_L, M_S pairs 1, 0 and 0, 0, however, each state of the $n^{\mathrm{i}}l^{\mathrm{i}}m_l^{\mathrm{i}}m_s^{\mathrm{i}}$ scheme passes into a linear combination of states belonging to different terms in the L, S scheme. The coefficients of these linear combinations can be found by matrix methods; most of the important cases have been tabulated.[14] For each of the states in the fourth line of table V, 3, e.g. the energy $W(\mathrm{a}, \mathrm{b})$ is to be taken as $\frac{1}{2}\,{}^1\mathrm{D}+\frac{1}{2}\,{}^3\mathrm{P}$, and for the state $(0\ \frac{1}{2})(0-\frac{1}{2})$ one finds $W(\mathrm{a}, \mathrm{b}) = \frac{2}{3}\,{}^1\mathrm{D}+$ $+\frac{1}{3}\,{}^1\mathrm{S}$.

Apart from this complication the method is straightforward; it gives useful rough estimates of term values if pair energies only are used and is capable of fair accuracy when triple or higher energies can be included. It has been generalised and extended by various authors.[8, 12, 15, 16]

8. The width of multiplet splitting

If we assume that, in addition to the electrostatic effects, magnetic interaction exists between the orbital motion and the spin of each individual electron, the vectors **L** and **S** are no longer individually constants of the classical motion, but only the total angular momentum **J**. In quantum mechanics the spin–orbit interaction is described by the operator

$$H_{\mathrm{mgn}} = \sum_{i} \xi(r_i)\mathbf{L}_i \cdot \mathbf{S}_i \qquad (17)$$

where $\xi(r_i)$ is defined as in (III, 53). The $\mathbf{L}_i$ and $\mathbf{S}_i$ will change in a complicated way during the classical motion, and therefore the matrix H_{mgn} will not be diagonal in the $l_i s_i$ scheme, and not even in the L, S scheme. Only **J** commutes with H_{mgn} so that the quantum number J can be used for labelling the states.

Fortunately, however, it is found that in many atoms the *magnetic interaction* can be regarded as *small* compared with the electrostatic forces, so that the non-diagonal elements in the L, S scheme can be neglected. The magnetic perturbation can then be applied to each term (L, S) individually. It is this approximation which we consider first and which leads to an interpretation of the structure of *normal* multiplets. Effects due to non-diagonal terms can, if small, be treated as perturbations due to the nearest term or terms.

In the most general case the magnetic interaction has to be treated on the same level as the electrostatic one and the specification of states by L and S has to be entirely abandoned.

The quantum-mechanical formulae can best be illustrated by reference to the classical motion. In the absence of magnetic interaction each vector $\mathbf{L}_i$ precesses rapidly, and in a complicated manner, about the direction of **L** which is fixed in space. Its component in the latter direction has a definite average value $|\mathbf{L}|_i \overline{\cos(\mathbf{L}, \mathbf{L}_i)}$. The same applies to all vectors $\mathbf{S}_i$ each which forms an average component $|\mathbf{S}|_i \overline{\cos(\mathbf{S}, \mathbf{S}_i)}$ in the direction of **S**. Weak magnetic interaction causes **L** and **S** to precess so slowly about **J** that the above average values can be used in further averaging over the slow precession to form $\overline{\cos(\mathbf{L}_i, \mathbf{S}_i)}$:

$$\begin{aligned}\sum_i \xi_i \overline{\mathbf{L}_i \cdot \mathbf{S}_i} &= \sum_i \xi_i \mathrm{L}_i \mathrm{S}_i \overline{\cos(\mathrm{L}_i, \mathrm{S}_i)} \\ &= \sum_i \xi_i \mathrm{L}_i \mathrm{S}_i \overline{\cos(\mathrm{L}, \mathrm{L}_i)}\, \overline{\cos(\mathrm{S}, \mathrm{S}_i)} \cos(\mathrm{L}, \mathrm{S}) \\ &= A' \cos(\mathrm{L}, \mathrm{S}).\end{aligned} \tag{18}$$

The corresponding matrix formula is

$$\begin{aligned}&\langle SLJM | \sum_i \xi_i \mathbf{L}_i \cdot \mathbf{S}_i | SLJM \rangle \\ &= \sum_i \langle L || \xi_i \mathrm{L}_i || L \rangle \langle S || \mathrm{S}_i || S \rangle \frac{J(J+1) - L(L+1) - S(S+1)}{2} \\ &= A(L, S) \; \frac{J(J+1) - L(L+1) - S(S+1)}{2}\end{aligned} \tag{19}$$

the analogy of which to (18) is obvious. The first two factors are the expectation values of $\xi_i \mathrm{L}_i$ and S_i in the L, S schemes, and the last factor is the eigenvalue of $\mathbf{L} \cdot \mathbf{S}$, apart from a factor $\hbar^2$.

The magnetic interaction energy described by (19) removes some of the remaining degeneracies of the terms: the different values of J define the different *levels* of each term. Since A is independent of J, the relative spacings of the levels are given by this formula and are found to agree with the Landé *interval rule* (III, 87):

$$W(L, S, J) - W(L, S, J-1) = \tfrac{1}{2} A[J(J+1) - (J-1)J] = AJ. \tag{20}$$

The *total splitting* $W(J_{\max}) - W(J_{\min})$ of a term is found to be

$$AS(L+1) \quad \text{for } L > S \quad \text{and } AL(S+1) \quad \text{for } L < S. \tag{21}$$

For comparison of widths of multiplet splitting in different terms arising from one configuration the dependence of the splitting factor A on L and S has to be studied. This involves, in general, a complicated transformation of matrices, but in many cases the result can be found quite simply, as in electrostatic interaction, by means of the *diagonal sum rule.*[17, 18] In order to apply this rule we have to write down the diagonal elements of the interaction matrix in the zero-order scheme n^{i}, l^{i}, $m_l{}^{\mathrm{i}}$, $m_s{}^{\mathrm{i}}$ and in the SLM_LM_S scheme. The sum must, in each case, be equal to the sum of the eigenvalues of the perturbation energy. In view of the special assumption that, to a first approximation, **L** and **S** can be regarded as commuting with the Hamiltonian, the sum can, just as in the case of electrostatic interaction, be restricted to states with any one given set of values

$$M_L = \sum_{\mathrm{i}} m_l{}^{\mathrm{i}} \quad \text{and} \quad M_S = \sum_{\mathrm{i}} m_s{}^{\mathrm{i}}.$$

Each diagonal element is the expectation value of the energy for the particular set of quantum numbers. In the zero-order scheme this is the sum of the energies of magnetic interaction for the individual electrons for specified values of the $m_l{}^i$ and $m_s{}^i$. This can, e.g. be found by means of the operators in table II, 1 from the relation $\mathbf{L} \cdot \mathbf{S} = \mathrm{L}_x\mathrm{S}_x + \mathrm{L}_y\mathrm{S}_y + \mathrm{L}_z\mathrm{S}_z$. The result is

$$\langle n^{\mathrm{i}}l^{\mathrm{i}}m_{\mathrm{l}}{}^{\mathrm{i}}m_{\mathrm{s}}{}^{\mathrm{i}}|\xi_{\mathrm{i}}\mathbf{L}_{\mathrm{i}} \cdot \mathbf{S}_{\mathrm{i}}|n^{\mathrm{i}}l^{\mathrm{i}}m_{\mathrm{l}}{}^{\mathrm{i}}m_{\mathrm{s}}{}^{\mathrm{i}}\rangle = a^{\mathrm{i}}m_{\mathrm{l}}{}^{\mathrm{i}}m_{\mathrm{s}}{}^{\mathrm{i}}. \tag{22}$$

It corresponds to the classical energy of interaction which would result if $\mathbf{L}_i$ and $\mathbf{S}_i$ were precessing about the z-axis forming average z-components $\mathrm{L}_{\mathrm{i}_z} = \mathrm{L}_{\mathrm{i}} \cos(\mathrm{L}_{\mathrm{i}}, z)$ and $\mathrm{S}_{\mathrm{i}_z} = \mathrm{S}_{\mathrm{i}} \cos(\mathrm{S}_{\mathrm{i}}, z)$ while all other components cancel out in the time-average. Only in enormously strong magnetic fields in the z-direction would this motion actually take place, and only then would the matrix in this scheme have exclusively diagonal elements.

In the LSM_LM_S scheme a corresponding argument gives the diagonal element

$$A(L, S)\langle LSM_LM_S|\mathbf{L} \cdot \mathbf{S}|LSM_LM_S\rangle = M_LM_SA(L, S). \tag{23}$$

The application of the sum rule may be illustrated by the example of the configuration $n\mathrm{p}n'\mathrm{p}$. For the pair of values $M_L = 2$, $M_S = 1$ the relation

$$\sum M_LM_SA(L, S) = \sum_i a^{\mathrm{i}}m_l{}^{\mathrm{i}}m_s{}^{\mathrm{i}} \tag{24}$$

has, according to table V, 3, only one term on the left and two, one for each electron, on the right side of eq. (24):

$$2A(^3\mathrm{D}) = \tfrac{1}{2}(a+a'),$$

therefore

$$A(^3\mathrm{D}) = \tfrac{1}{4}(a+a'),$$

where a and a' refer to the electrons $n\mathrm{p}$ and $n'\mathrm{p}$.

From the third line of table V, 3 follows similarly

$$A(^3\mathrm{D})+A(^3\mathrm{P}) = \tfrac{1}{2}(a+a'),$$

$$A(^3\mathrm{P}) = \tfrac{1}{4}(a+a') = A(^3\mathrm{D}),$$

The total width of splitting for the two terms is found in the ratio

$$\Delta^3\mathrm{D}/\Delta^3\mathrm{P} = 5/3. \tag{25}$$

The procedure can clearly be applied to other configurations and leads to expressions for the splitting factors A in terms of the factors a^i for the individual electrons. Again, the restriction characteristic of the sum rule applies: if a configuration contains several terms of the same kind, only the sum of the splitting constants is found. Equation (24) used to be known as the Γ sum rule.

Table V, 9 gives the splitting factors for some important configurations. For n equivalent electrons it is found that

$$A_{N-n} = -A_n \tag{26}$$

where $N = 2(2l+1)$ is the number of electrons in the complete sub-shell. A is usually positive for $n < 2l+1$ and negative for $n > 2l+1$, so that *regular multiplets* are usually found in atoms or ions in which the *first half* of a sub-shell is being filled in, and *inverted multiplets* for those in which the *second half* is being filled. A is always negative for a configuration of one hole in a complete sub-shell, such as an X-ray term. The cause of this has been explained in IV.C.3. For a half-filled shell, such as p^3 or d^5, A is always zero, because all spins are parallel and the sum (24) vanishes since $\Sigma\, m_l{}^{\mathrm{i}} = 0$.

The results of the theory are generally found to agree well with observed multiplet splittings provided these are considerably smaller than the differences between terms, in accordance with the assumption of the magnetic interaction as weak compared with the electrostatic interactions. Examples will be given in section B.

The relations discussed above not only assume that the L, S, J scheme—or Russell–Saunders scheme—is a good approximation, but also that the only magnetic interaction is that of each spin with its own orbit. Spin-other-orbit interaction appears to be generally small in conditions of Russell–Saunders coupling. It would not affect the validity of the interval rule but the values of the coefficients A.

More important is the spin–spin interaction which is prominent in He and the helium-like ions. In all these, the Russell–Saunders scheme is an extremely good approximation as shown by the smallness of the multiplet splitting compared with the singlet–triplet

TABLE 9

Multiplet splitting factors

	term	A		term	A
p, p^5	^2P	$\pm a$	d^4, d^6	^5D	$\pm \frac{1}{4}a$
				^3H	$\pm \frac{1}{10}a$
p^2, p^4	^3P	$\pm \frac{1}{2}a$		^3G	$\pm \frac{3}{20}a$
p^3	^2D, ^2P	0		^3F	sum =
d, d^9	^2D	$\pm a$		^3F	$\pm \frac{1}{12}a$
d^2, d^8	^3F, ^3P	$\pm \frac{1}{2}a$		^3D	$\mp \frac{1}{12}a$
d^3, d^7	^4F, ^4P	$\pm \frac{1}{3}a$		^3P	sum =
	^2H	$\pm \frac{1}{5}a$		^3P	$\pm \frac{1}{2}a$
	^2G	$\pm \frac{3}{10}a$	d^5	all terms	0
	^2F	$\mp \frac{1}{6}a$	ps, (ds, ...)	^3P, (^3D) ...	$\frac{1}{2}a_p(\frac{1}{2}a_d$ ...)
	^2D	sum =	p^2s	^2P	$\frac{2}{3}a_p$
	^2D	$\pm \frac{1}{3}a$		^4P	$\frac{1}{3}a_p$
	^2P	$\pm \frac{2}{3}a$			
				^2D	0
			p^3s	all terms	0
			$npn'p$	^3P, ^3D	$\frac{1}{4}(a+a')$
			pd	^3P	$-\frac{1}{4}a_p+\frac{3}{4}a_d$
				^3D	$\frac{1}{12}a_p+\frac{5}{12}a_d$
				^3F	$\frac{1}{6}a_p+\frac{1}{3}a_d$
			dd	^3P, ^3D	$\frac{1}{4}(a+a')$
				^3F, ^3G	

difference and by the validity of the intensity rules. But there is a complete breakdown of the interval rule owing to spin–spin interaction, and there is much evidence to show that spin–spin interaction is by no means negligible in several of the lighter atoms (see p. 188). The effect is difficult to calculate, and opinions still differ on its importance in heavier atoms.[19] Araki[20] is inclined to make it responsible for many deviations from the interval rule, but certainly in some cases other factors have been shown to be more important (ref. III.146).

9. Selection rules and line strengths

It is clear from the discussion in II.D that two of the selection rules for dipole radiation must be expected to hold rigorously for atoms of any number of electrons with any kind of interaction between them. In the absence of external fields the total angular momentum of the atom is strictly constant in the classical model, so that J is a quantum number which can be rigorously associated with the total angular momentum. This fact and the general properties of the dipole radiation field lead to the selection rules

$$\Delta J = 0, \pm 1, \quad \text{but } J = 0 \rightarrow J = 0 \text{ is forbidden.} \tag{27}$$

The effect of the nuclear spin on this selection rule will be discussed in VI.

Another property which is rigorously defined for every atomic state is the *parity*. When it is a meaningful approximation to ascribe quantum numbers to individual electrons, the parity depends merely on the configuration and is *even* or *odd* according to whether $\Sigma\, l_i$ is even or odd. But even if the configuration cannot be defined at all, the distinction between even and odd atomic states is still completely valid. In a sense which will be further explained in section 12 one can say that states of different parity do not mix or perturb one another. The *Laporte rule* forbidding dipole transitions between terms of the same parity thus holds without restriction for free atoms.

All other selection rules, and all formulae for ratios of strengths of lines are derived from assumptions that certain angular momenta are constant to some degree of approximation, in the sense that the rate of their precession is slow compared with the frequency of any motion or precession about their own direction. In quantum mechanics, one expresses these conditions by assuming the validity of certain commutation relations. These lead, by methods of matrix algebra, to the result that certain elements of the dipole matrix vanish.

In the *zero-order approximation* an atomic state is described by independent states of motion of the individual electrons, each moving in a central force field. In the radiation of such a system of independent particles only the quantum numbers of one electron will change. To this approximation, the following selection rules will hold:

$$\Delta l_i = \pm 1, \qquad \Delta l_{j,k\,--} = 0. \tag{28}$$

Simultaneous changes in quantum numbers of two electrons correspond to sum- or difference-frequencies due to coupling between them. Transitions violating the above selection rule can therefore be expected to be observed only when configuration interaction is appreciable. In configurations containing d-electrons the energies of the single electron states nd and $(n-1)$s are often not very different and this fact, in view of their equal parity, can lead to strong configuration interaction; Transitions $\Delta l_i = \pm 1$, $\Delta l_j = \mp 2$ are then frequently observed.

In *perfect Russell–Saunders coupling*, the vectors **L** and **S** are constant and independent of one another. **S** is then not coupled at all to the orbital motions of the electrons and cannot be affected by the dipole radiation. This leads to the selection rule

$$\Delta S = 0. \tag{29}$$

The selection rules for L can be derived from the correspondence principle if it is remembered that the approximate constancy of $\mathbf{L}^2$ is connected with the existence of a cyclic variable and a uniform precession about the direction of **L**. The radiation is determined by the variable electric dipole moment

$$e\mathbf{P} = e\sum_i \mathbf{r}_i$$

due to all electrons, and in particular its components *parallel* and at *right angles* to **L**. The latter will contain the precessional frequency but the former will not. These two components will thus give rise to the two kinds of transitions

$$\Delta L = \pm 1, 0. \tag{30}$$

In contrast to the motion of a single electron, that of several electrons is not confined to a plane normal to **L**, so that $\Delta L = 0$ is allowed.

We can expect the selection rules for S and L to be fulfilled only as long as the magnetic interaction is small, i.e. as long as the width of the multiplet splitting is small compared with the term differences. Intercombination lines ($\Delta S > 0$) are in fact the stronger the less this condition is fulfilled; they are extremely weak in the spectra of the light elements and fairly strong in the heavy elements (see III.B.6).

Beyond the statement that certain transitions do not occur at all in dipole radiation, the correspondence principle has led to conclusions on the relative strengths of lines within one set of configurations and especially within one multiplet. The influence of the spin vector **S** on the orbital motion, for extreme Russell–Saunders coupling, consists merely in forcing the vector **L** into a certain orientation relative to **J**. The total rate of radiation can obviously not be affected by this, and the life time of the excited state must be independent of J. Together with the fact that the number of atoms in a given level is proportional to its statistical weight, this leads to the *sum rule* which had been stated for the special cases of doublets and triplets (p. 189) and which holds for any multiplets as rigorously as the conditions of Russell–Saunders coupling are fulfilled.

With the exception of a few simple cases the sum rule does not suffice for a determination of the intensity ratios, and more extensive use has to be made of the correspondence principle. For any given S and L, the value of J determines the angles between the three angular momenta. Let us suppose we are dealing with a line for which $\Delta L = 0$; the electric dipole for this transition is parallel to **L**. Its components parallel and normal to **J** then give, according to the correspondence principle, the amplitudes of the radiation for the transitions $\Delta J = 0$ and $\Delta J = \pm 1$ respectively. An uncertainty arises if J or L differ for the two states involved, and some average values for the angles have to be used. By combining this method with the sum rule however, Kronig,[21] Hönl and Sommerfeld[22] and Russell[23] succeeded in deriving formulae whose completely rigorous validity was later confirmed by quantum-mechanical methods.[24] The results are contained in formulae (III.88) and table App. 1.

In each multiplet transition, the strongest lines are those in which the change of J is the same as that of L, the so-called *principal lines*. Among these, the strength is the greater the greater J.

Quantum mechanics has provided a mathematical procedure describing the same physical situation in a more consistent though more abstract way. The precessional motion of a vector about a resultant **J** is expressed by the commutation rule (II.80). It applies to each of the vectors **L** and **S** with regard to **J** in Russell–Saunders coupling, and also to the radius vector $\mathbf{r}_i$ of any electron with regard to $\mathbf{L}_i$ and also with regard to **L** and **J**. This commutation rule leads directly to the result that all matrix elements $\langle \alpha, J, m|\mathbf{P}|\alpha', J', m'\rangle$ vanish except those for which $J'-J = 0$ or ± 1. The symbol α stands for any further, unspecified quantum numbers. The selection rules for J and, in case of Russell–Saunders coupling, for L thus follow directly from the definition of electric dipole radiation.

The dependence of the matrix elements of **P** on m expresses the dependence on the direction in space which becomes observable only in the presence of external fields. It is found that P_z vanishes except for $m = m'$, in agreement with the selection rule for the π-components in the Zeeman effect, a rule which holds for complex as well as simple spectra. Those components of **P**, which are associated with the σ-components (see p. 102), are finite only for $m'-m = \pm 1$.

Each matrix element of **P** can be written as the product of a factor which is independent of m and is denoted by $\langle \alpha, j || \mathbf{P} || \alpha' j' \rangle$ and a numerical factor which depends on j and m in a simple way and can be taken from tables. The sum relation

$$\begin{aligned}\sum_m |\langle \alpha, J, m|\mathbf{P}|\alpha', J', m\rangle|^2 &= \sum_m |\langle \alpha, J, m|\mathbf{P}|\alpha', J', m+1\rangle|^2 \\ &= \sum_m |\langle \alpha, J, m|\mathbf{P}|\alpha', J', m-1\rangle|^2\end{aligned} \tag{31}$$

ensures the unpolarised nature of the total radiation emitted.

The assumption that coupling between spins and orbital momenta is negligible is now introduced by the statement that **S** commutes with **L** and with **P**. This is sufficient for deriving, by pure matrix algebra, the dependence of the matrix elements

$$\langle \alpha, S, L, J || \mathbf{P} || \alpha', S', L', J' \rangle$$

on J and J'. Their squares, multiplied by the statistical weight $2J+1$ of each level, give the relative strengths of the multiplet transitions, as contained in the formulae (III.88) and the table App. 1.

The selection rule for L follows, exactly as that for J, from the validity of the commutation rule for **L** and **P**. The dependence of the matrix elements of **P** on L and S determines the relative strengths of the different multiplets arising from the transitions between all the terms of two configurations, a so-called transition array. In simple cases, matrix methods can be used for deriving this dependence; this applies specially to *super-multiplets* (see p. 264), in which the relative strengths of the different multiplets is given by the formula (III.88) if S, L, J are replaced by L_1, l, L, where l is the orbital quantum number of the electron whose quantum numbers are changing in the transition.

In particular, a sum rule can be shown to hold stating that the sum of the strengths of all multiplet transitions from any term of one configuration to all terms of another configuration is proportional to the statistical weight $2L+1$ of the term and independent of S. For the example given on p. 264 each of the sums has only one value, so that the rule yields directly the intensity ratio

$$\mathrm{a} : \mathrm{b} : \mathrm{c} = \mathrm{a}' : \mathrm{b}' : \mathrm{c}' = 3 : 5 : 7.$$

In other cases, intensity ratios can be calculated by means of the diagonal sum rule. The experimental knowledge of relative strengths in transition arrays is very scanty.

10. Different coupling conditions

In all the lower terms of the lighter elements and in many low terms of the heavier elements, the Russell–Saunders coupling scheme is found to be a good approximation. In view of the marked tendency of the width of multiplet splittings to increase with atomic numbers, it is not surprising that in many configurations of heavier elements the magnetic interaction can no longer be regarded as small. The interval rule or even the qualitative arrangement of levels in multiplets will then break down, and the quantum numbers L and S will lose their significance.

a. The j, j coupling scheme

In the extreme case in which the magnetic l, s-interactions are large compared with the angle-dependent part of the electrostatic interactions, the coupling scheme becomes simple again: in the first approximation, spin and orbital momentum of each electron are coupled together and form a resultant angular momentum with quantum number j_i:

$$\mathbf{L}_i + \mathbf{S}_i = \mathbf{J}_i \qquad (j_i = l_i \pm \tfrac{1}{2}). \tag{32}$$

In the second approximation, electrostatic forces cause the different $\mathbf{J}_i$ to be coupled and to form the resultant $\mathbf{J}$:

$$\mathbf{J} = \sum \mathbf{J}_i \qquad (J = \sum j_i, \sum j_i - 1, \ldots). \tag{33}$$

Spin–spin and spin–other orbit interactions, and also magnetic orbit–orbit interactions are again usually disregarded.

This type of coupling is known as j, j-coupling. In its extreme form, it is much less frequently found than L, S-coupling and is restricted to certain types of configurations. The transition from L, S-coupling to j, j-coupling and the conditions which favour the latter can best be demonstrated on some examples.

We consider the simple configurations $ns\,n'$p in which the only magnetic interaction is that of the p-electron with its spin. We compare, at first, elements in one column of the periodic table, taking the lowest possible values of n and n' for each element. If we choose the examples so that the p-electron has a higher energy than the s-electron, i.e. $n' \geqq n$, the magnetic interaction will be comparatively small, making conditions unfavourable for j, j-coupling. One finds, in fact, that in the elements Mg, Ca, Sr, Ba the configurations 3s3p, 4s4p, 5s5p, 6s6p, and also 3s4p, 4s5p, 5s6p, 6s7p form triplet- and singlet-terms with levels in the normal order and with only small deviations from the interval rule, apart from some irregularities which can be readily explained as perturbations.

In configurations, however, in which the p-electron has lower energy than the s-electron, the magnetic l, s-interaction is comparatively strong and the coupling is found to pass into practically complete j, j-coupling in the heavier elements. This is shown in fig. V, 4 where the term values arising from the configurations 2p3s, 3p4s,

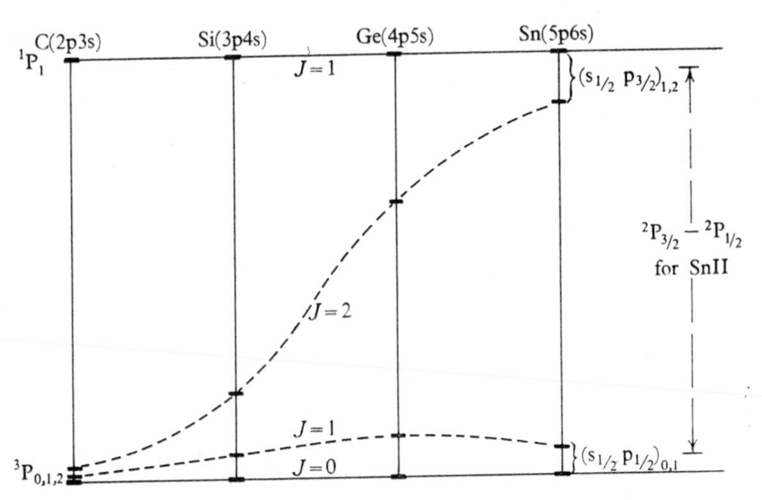

FIG. V, 4. Transition from L, S- to j, j-coupling in configurations np(n+1)s.

4p5s, 5p6s in the elements of the fourth column are plotted. In each case the scale has been so chosen that the difference between the lowest and the highest level appears the same. The regrouping of the levels in the sequence from C to Sn is due to increasing penetration effects causing increasing predominance of the magnetic l, s-interaction in the p-electron compared with the electrostatic interaction which does not depend very markedly on penetration effects. This interpretation is supported by the fact that the difference between the two groups of levels for Sn agrees closely with the difference 5p $^2P_{\frac{3}{2}}$ – 5p $^2P_{\frac{1}{2}}$ of SnII as shown in the figure.

For j, j-coupling the levels have to be described by different symbols, such as $(5p_{\frac{3}{2}}6s_{\frac{1}{2}})_1$ for the highest level in fig. V, 4, where the suffixes in the bracket give the values j_1 and j_2, and that outside the values of J.

Similar transitions to j, j-coupling can be found within the spectrum of one element in the sequence of configurations nsn'p, $(n+1)$s n'p, $(n+2)$s n'p . . . The magnetic interaction can be expected to remain almost constant in this sequence, while the electrostatic interaction becomes weaker as the s-electrons increases its quantum number. The limit $n \to \infty$ leads to a free electron and an ion in either of the states $^2P_{\frac{1}{2}}$ or $^2P_{\frac{3}{2}}$; the different levels must therefore converge to two limits.

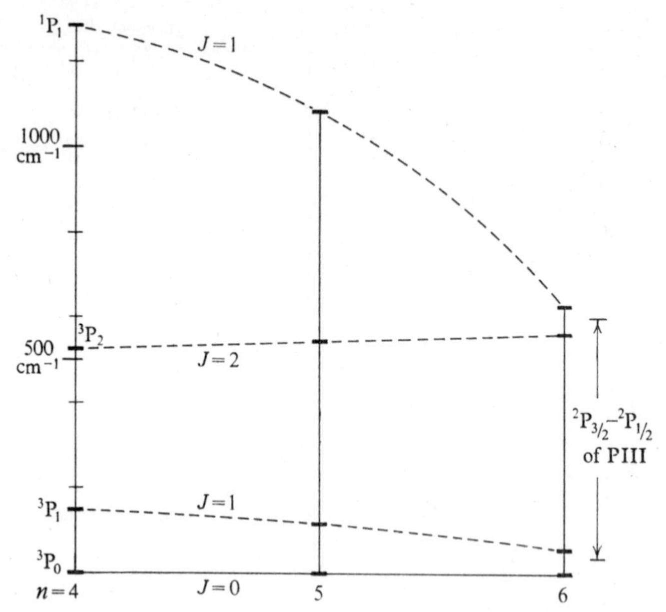

FIG. V, 5. Transition from L, S- to j, j-coupling with increasing n in terms nsn'p of PII.

Unfortunately perturbations become more frequent in higher terms, e.g. perturbations of ns n'p by configurations n''d n'p, and these tend to confuse the conversion to j, j-coupling. An example which is fairly free from perturbations is shown in fig. V, 5, which refers to the Si-like spectrum of PII. For $n = 6$ the wave-number difference between the doublets is seen to be equal to the difference 3p $^2P_{\frac{3}{2}}$ – 3p $^2P_{\frac{1}{2}}$ of PIII.

Whenever the process of ionisation produces an ion with a multiple ground term, the levels of the atom converge to several limits. This involves rearrangement of levels and change in the coupling conditions, though the result is not always j, j-coupling. Another form of coupling, typical of high quantum numbers, will be described on p. 286.

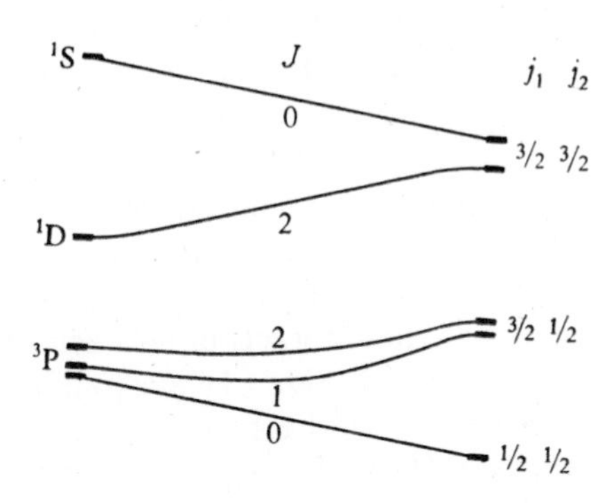

FIG. V, 6. Diagram of transition from L, S- to j, j-coupling in configuration p².

For configurations $n\mathrm{p}^2$ the transition to j, j-coupling is shown diagrammatically in fig. V, 6. It can be proved that in this and in any kind of transition from one coupling scheme to another the curves belonging to levels of equal J do not cross. This is due to the fact that the total angular momentum remains rigorously constant in all coupling conditions. It is analogous to the similar rule applying to the magnetic quantum number M in magnetic conversion processes.

The quantum-mechanical treatment of energy levels in j, j- and intermediate coupling may be only briefly indicated. In complete j, j-coupling, the electrostatic energy is neglected in the first-order approximation, and the magnetic energy consists simply in the sum of the l, s-interaction energies for all the electrons taken individually:

$$E_0 = \sum_i a_i \frac{j_i(j_i+1) - l_i(l_i+1) - s_i(s_i+1)}{2} \tag{34}$$

where a_i is defined by (III, 83) and depends only on n_i and l_i. In the examples of configurations $n\mathrm{s}\ n'\mathrm{p}$ the sum reduces to one term.

In the second approximation, the electrostatic energy has to be added as a perturbation acting on a system which is still highly degenerate since E_0 is independent of J and m. The resulting energy can be expressed in terms of the radial integrals F_k and G_k and is to be added to E_0. As an example, the expressions for the configuration $n\mathrm{p}^2$ may be quoted:

$$\begin{array}{llll} j = \frac{3}{2} & j' = \frac{3}{2} & J = 2: & E = a_p + F_0 - 3F_2 \\ & & J = 0: & E = a_p + F_0 + 5F_2 \\ j = \frac{3}{2} & j' = \frac{1}{2} & J = 2: & E = -a_p/2 + F_0 - F_2 \\ & & J = 1: & E = -a_p/2 + F_0 - 5F_2 \\ j = \frac{1}{2} & j' = \frac{1}{2} & J = 0: & E = -2a_p + F_0. \end{array} \tag{35}$$

The relative strengths of lines connecting two configurations can be calculated by methods similar to those used in L, S-coupling.

The selection rule that only one j_i can change in dipole radiation follows from the fact that the interaction between the electrons only appears in the higher approximation. In a configuration of two electrons, the relative strengths of the transitions j_1, j_2, J to j_1, j_2', J' is given by the formulae (III.88) if j_1, j_2, j_2' are substituted for S, L, L'.

In order to derive the energy levels in the general case of intermediate coupling where electrostatic and magnetic interactions are of the same order of magnitude, one has to use the matrix of the sum of electrostatic and magnetic energies and solve the secular equation.[25, 26] The matrix elements can be written either in the L, S, J scheme or the j_1, j_2 . . scheme. In the former, the electrostatic energy contributes only diagonal terms, namely the same as in Russell–Saunders coupling. The magnetic l_i, s_i-interaction is diagonal in J and M; it connects different values of L and S, but only those differing by one unit. For those values of J which appear only in one level, the perturbation energy is simply equal to the corresponding diagonal element. For the values of J which occur n times, an equation of the nth degree has to be solved.

This may be illustrated by the example of the configuration sl, where l stands again for one of the symbols p,d,f, . . The complete interaction matrix is

	$^3L_{l+1}$	^3L_l	^1L_l	$^3L_{l-1}$
$^3L_{l+1}$	$F_0 - G_0 + \frac{1}{2}a_l l$			
^3L_l		$F_0 - G_0 - \frac{1}{2}a_l$	$\frac{1}{2}a_l\sqrt{[l(l+1)]}$	
^1L_l		$\frac{1}{2}a_l\sqrt{[l(l+1)]}$	$F_0 + G_0$	
$^3L_{l-1}$				$F_0 - G_0 - \frac{1}{2}a_l(l+1)$

(36)

The diagonal elements contain the electrostatic energies from table V, 8, and the magnetic energies derived from table V, 9. The off-diagonal elements are calculated by the general rules of operator algebra as matrix elements of the scalar product $\mathbf{L} \cdot \mathbf{S}$ of two commuting vectors each of which obey the rule (II.80) with regard to the resultant $\mathbf{J}$. The diagonal elements for the values of J occurring only once, namely $l+1$ and $l-1$, have been calculated exactly as in pure Russell–Saunders coupling; if they are calculated in the j,j-coupling scheme, the same expressions are found. For the magnetic energy this can be verified immediately from (31). Quite generally any diagonal element whose J occurs only once is the same in any representation and under any coupling condition which leaves J a good quantum number. This is analogous to the similar rule with regard to M (see p. 207) which was found to hold in conditions where M remained a good quantum number for all strengths of magnetic fields.

In the present case, the matrix elements are independent of M which has therefore been omitted in the numbering of the rows and columns. For the values $J = l$ the secular equation is quadratic, and the two solutions are included in the following list of all the energy values for arbitrary coupling conditions:

$$\begin{aligned} E_1 &= {}^1L_l = F_0 - \tfrac{1}{4}a_l + \sqrt{[(G_0 + \tfrac{1}{4}a_l)^2 + \tfrac{1}{4}a_l^2 l(l+1)]} \\ E_2 &= {}^3L_{l+1} = F_0 - G_0 + \tfrac{1}{2}a_l l \\ E_3 &= {}^3L_l = F_0 - \tfrac{1}{4}a_l - \sqrt{[(G_0 + \tfrac{1}{4}a_l)^2 + \tfrac{1}{4}a_l^2 l(l+1)]} \\ E_4 &= {}^3L_{l-1} = F_0 - G_0 - \tfrac{1}{2}a_l(l+1). \end{aligned} \tag{37}$$

The relative position of the levels is determined solely by the parameter a/G_0 as can be seen when the equations are divided by G_0. The following relation which results from the four equations (37) is convenient for comparison with experimental results:

$$\frac{E_3 - E_4}{E_2 - E_3} + \frac{l}{l+1}\,\frac{E_2 - E_4}{E_1 - E_4} = \frac{l}{l+1}. \tag{38}$$

For the configuration sp the value of l is 1 and the left-hand side becomes equal to 1/2. The experimental values of this expression for the levels involved in figs. V, 4 and V, 5 are shown in table V, 10.

TABLE 10

Interval ratio (38) in intermediate coupling

C	Si	Ge	Sn	P II	P II	P II	theor.
2p3s	3p4s	4p5s	5p6s	3p4s	3p5s	3p6s	$npn's$
0·49	0·50	0·50	0·51	0·58	0·51	0·55	0·50

In this and numerous other appropriate cases, the agreement with the theoretical value 0·5 is satisfactory. The formula also holds for configurations such as p^5s, d^9s, $f^{13}s$.[26,27,28]

b. The j, l-coupling scheme

In configurations containing one or several electrons with $l > 0$ outside closed shells it can happen that some of the magnetic interactions are small and others large compared with the electrostatic effects. In configurations consisting of a shell with one hole and of one electron of large l, the levels often occur in two groups, and each level within a group consists of an often unresolved doublet. This has been found in CuII[29] and in rare gases and has been explained in the following way.[30,31]

Taking the example of the configuration $3d^95g$ in Cu II, we find that the parent ion forms an inverted doublet $d^9\,{}^2D_{\frac{5}{2},\frac{3}{2}}$ whose splitting is due to spin–orbit interaction which is fairly strong and is known from the spectrum of Cu III. It accounts for the spacing between the two widely separated groups (fig. V, 7). Owing to the considerably weaker electrostatic interaction, the **L** of the 5g-electron couples with the **j** of the core, forming a resultant angular momentum **K** whose quantum number K can assume 6 and 4 values for the values $j = 5/2$ and $3/2$ respectively. The values of the spacing are shown in fig. V, 7 in terms of the Slater integral F_2. Each of these levels is double, with extremely small splitting, owing to the magnetic coupling between **K** and the spin of the g-electron, resulting in the quantum numbers $J = K \pm \frac{1}{2}$. The name j, l-coupling has been suggested for this phenomenon[31] and the following nomenclature: $(j)l[K]_J$ such as $({}^2D_{\frac{3}{2}})5g[4\frac{1}{2}]_5$ for one of the levels in the example discussed above.

In the given example, the experimental values agree so well with the calculations that discrepancies would not be visible on the scale of the figure. Examples of j, l-coupling in N II have been studied by Eriksson.[32]

c. General sum-rules

Two sum rules for the strengths of transitions between the levels of two configurations can be shown to hold for any kind of coupling, provided that configuration interaction can be neglected.

The *J-group sum rule*[33] states that the sum of the strengths of all transitions from all levels of given J of one configuration to all levels of given J' of another configuration is independent of coupling; its value can therefore be calculated for Russell–Saunders coupling and will hold for any kind of coupling.

The *J-file sum rule*(34) states that the sums of the strengths of all transitions from any one level of given J to all levels of another configuration are in the ratio of the statistical weights $2J+1$. The rule holds only if the latter configuration contains no electron which is equivalent to the electron whose quantum numbers are changing; it holds for transitions *to* any level in the same way as for transitions *from* any level.

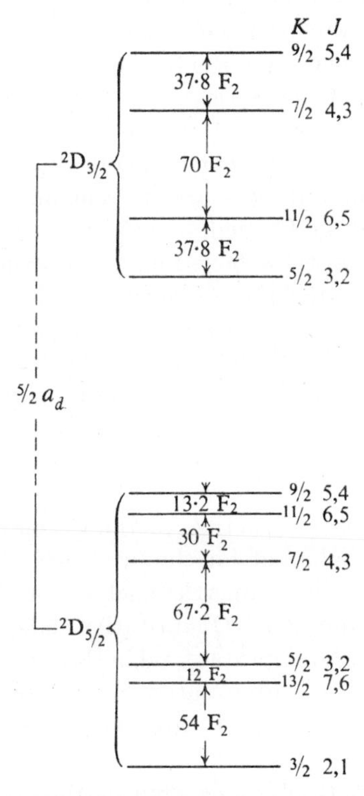

FIG. V, 7. j, l-coupling in Cu II; the levels of ($3d^9\ ^2D_{5/2,\ 3/2}$) 5g, with $a_d = 828{\cdot}6$, $F_2 = 0{\cdot}315$ cm^{-1}. (G. H. Shortley & B. Fried, ref. 30.)

11. Zeeman effects in complex spectra

In practically all experiments on Zeeman effects in complex spectra, the wave-number changes caused by the field are small compared with the differences between levels of the free atom, so that the field can be regarded as *weak*. Under these conditions the *anomalous* Zeeman effect observed is very similar to that in alkali-like spectra and shows the following features, regardless of coupling conditions.

1. The patterns are symmetrical with regard to wave-number, intensity and polarisation of the components. The π-components form a central group, the σ-components two symmetrically displaced groups, and the components are equidistant in each. π- and σ-components are defined in the same way as in the normal effect by their polarisation in transverse and longitudinal observation. The sum of the intensities of the π-components is equal to that of the σ-components, in transverse observation.

2. Different lines of one series show the same Zeeman pattern, and even the spacings are the same as long as the coupling conditions remain the same. The pattern only depends on the angular momentum quantum numbers and it remains the same if initial and final levels are interchanged.

The interpretation of these facts follows the same lines as in alkali-like spectra. In the free atom, every level is characterized by a value of J and a factor g describing its magnetic properties which depend on the relative extent to which spins and orbital momenta contribute to the vector $\mathbf{J}$. In a magnetic field, each level forms $2J+1$ equidistant levels of energy $Mg\mu_0\mathscr{H}$, where $M = J, J-1 \ldots -J$. The structures of the patterns just described then follow from the selection rules for M.

While the qualitative structure of the Zeeman pattern of a line depends only on the values of J in the two levels, the spacings depend on g and thereby on other properties such as coupling conditions and on other quantum numbers. Quantitative data on Zeeman effects are therefore not only an important aid in the term analysis of spectra but also in the detailed study of atomic states and their interpretation. Landé g-values, if known, are usually included in tables of atomic energy states.

From a completely resolved Zeeman pattern the g-values of both levels can be immediately derived; thus the term diagram fig. V, 8 shows that the spacing of the components within any of the groups of π- and σ-components is proportional to the difference of the g-values of the two levels.

For Russell–Saunders coupling, the value of g depends only on L, S and J and is given by (III.95). In view of this generalised use of the formula then derived for alkali-spectra, capital letters S and L had been used. On the basis of the vector treatment, this generalisation can easily be justified. Since $\mathbf{L}$ is the resultant of orbital angular momenta and **S** that of spin momenta, the ratio of magnetic moment to angular momentum must be the same, in each case, as that for one electron. In terms of matrix algebra the argument is analogous: the resultant momenta **L** and **S** obey the same commutation rules with regard to one another and **J** as the corresponding single electron momenta, and also their g-values are the same.

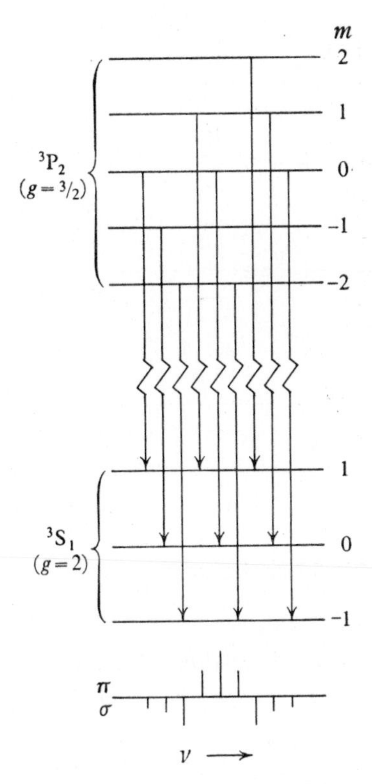

FIG. V, 8. Anomalous Zeeman effect in triplet transition $^3S_1 - ^3P_2$.

Table V, 11 gives a list of g-values for Russell-Saunders coupling and plate 12 shows, as an example, the Zeeman pattern of the line 4254 A of Cr, $^7S_3 - ^7P_4$, which belongs to a normal multiplet; the equal spacings in the π- and σ-group can be clearly seen.

One might intuitively expect the value of g to lie between 1 and 2, the limits of pure orbital and pure spin effects. But the semi-classical vector diagram shows that this is not true for small J. The approximate formula (p. 201) can be written $g = 1 + \frac{1}{2} + (S^2 - L^2)/2J^2$ which can have arbitrarily large absolute values for small J, and they can be positive or negative according to the sign of the bracket. In fact, the table, which has been calculated from the rigorous formula (III.95), contains values > 2 and < 0.

TABLE **11**

Landé g values

S	J	$L = 0$	1	2	3	4
1/2	$L-1/2$	—	2/3	4/5	6/7	8/9
	$L+1/2$	2	4/3	6/5	8/7	10/9
1	$L-1$	—	0/0	1/2	2/3	3/4
	L	—	3/2	7/6	13/12	21/20
	$L+1$	2	3/2	4/3	5/4	6/5
3/2	$L-3/2$	—	—	0	2/5	4/7
	$L-1/2$	—	8/3	6/5	36/35	62/63
	$L+1/2$	—	26/15	48/35	26/21	116/99
	$L+3/2$	2	8/5	10/7	4/3	14/11
2	$L-2$	—	—	0/0	0	1/3
	$L-1$	—	—	3/2	1	11/1
	L	—	5/2	3/2	5/4	23/20
	$L+1$	—	11/6	3/2	27/20	19/15
	$L+2$	2	5/3	3/2	7/5	4/3
5/2	$L-5/2$	—	—	—	−2/3	0
	$L-3/2$	—	—	10/3	16/15	6/7
	$L-1/2$	—	—	28/15	46/35	8/7
	$L+1/2$	—	12/5	58/35	88/63	14/11
	$L+3/2$	—	66/35	100/63	142/99	192/143
	$L+5/2$	2	12/7	14/9	16/11	18/13

For complete j, j-coupling, the Landé g-factors can be derived in a way similar to that for Russell–Saunders coupling. For each electron, the magnetic energy of spin and orbital momentum in the external field is found by first projecting the vector $\mathbf{S}_i$ (or $\mathbf{L}_i$) on the direction of $\mathbf{J}_i$, and then projecting this component on the direction of $\mathbf{J}$, and finally this component on the direction of $\mathscr{H}$. Using angles in the vector diagram, we can, e.g. write for the spin

$$\overline{\cos(\mathrm{S}_i, \mathscr{H})} = \cos(\mathrm{S}_i, \mathrm{J}_i)\cos(\mathrm{J}_i, \mathrm{J})\cos(\mathrm{J}, \mathscr{H});$$

since

$$J_i g(J_i) = 2\,S_i\cos(\mathrm{S}_i, \mathrm{J}_i) + L_i\cos(\mathrm{L}_i, \mathrm{J}_i)$$

and the magnetic energy is given by

$$\mathscr{H} J g(J) \mu_0 \cos(\mathbf{J}, \mathscr{H}),$$

we find:

$$Jg(J) = \sum_i J_i g(J_i) \cos(\mathbf{J}_i, \mathbf{J}), \tag{39}$$

where $g(J_i)$ is the single-electron g-value. In order to get the same result as by rigorous matrix methods, one has to make the usual substitution of $j(j+1)$ for j^2 for all angular momenta. For two electrons the result is

$$g(J) = g(J_1)\frac{J(J+1)+J_1(J_1+1)-J_2(J_2+1)}{2J(J+1)} + \\ +g(J_2)\frac{J(J+1)+J_2(J_2+1)-J_1(J_1+1)}{2J(J+1)}. \tag{40}$$

For intermediate coupling the g-values of the levels are generally not easy to calculate. But the quantum-mechanical operator $\mathbf{L} \cdot \mathbf{J} + 2\mathbf{S} \cdot \mathbf{J}$ (see p. 205) can be expanded in terms of its values in the Russell–Saunders scheme, of the same J, and this leads to a sum rule which often allows the g-values for intermediate coupling to be derived from g-values in Russell–Saunders coupling.

This *g-sum rule* states that for any given configuration the sum of the g-values of all levels with the same J is independent of the coupling scheme. The value of this sum can therefore be established from the formula for Russell–Saunders coupling (or j, j-coupling). If there is only one level of a given J, its value is independent of coupling and is given by (III.95).

With increasing strength of the field, a transformation of the coupling conditions similar to that described in III.F.4 must be expected to occur. With the field strengths obtainable in practice, this *Paschen–Back-effect* can hardly ever be fully realised, but its initial stages have often been observed as unsymmetries of Zeeman patterns in strong fields.

The theory of the transformation effect is similar to that for alkali spectra. M is a good quantum number in fields of any strength, and the secular equation has to be solved for every value of M; the number of states of given M determines the degree of the equation. Such calculations are of much greater importance in the closely similar case of hyperfine structures which will be discussed in VI.

One theoretical aspect is, however, worth mentioning in the present context, namely the application of the diagonal sum rule to all states of a given M in the $n^il^im_l^im_s^i$ representation. The field strength can be imagined to become so great that not only J but even L and S cease to be good quantum numbers. The summation can therefore not be restricted to levels of a given J or even of a given term.

The result is the *g-permanence rule*(35): the sum of the g-values of all states of a given M, for any given configuration is independent of the field strength. It arises from the fact that the sum of the magnetic energies $\Sigma\,\mathscr{H}M\mu_0 g$ for all states of the same M is independent of the coupling conditions. These energy values are the diagonal elements of the sub-matrix, for the given M, of the magnetic interaction energy

$$\sum_i \gamma_0\mathscr{H}(L_z{}^i+2S_z{}^i).$$

Only in the extremely strong fields mentioned is this matrix diagonal, but in all fields the diagonal sum must be equal to the sum of the eigenvalues.

The permanence rule allows the Landé g-factors to be derived simply and rigorously. Equating the energy sums in extremely strong and in weak fields:

$$\sum \mathscr{H}\mu_0(m_l{}^i+2m_s{}^i) = \sum \mathscr{H}\mu_0 Mg(J) \tag{41}$$

where $m^i = m_l{}^i+m_s{}^i$ and the sum is to be taken over a constant $M = \Sigma\, m^i$, we find

$$\sum g(J) = \sum \frac{m_l{}^i+2m_s{}^i}{m_l{}^i+m_s{}^i}. \tag{42}$$

As an example, the g-values of the levels of the term sp ^{3}P may be derived from (42). There is only one state with $M = 2$, namely that which is described in weak fields by the symbol sp $^3P_2$, $M = 2$ and in very strong fields by sp, $m_l{}^1 = 0$, $m_s{}^1 = \frac{1}{2}$, $m_l{}^2 = 1$, $m_s{}^2 = \frac{1}{2}$. Equation (42) thus gives

$$g(2) = \frac{1+2(\frac{1}{2}+\frac{1}{2})}{2} = \tfrac{3}{2}.$$

For $M = 1$ there are two states, sp $^3P_2$, $M = 1$ and sp $^3P_1$, $M = 1$, or sp, $m_l{}^1 = 0$, $m_s{}^1 = \frac{1}{2}$, $m_l{}^2 = 1$, $m_s{}^2 = -\frac{1}{2}$ and sp, $m_l{}^1 = 0$, $m_s{}^1 = \frac{1}{2}$, $m_l{}^2 = 0$, $m_s{}^2 = \frac{1}{2}$, so that

$$g(2)+g(1) = \frac{0+2\times\frac{1}{2}+1-2\times\frac{1}{2}}{1}+\frac{0+2\times\frac{1}{2}+0+2\times\frac{1}{2}}{1} = 3,$$

giving $g(1) = 3-g(2) = 3/2$.

The importance of Zeeman effects in the identification of the nature of forbidden lines has been mentioned in III.F. Some interesting and apparently contradictory results on forbidden He lines(36, 37) have been explained by a full theoretical analysis of the positions and intensities of the components.(38)

12. Perturbation and auto-ionisation

The regular arrangement of lines in a spectral series is sometimes found to be disturbed; one or several lines are displaced, or their intensities are irregular. Also in multiplets the ratio of the intervals

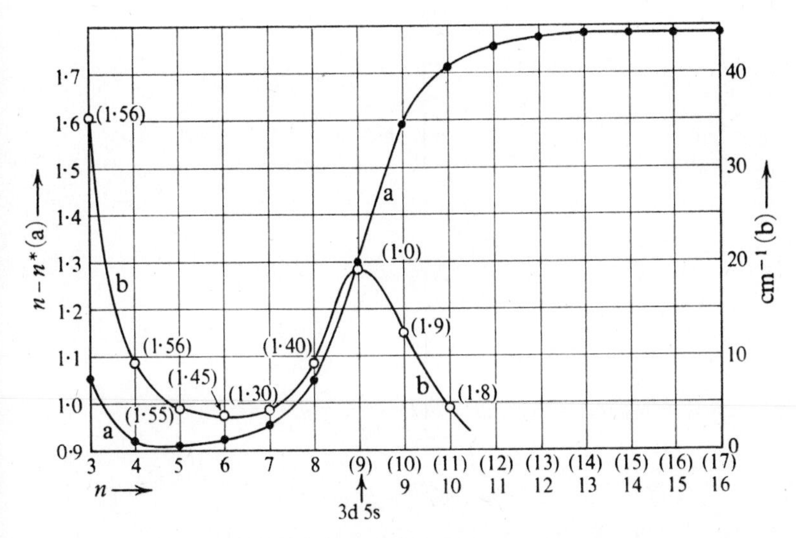

FIG. V, 9. Perturbed series 4s*n*d ^{3}D in Ca; curve (a); term defect $n-n^*$; (b) width of triplet splitting; numbers in brackets: interval ratio.

and intensities of the components may differ from the prediction of the theory. The failure of a term to be in the position to be expected by interpolation or simple theory and to form transitions of the expected strengths is known in spectroscopy as *perturbation*. This use of the term differs in one respect from that applied to changes due to external fields where we can observe the unperturbed state and then apply the perturbation. In contrast to this, the unperturbed state of the spectroscopic perturbation is a hypothetic one which would exist if certain theoretical assumptions were strictly valid. In this case the perturbation potential is that potential energy which has to be added in order to describe the system more correctly. The theoretical treatment is, however, similar in both cases and the qualitative features of perturbation phenomena can be well understood with reference to the mathematical theory of perturbations (II.C.3).

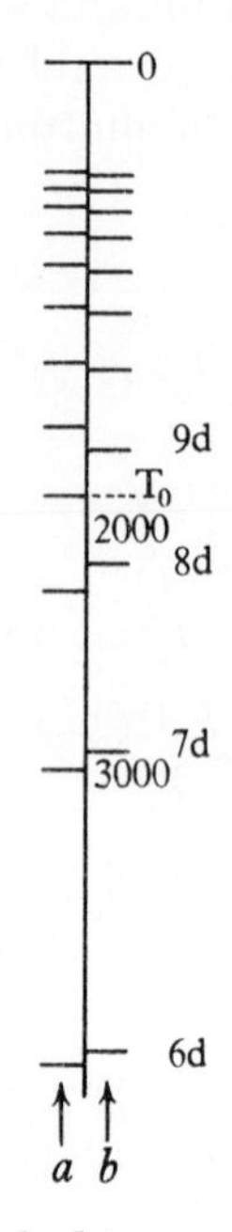

FIG. V, 10. Perturbed term series 4s*n*d ³D in Ca.

One not uncommon type of perturbation occurs as irregularity in the spacings of the terms within a series. Figure V, 9 shows a plot of the term defects n–n^* in the series 4snd ³D of Ca.[39] Instead of being nearly constant, the values show a rapid change by almost 1 between $n = 6$ and 11. In this range of term values the regular behaviour is thus interrupted with the result that one excess term has been fitted in by re-arrangement of the others. This is also shown in fig. V, 10 where the actual term values are plotted on the left, and the hypothetical, *unperturbed* terms on the right.

The observed effects can be explained in the following way. A term 3d5s ³D (T_0 in fig. V, 10) would be expected to lie at about

2000 cm^{-1}, where the centre of the perturbation is observed; it has the same parity and its levels have the same J-values as those of the terms 4snd ^{3}D. It also has the same L and S. Under these conditions of several levels of similar type lying close together, the specification of the configuration loses its meaning and each of the neighbouring states has to be regarded as a mixture of both configurations.

The regular spacing within a term series arises from the successive change of one quantum number in a definite configuration, causing corresponding changes in the radial wave-function. More generally, the terms of a configuration had been derived by applying electrostatic perturbation of the first order to the degenerate state specified by the configuration. A second-order perturbation would have introduced terms of the type (II.86) linking several configurations. Such terms are no longer negligible when the energy difference in the denominator becomes very small; their effect is always a mutual *repulsion* between the two states involved, as a discussion of the sign easily shows. The non-diagonal matrix element, in the numerator of (II.86), vanishes for configurations of different parity and for states of different J. For extreme Russell–Saunders coupling it also vanishes for levels differing in either L or S.

In the present example, the matrix elements connecting two closely adjacent states 3d5s $^3D_{1(2,\,3)}$ and 4snd $^3D_{1(2,\,3)}$ for $n = 8$ or 9, are not negligible and cause mutual repulsion, making the difference between the latter two terms anomalously large. These two terms of the perturbed series are no longer correctly described by the configuration symbols 4s8d and 4s9d but partly by 3d5s. The same applies to a smaller extent to the terms with $n = 7$ and 10. For $n > 7$ the numbering becomes ambiguous, according to whether the perturbing term is included or not, but the total number of terms and levels is never changed by perturbation.

The perturbation effect also manifests itself in the change of the width of the multiplet splitting with n. Figure V, 9(b) shows the overall width of the triplet splitting as function of n. Instead of the regular decrease to be expected for an unperturbed series, the splitting is found to rise to a maximum at the centre of the perturbation. Also the ratio $(D_3-D_2)/(D_2-D_1)$ which is given in brackets in the figure, shows the greatest deviation from the theoretical value 1·5 for the most perturbed terms.

The perturbation within a series, as function of n, is mainly determined by the denominator of (II.86) and is closely similar to an anomalous dispersion function. A number of similar perturbations have been studied, e.g. in Ba, Hg, Cu and Al II.[40–43]

The curve a in fig. V, 9 shows that even for fairly distant terms the perturbing effect has to be taken into account if accurate term values are required. Such weak perturbation is usually described by the general term *configuration interaction*. To identify effects of this kind with certainty one has to calculate their absolute magnitude which is unfortunately rather difficult.

One of the few cases where a quantitative estimate has succeeded in establishing the nature of a perturbation quite convincingly is that of an anomaly found in the spectrum of Mg. In this otherwise simple and regular spectrum it is found that in all configurations 3snd the term ^1D lies below ^3D, in contradiction to the general experience and the theoretical result that the triplet term is the lower in all configurations sl. Bacher (ref. III. 106) has ascribed this anomaly to the perturbing action of the configuration $3p^2$. This forms the terms ^3P, ^1D and ^1S, of which only the first has been actually identified; the other two are expected to lie a little above the ionisation limit. None of these can perturb the terms 3snd ^3D because the smallness of the magnetic interactions ensures that the coupling is close to the Russell–Saunders limit so that the selection rules for L and S will hold. The 3snd ^1D terms, however, are depressed by repulsion due to $3p^2\,{}^1D$. Bacher calculated the magnitude of the effect by means of rather crude, hydrogen-like wave functions and found it to be of the right order of magnitude.

Configuration interaction can have a marked effect on the multiplet splitting of a term. Even a very small admixture of one term, with a large splitting factor, in another with a much smaller splitting factor, can radically change the splitting in the latter. The inverted order of the ^2D- and ^2F-terms in some alkali atoms have been explained in this way (ref. III. 105, 106). Excitation of one of the p-electrons of the core leads to a term of very high energy—in fact an X-ray term—but of very wide, inverted doublet structure. Its perturbing influence on those terms of the optical spectrum which have a small splitting might explain the anomalies observed in the latter. In K, e.g. the term $3p^54p4s\,{}^2D$ is an even, inverted term and can perturb the terms $3p^6nd\,{}^2D$ and cause them to be inverted. Other explanations have, however, been suggested for these anomalies (see ref. III, 107).

Altogether, numerous irregularities are found in the higher terms of spectra, but it is usually difficult to establish their cause with certainty, for lack of a simple method of estimating absolute values of perturbation effects. Only one arbitrarily chosen example may be mentioned. In the 4snp ^3P terms of Ca the interval rule is found to be fairly well observed for $n = 4$ and 6, but not for $n = 5$ where the interval ratio is 2·9 instead of 2. An inspection of the tables shows that the term 3d4p $^3F_{2,3,4}$ lies closely below 4s5p $^3P_{0,1,2}$. Only the level $^3P_2$ of the latter can be perturbed by the former, namely by the level $^3F_2$; parity and J-value will allow this perturbation. The difference in L will make the effect small but not zero because magnetic interactions are not very small and cause Russell–Saunders coupling to be only a moderately good approximation. The mutual repulsion of the two levels with $J = 2$ may therefore well explain the anomaly. The interpretation could be tested if the Landé g-values of the levels were known. One would expect that of the level $^3P_2$ to be anomalous, but not those of $^3P_1$ and $^3P_0$.

Intensities are also affected by perturbations, but few experimental data are available. Such effects are brought about by the mixing of radial wave functions, either due to electrostatic interactions or magnetic spin-orbit interactions. An example of the latter is the anomaly in the intensity ratio of Cs doublets described in III.C.4.

Spectral lines whose upper term lies above the normal ionisation limit are sometimes observed to be abnormally broad, and their intensities show anomalies which can be ascribed to a very short life time of the upper state. Such effects were first observed by Shenstone[(44)] and ascribed by him and others[(45–47)] to the process of *auto-ionisation*. It is the same which gives rise to the *Auger effect* in X-ray spectroscopy where it manifests itself by the emission of electrons (IV.C.5).

If an electron other than the most loosely bound one is excited in an atom, or if two electrons are excited simultaneously, the energy of the resulting state is often higher than the lowest excitation energy. An atom in such a state has enough energy to pass into an ion in its ground state and a free electron with excess kinetic energy. The latter state can be regarded as a continuous eigenstate or rather as a superposition of such states in which the quantum number l of the free electron can have various values 0, 1, 2, . . and the quantum number J of the whole atom also has certain definite values.

If the Hamiltonian connects the discrete state in question with one of the possible continuous states, the eigenfunctions and properties of the two are mixed. The excited state is then no longer truly stationary, the energy value becomes indistinct and the lines connected with the state appear broadened. One can also describe the situation by saying that an atom in this state has a certain probability of passing, by a radiationless process, into the ionised state, with the electron leaving the ion with excess energy. The shortened life time leads to a greater width of the energy level, in accordance with the uncertainty principle.

The selection rules for radiationless transitions are the same as for perturbations: the terms must have the same parity and J-value, and in Russell–Saunders coupling they must have the same L and S.

To what extent the observed effects can be understood may be illustrated by the ^{3}P-terms due to the configurations np^2 in Zn, Cd and Hg, where n is 4, 5 and 6 respectively.[48] They lie above the limit ns of the normal spectrum. Auto-ionisation can be caused by the continuous states of ns∞s ^{1,3}S and ns∞d ^{1,3}D, but not by those of ns∞p ^{1,3}P which are of odd parity. For strict Russell–Saunders coupling, none of the former could cause auto-ionisation, owing to the L-selection rule.

Lines connecting with the p^2 ^{3}P levels have been identified with some certainty, and it has been found that combinations with $^3P_0$ and $^3P_1$ are sharp while combinations with $^3P_2$ are slightly diffuse in Zn, very wide in Cd (about 176 cm^{-1}) and possibly even wider in Hg where the observation is rather uncertain. The possibility of auto-ionisation of $^3P_2$ is easily understood by the deviation from Russell–Saunders coupling which increases from Zn to Hg. The state p^2 $^3P_2$ is mixed with p^2 $^1D_2$ which mixes with the continuous states of ns∞d $^1D_2$. No similar perturbation can occur to p^2 $^3P_1$ which is the only level of p^2 with $J = 1$ and for which the Russell–Saunders coupling scheme therefore always remains valid. The level p^2 $^3P_0$ would be expected to be mixed with p^2 $^1S_0$ and therefore also liable to autoionisation due to ns∞s $^1S_0$. The apparent absence of broadening in the lines connected with this level can only be explained by the assumption *ad hoc* that the probability of the radiationless transition is very much smaller.

The theoretical calculation of autoionisation probabilities is unfortunately difficult and only a few cases have so far been treated.[49]

The existence of a level with strong autoionisation, of energy E above the normal ionisation potential implies that the radiative recombination process of an electron of kinetic energy E with the ion has a greatly enhanced probability; this follows immediately from an application of the principle of detailed balancing to a system of atoms, ions, electrons and radiation. It has, in fact, been observed that high current densities in discharges enhance the lines which show autoionisation, compared with other lines, an effect which is evidently due to recombination followed by emission.

B. TYPES OF SPECTRA

In spite of our still very incomplete knowledge of the spectra of most elements, the existing numerical material is far too large to find room in a textbook. Works of tables giving wavelengths of lines and values of terms have been referred to in I.3. The following chapter is intended as a brief survey, stressing features of special importance and providing examples of the application of the theory to individual spectra.

1. The "displaced" terms in two- and three-electron spectra

The spectra of the alkaline earth elements and their iso-electronic ions show some prominent multiplets which do not fit into the scheme of a simple spectrum whose structure can be explained by the excitation of a single electron. These anomalous multiplets are due to transitions from a *dashed* or *displaced* term to one of the normal triplet terms. The interpretation, given mainly by Russell and Saunders,(50,51) of these displaced terms as due to the excitation of two electrons formed the starting point for the understanding of complex spectra.

While the terms of the simple spectra of Be, Mg, Ca . . are based on the ground configuration of the corresponding ion, 2s, 3s, 4s, . ., these anomalous terms are based on excited ionic configurations, such as 2p in Be, 3p in Mg, and converge towards a series limit which is displaced above the normal limit by the excitation energy of the ion: 2s – 2p, 3s – 3p, . .. It is for this reason that they are called *displaced* terms.

A particularly striking illustration is the multiplet at 2780 A in Mg (plate **13**). It forms a group of five lines, the central one of which is a blend of two lines. The almost perfect symmetry of spacings and intensities, resembling that of a Zeeman pattern, suggests the interpretation as a transition between two terms having the same values of S and L and nearly the same width of splitting. Figure V, 11 demonstrates this interpretation as a so-called PP′ transition. The intervals have been drawn according to the interval rule and the intensities would obey to formulas (III.88). From the values of the splitting, the lower of the terms can be identified with the normal term 3s 3p ^{3}P^0, the upper term of the resonance triplet. The lowest configuration which has odd parity and can therefore combine with this term is 3p^2, and of the three terms ^{3}P, ^{1}D and ^{1}S arising from this configuration only the first will combine at all strongly, on account of the selection rule $\Delta S = 0$.

The identification of the anomalous term P′ as $3p^2\ {}^3P$ is confirmed by the fact that the width of splitting is nearly equal to that of the term $3s3p\ {}^3P^0$: according to table V, 9 both terms have the splitting factor $\frac{1}{2}a_p$. The value of a_p itself will be slightly affected by the screening effect due to a 3s electron is one case and a 3p electron in the other, but the effect is small. A similarly close agreement of the splitting factors is found for the corresponding terms of AlII, SiIII, PIV, SV, ClVI and AVII.

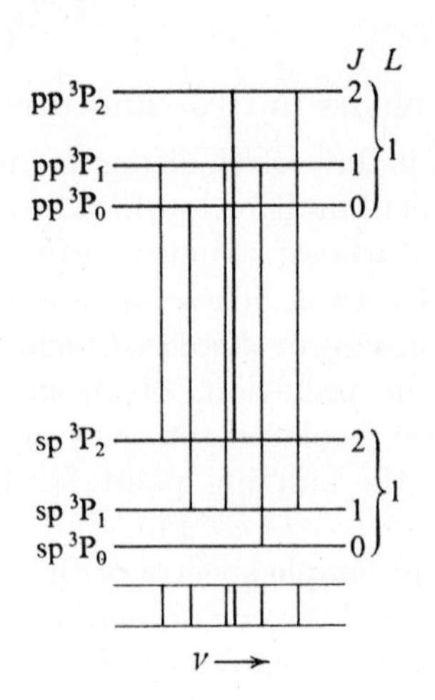

FIG. V, 11. Term diagram of PP′-triplet.

The strong doublet visible on the left in plate **13** is the resonance doublet $3s\ {}^2S - 3p\ {}^2P$ of MgII (λ2796, 2803). The approximate agreement of its frequency with that of the anomalous triplet further confirms the identification of the latter: the presence of an additional 3p electron in both upper and lower term would be expected to have little effect on the term difference. This means that not only the hypothetical series limit of the anomalous terms of Mg, i.e. of the configurations $3pnp$ ($n \to \infty$) but the lowest term itself is "displaced", at least approximately, by the term difference $3s - 3p$ of the ion.

Entirely similar relationships are found for the displaced terms $2p^2\ {}^3P$ of Be and its iso-electronic ions, and for the so-called $P - P'$ transitions to the normal ^3P terms.

The interpretation of the displaced terms in Ca and the other alkaline earth elements is a little less simple. Though it is again obvious that they arise from the excitation of two electrons, the assignment of quantum numbers is more difficult. The lowest excited configuration of Ca is 4s4p, but in Ca^+ it is found that 3d is lower than 4p. There are thus two possible even configurations due to excitation of two electrons in Ca: $4p^2$ and $3d^2$, and it is not immediately obvious which of them is lower. In fact two even ^3P terms are known, and by arguments similar to those given above, the lower of them has to be ascribed to $4p^2$: the total splitting $^3P_0 - {}^3P_2$ is found to be 134 cm^{-1}, in sufficient agreement with the splitting of 158 cm^{-1} of the normal term 4s4p ^3P.

The total splitting of the higher of the two even ^3P terms is 39·5 cm^{-1}; if it is interpreted as $3d^2\,{}^3P$, its total splitting can be compared with the doublet splitting of 3d ^2D of Ca^+. From table V, 9 and eq. (21) we find for the former $\Delta\nu_{total} = a$, and for the latter $\Delta\nu_{total} = (3/2)a$. The total splitting of 3d ^2D is 60·7 cm^{-1} which compares very well with (3/2)39·5 cm^{-1}.

Apart from the terms of the configuration $3d^2$, those of the series 3d4d, 3d5d, and 3d6d are known in Ca and allow the displaced series limit to be extrapolated.

Figure V, 12 shows the relation between the single terms, displaced terms and terms of the ion; the multiplet splitting is not shown. The three terms of the configuration $4p^2$ lie in the order predicted by the theory, though the ratio $({}^1S - {}^1D)/({}^1D - {}^3P)$ is much smaller than the theoretical value 1·5. In the 3d4p configuration, the D- and F-terms show the order of singlets and triplets as predicted (see table V, 8), but the order of the P-terms is reversed. This may be due to perturbation by terms of the n_0snp configurations which could also account for the fact that the P-term, taken as the mean of ^1P and ^3P is slightly lower than F, in contradiction to the theory.

For the configurations $n\mathrm{p}^2$ the other two terms ^1D and ^1S have also been found in some of the other alkaline earth spectra, but the relative positions of the three terms ^3P, ^1D and ^1S appear to have little relation to the predictions of the theory; even the order is not always correct. How far these discrepancies are due to incorrect identifications and how far to perturbations is difficult to say.

In these $P - P'$ transitions it is to be noted that only one electron changes its quantum number during the transition. This accounts for their comparative strength. The excitation of the P' term from the ground term requires a double electron jump. This process may not be very frequent in a single electron collision of a gas discharge, but the long life time of the metastable states offers the possibility of excitation in two steps.

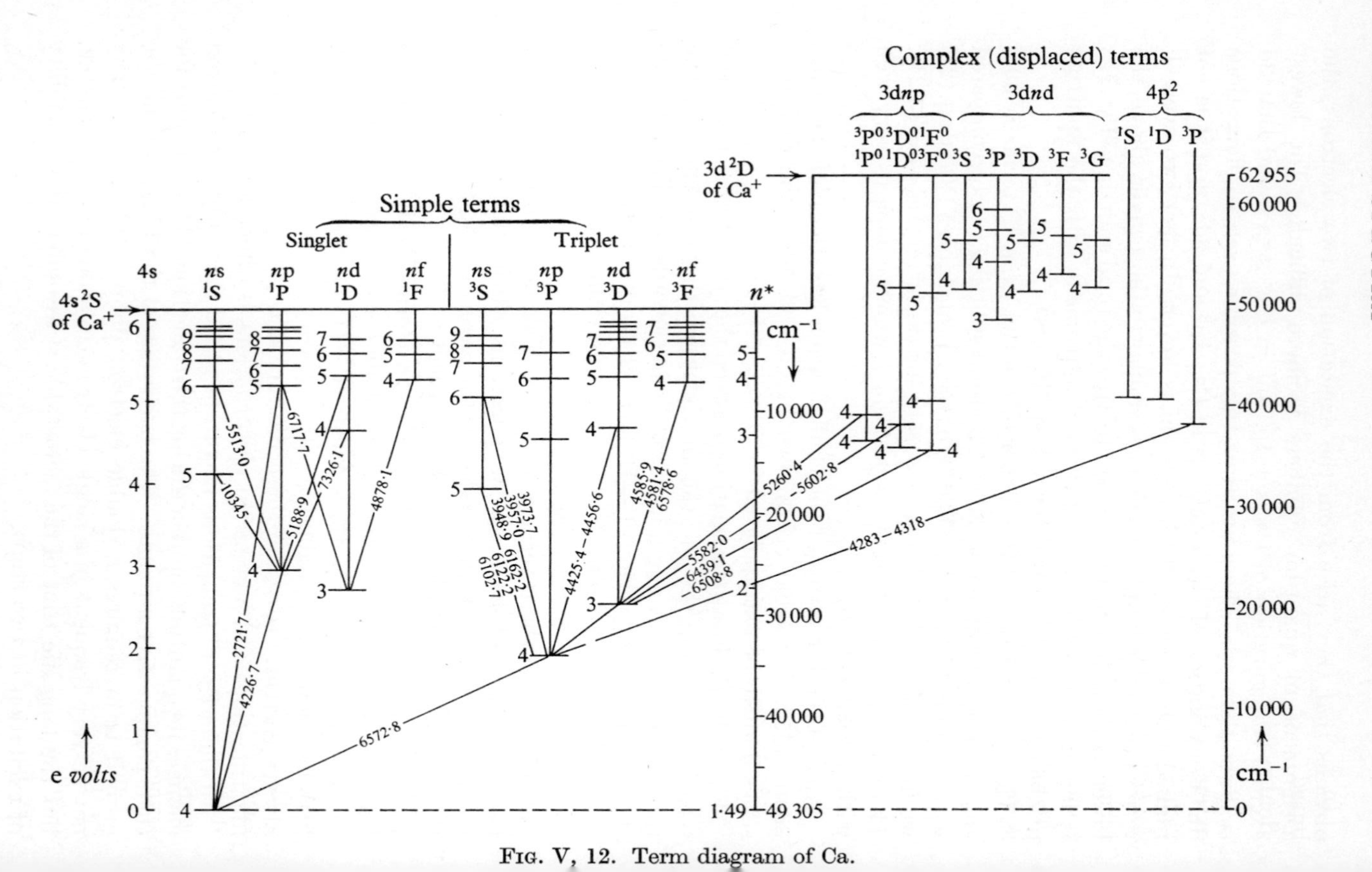

FIG. V, 12. Term diagram of Ca.

In He, a number of lines in the far ultra-violet have been attributed to terms involving the excitation of two electrons, but the identifications are uncertain.[52]

In alkaline-earth-like spectra of ions, $P - P'$ multiplets have been extensively studied, especially by Edlén.[53] Pairs of terms such as 2s2p ^3P and $2p^2\ ^3P$ can be regarded as *screening doublets* or *irregular doublets* since only the value of l of one electron differs in the two terms. Comparison of such term differences within an iso-electronic sequence are, in fact, found to obey the law of this type of doublets: the wave-number differences are linear functions of Z. This property is very useful for the identification of displaced terms.

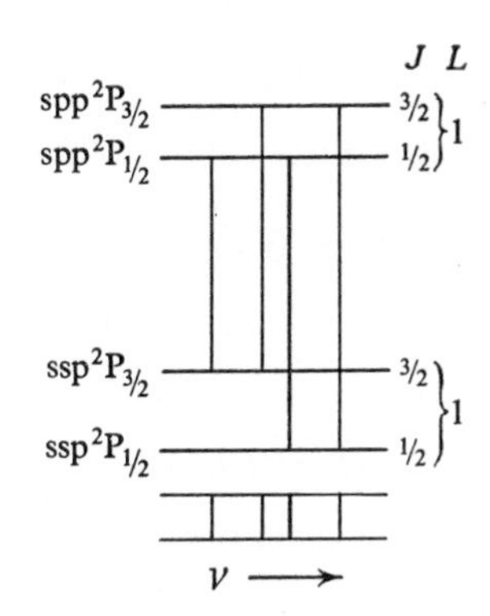

FIG. V, 13. Diagram of PP′-doublet.

Also the spectra of the earth-element type, with three electrons outside closed shells, show anomalous groups of lines, such as $^2P - {}^2P'$ transitions (see fig. V, 13). The upper term is again to be explained as a *displaced* term, but it can be produced by the excitation of only one electron from the ground state, namely one from the closed subshell ns^2.

This may be illustrated by the example of boron, whose ground term is $2s^22p\ ^2P$. Excitation of the 2p electron leads to the terms of the simple spectrum. If, instead, one of the 2s electrons is excited, configurations such as 2s2pnp, 2s2pnd converging to the excited configuration 2s2p of the ion can result.

Unfortunately the spectrum of B is very incompletely known; fig. III. 29 shows all the known terms. All the four possible terms of $2s2p^2$ have been found, though the position of the quartet term relative to the doublet terms had to be estimated, by means of the irregular doublet rule, from the terms of the iso-electronic ions where intercombinations have been observed. This kind of uncertainty is quite common.

In order to compare the terms of $2s2p^2$ with the theoretical predictions in table V, 8, we have to eliminate G_1. This can easily be done by means of the definition $\bar{P} = 1/3(2\,{}^4P + {}^2P)$ which causes G_1 to disappear from the term differences. The same relation is then found as for p^2: $({}^2S - {}^2D)/({}^2D - \bar{P}) = 1{\cdot}5$.

The experimental values of this ratio for the sequence B to Ne VI are shown in table V, 12. The order of the terms is correct in all cases, but the agreement of the ratio with theory is poor, except for the highly charged ions.

TABLE 12

Relative positions of terms of $2s2p^2$

	B	C II	N III	O IV	F V	Ne VI	theor.
$\dfrac{{}^2S - {}^2D}{{}^2D - \bar{P}}$	3·5	2·3	2·1	1·8	1·9	1·8	1·5

TABLE 13

Ratio of total splittings of ${}^4P$ and ${}^2P$

						theor.
$2s2p^2$	B	C II	N III	O IV	F V	
	0·92	1·22	1·28	1·30	1·32	1·25
$3s3p^2$	Al	Si II	P III	S IV	Cl V	
	1·45	1·39	1·42	1·43	1·44	

The terms of the configurations $nsnp^2$ offer an opportunity for testing the formulae for the width of the multiplet splitting of different terms of the same configuration. From the values A of table V, 9 and the formulae (21) one derives directly the ratio of the total splittings

$$\Delta\,{}^4P/\Delta\,{}^2P = 5/4,$$

and $\Delta\,{}^2D = 0$. Table V, 13 shows the experimental values for $2s2p^2$ to be in good agreement with the theory, except for B where the measurements are probably not accurate enough. For $3s3p^2$, the agreement is poor, in spite of the consistency within the sequence. The width of the splitting of ${}^2D$ is in most cases about ten times smaller than that of the P terms, in qualitative agreement with the theory.

2. Absorption spectra due to inner electrons

A type of spectrum which is related to displaced spectra and can be regarded as an intermediate between optical and X-ray spectra has been described by Beutler and others[54] in a series of papers. They observed large numbers of absorption lines caused by vapours of metals such as K, Rb, Cs, Hg, Cd, Zn and Tl in the spectral range of 1200–600 A.

They resemble X-ray spectra in so far as they arise from excitation of electrons in closed shells, but they are optical spectra in the sense that, in the final state, the electron is in a well-defined, "optical" level.

In Cs, e.g. Beutler observed a strong doublet, 1007·5 and 918·8 A, to be interpreted by the transitions

$$5s^25p^66s\ {}^2S_{\frac{1}{2}}–5s^25p^5({}^2P_{\frac{3}{2}})6s^2\ {}^2P_{\frac{3}{2}},$$

$$5s^25p^66s\ {}^2S_{\frac{1}{2}}–5s^25p^5({}^2P_{\frac{1}{2}})6s^2\ {}^2P_{\frac{1}{2}}.$$

The doublet splitting (9810 cm^{-1}) is of similar magnitude as that in the ground term $5s^25p^5\ {}^2P_{\frac{3}{2},\frac{1}{2}}$ of Xe^+ (10537 cm^{-1}) or as the difference $5s^25p^5({}^2P_{\frac{1}{2}})6s[1/2] - 5s^25p^5({}^2P_{\frac{3}{2}})6s[\frac{3}{2}]$ in Cs^+ (15400 cm^{-1}).

The absolute term value of $5s^25p^56s^2$, referred to the limit $5s^25p^56s$ is not very different from the corresponding value in Ba: the electron configuration $5s^25p^56s$ with $Z = 55$ has a similar effect as the configuration $5s^25p^66s$ with $Z = 56$, with regard to the addition of a further 6s electron.

A large number of further absorption lines of shorter wavelength were found in Cs, converging to the limits of the configuration $5s^25p^56s$ of Cs^+, and similar comparisons can be made for these. Their multiplet structure is rather complex, and some of the original identifications had to be modified (ref. I. 15, vol. III). There is no doubt, however, that they arise from a multitude of states of a *valency* electron moving in the field of a *core* $5s^25p^56s$ which is itself in an excited state.

The first absorption doublet appears to be of great strength, with f-value of the order of 1.

In the other elements mentioned above, the spectra were of similar type, and the interpretation follows similar lines.

Many lines were observed to be broadened, owing to autoionisation.

3. The elements of the p groups (short periods)

It is convenient to consider as one group all those elements whose electrons outside closed shells or sub-shells are p electrons, when the atom is in the ground state. Iso-electronic ions are, of course, always included; in many cases their spectra are more completely known than those of the neutral atoms. This group contains the following elements:

3	4	5	6	7	8
B	C	N	O	F	Ne
Al	Si	P	S	Cl	A
Ga	Ge	As	Se	Br	Kr
In	Sn	Sb	Te	J	X
Tl	Pb	Bi	Po	At	Rn.

Column 3 has been dealt with, partly in III and partly in the immediately preceding section.

The elements of the *4th column*:

C, Si, Ge, Sn, Pb

have the ground configuration ns^2np^2 ($n = 2, 3, 4, 5, 6$) whose terms are predicted by the theory as ^3P, ^1D and ^1S, with the ratio of the spacings $(^1S-^1D)/(^1D-^3P) = 1{\cdot}5$ (fig. V, 16).

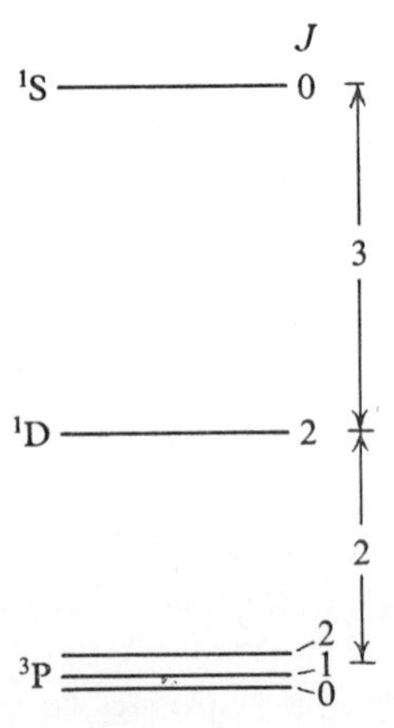

Fig. V, 16. Terms of configuration p^2.

For $2p^2$, one finds that the ratio lies between 1·12 and 1·14 for all atoms of the sequence C I to Mg VII. The close agreement among these values is especially remarkable in view of the disagreement with the theory; this has no obvious explanation.

For the configuration $3p^2$ in the atoms Si, P II, S III and Cl IV, the values of the ratio are 1·48, 1·47, 1·48, 1·47 in very good agreement with one another and with theory.

For Ge, Sn and the iso-electronic ions, the ratio agrees well with the theory, in spite of the considerable deviation from Russell–Saunders coupling. It has the values 1·51 for Ge and As II, and 1·42 and 1·31 for Sn and Sb II respectively.

The known terms of *carbon* are shown in fig. V, 15 as a typical example of the elements of the fourth column; only a few of the very high series members have been omitted.

The terms on the left contain the configurations produced by the excitation of one of the two 2p electrons of the ground configuration $2s^22p^2$. They are analogous to the simple spectra of the first three columns of the periodic table and converge towards the lowest ionisation limit for which the ion C^+ is left in the ground term $2s^22p\ ^2P$, which corresponds to the ground state of B (fig. III, 29).

The next higher terms arise from the excitation of one of the 2s electrons resulting in the configurations $2s2p^2nl$. They can be regarded as displaced terms converging to the terms $2s2p^2$ of the ion. The lowest of these configurations is $2s2p^3$. Three of the possible six terms are known; the lowest of these, ^5S, is of special interest for the study of carbon bonds. It was found experimentally[(55)] after several attempts at calculating its value had led to widely divergent results.

The wavelengths of some of the transitions have also been entered in fig. V, 15. Intercombination lines, such as the multiplet $^3P-{}^5S$ at 2965 A, are weak but important as a means of fixing the relative positions of terms of different multiplicity.

The structure of the term diagrams of Si, Ge, Sn and Pb is similar to that of C, but the increase of the spin–orbit interaction in this sequence causes the transition from Russell–Saunders coupling to j, j-coupling which has been discussed in V.A.

A feature which is typical of most complex spectra is the existence of low-lying terms which have the same configuration as the ground term. They are metastable, since electric dipole transitions to the ground term are forbidden by the parity rule. In stellar nebulae and in the solar corona where the intensities of lines are not determined by the values of the transition probability (see p. 82) transitions $^1D-{}^1S$ and $^3P-{}^1D$ within one and the same configuration ns^2np^2 appear as strong *nebula* or *Coronal* lines (ref. II. 15, 16, 17). In

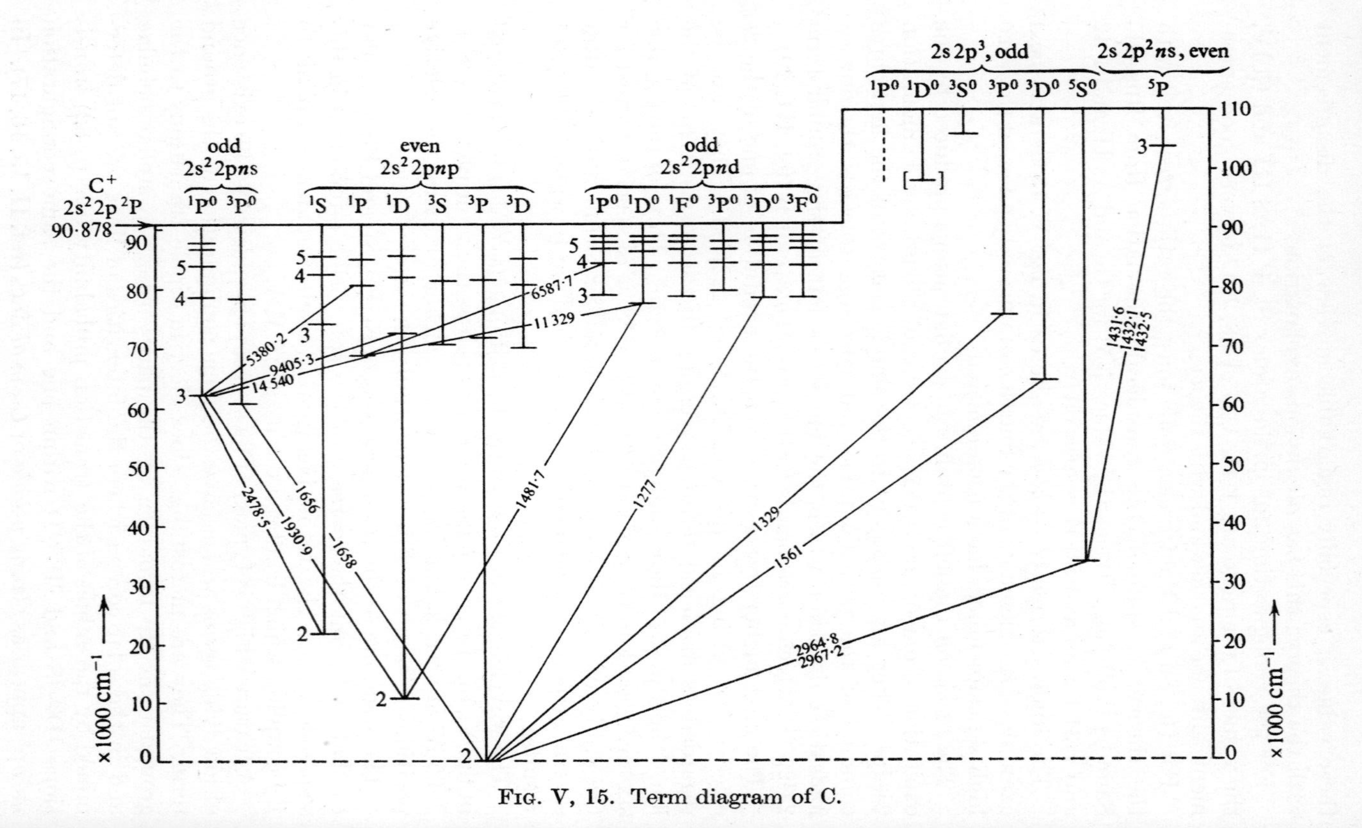

Fig. V, 15. Term diagram of C.

the configuration $2s^2 2p^2$ such transitions have been observed as nebular lines of N II and O III and as coronal lines of the highly stripped atom Ca XV, and in the configuration $3s^2 3p^2$ in the coronal lines of Fe XIII and Ni XV. Even transitions between the levels of the ground triplet ^3P have been identified in the latter two ions as coronal lines; they must be ascribed to magnetic dipole radiation.

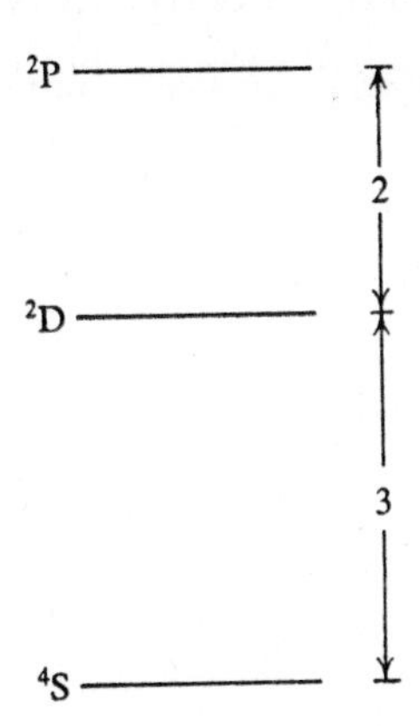

FIG. V, 16. Terms of configuration p^3.

The elements of the *fifth column*

N, P, As, Sb, Bi

have the ground configuration $ns^2 np^3$, forming the terms ^4S, ^2D and ^2P (fig. V, 16) with the theoretical ratio $(^2P - {}^2D)/(^2D - {}^4S) = \frac{2}{3}$ (see table V, 8). The observed values are shown in table V, 14.

As in the configurations $ns^2 np^2$, the agreement within an iso-electronic sequence is remarkably good, but the agreement with theory is poor for $2s^2 2p^3$ where it would have been expected to be best. The values for $n = 4$, 5 and 6 must be increasingly affected by the breakdown of Russell–Saunders coupling; they can hardly be regarded as a test for the theory.

The multiplet splitting vanishes, according to the theory (table V, 9) for both terms ^2P and ^2D of the configurations $ns^2 np^3$; the observed splittings must be ascribed to perturbations.

The term structure of *nitrogen* (fig. V, 17) may be discussed in more detail as typical of all the elements of this column. Starting from the ground configuration $2s^2 2p^3$ whose terms ^4S, ^2D and ^2P are all known, we consider at first one of the 2p electrons to be excited. This results in series of configurations $2s^2 2p^2 ns$, $2s^2 2p^2 np$, $2s^2 2p^2 nd$.

TABLE 14

Values of $(^2P - ^2D)/(^2D - ^4S)$

						theor.
$2s^22p^3$	N	O II	F III	Ne IV		
	0·500	0·508	0·514	0·518		
$3s^23p^3$	P	S II	Cl III	A IV	K V	
	0·645	0·650	0·653	0·652	0·653	
$4s^24p^3$	As					
	0·715					0·667
$5s^25p^3$	Sb					
	0·908					
$6s^26p^3$	Bi					
	1·121					

The ten possible terms of the configuration $2s^22p^2np$ can be divided by their *genealogy* into three groups of terms based on the terms ^3P, ^1D and ^1S of the ion (see p. 306). This means that we consider first the interaction between the four most strongly bound electrons of $2s^22p^2$ and then add the more loosely bound np electron ($n > 2$). For the configuration $2s^22p^23p$ we shall thus expect a low-lying family of terms which can be written $2s^22p^2(^3P)3p$, consisting of the six terms ^2S, ^4S, ^2P, ^4P, ^2D, ^4D and two families of higher doublet terms of the type $2s^22p^2(^1D)np$ and $2s^22p^2(^1S)np$. The term structure of the elements of this column follows, in fact, this qualitative pattern very well, but for the calculation of the term differences the matrix of the electrostatic interaction has to be extended over all states of the entire configuration. The diagonal sum method does not lead to a complete solution, because there are three ^2P terms and two ^2D terms, but the problem can be solved by means of the Racah method.[56] It is found that

$$(^4P - ^4D)/(^4D - ^2S) = 2/3;$$

the experimental values are: 0·52 and 0·51 for $2p^23p$ and $2p^24p$ of N; 0·52 for $2p^23p$ in O II and 0·60 for $3p^24p$ in S II.

For the other terms no simple ratio of differences can be formed in such a way that the values of the Slater integrals cancel out; no similar comparison with theory can therefore be made.

The terms arising from the parent term ^1D, such as $2p^2(^1D)np$, can be regarded as displaced terms converging to the second ionisation limit ^1D of N^+. Only the two terms ^2P and ^2D of $2p^2(^1D)3p$ are known.

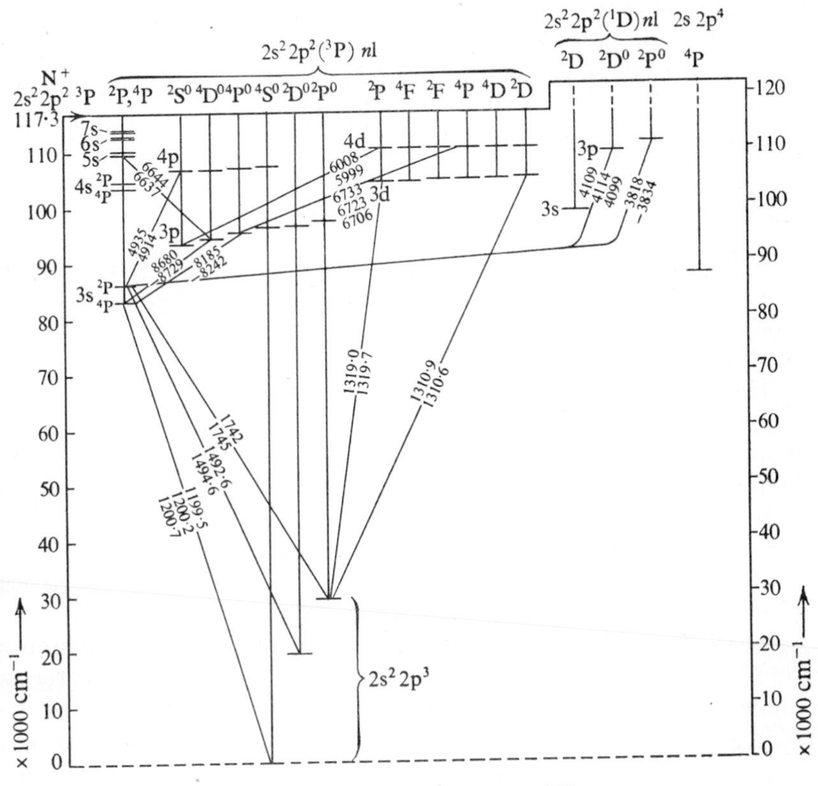

FIG. V, 17. Term diagram of N.

In view of the fact that in C and N II the term $2s2p^3\ ^5S$ lies fairly low, it is not surprising that in N the configuration $2s2p^4$ has at least one fairly low term ^4P. In O II and in the corresponding configuration $3s3p^4$ of P, all four possible terms ^4P, ^2P, ^2D and ^2S are known.

In order not to confuse the term diagram, few transitions have been entered in fig. V, 17; all lines connected with the three terms of the ground configuration lie in the vacuum ultra-violet.

The ratios of the widths of the multiplet splitting of different terms arising from any one of the configurations $2s^22p^2ns$ agree well with the values from table V, 9 for low n. Disagreements in the higher terms are presumably due to perturbations which generally become much more frequent in higher terms.

In the *sixth column*, the ground configuration ns^2np^4 leads to the same three terms as in the 4th column, but the levels of the ground term ^3P are inverted. Figure V, 18 shows the term diagram of *oxygen* as a typical example. The following terms have been identified, some in several members of a series, as arising from addition of an s-, p-, d-electron to the terms of the ground configuration of O II:

	*n*s		*n*p				*n*d				
$2s^22p^3(^4S)$	^5S	^3S		^5P	^3P			^5D	^3D		
$2s^22p^3(^2D)$	^3D	^1D	^3D	^1D	^3F	^1F	^3P	^3F	^1F	^3G	^1G
$2s^22p^3(^2P)$	^3P	^1P	^3D	^1S	^1P	^1D					

The terms based on ^4S have been shown on the left, in the figure. These form all the lower terms; it can be seen from the diagram that the strong lines in the spectrum of *O* I lie mostly in the far ultra-violet and the near infra-red.

The quadrupole transitions between the terms of the ground configuration form the strong *auroral line* λ5577, one of the most prominent lines in northern lights, and two strong, red *nebular lines*. In laboratory sources, all these are extremely weak and difficult to observe.

The line λ1302 can be regarded as the resonance line of the atom. Most of the corresponding terms are known in F II, and many of them also in the ions from Ne III to S IX.

In the second row of the periodic table, the spectrum of S and those of the iso-electronic ions up to Fe XI have been extensively analysed.

In Se and Te the terms of the ground configuration are known, but otherwise the spectra have been only partially analysed. In both these atoms the d-electrons in closed shells have fairly low excitation potentials and lead to a great complexity of excited terms; these often show so much configuration interaction that description by one configuration has little meaning.

The seventh column is formed by the *halogen* atoms whose spectra are very imperfectly known. This applies especially to *fluorine*, of which all the known terms are shown in fig. V, 19.

In the ground configuration $2s^22p^5$ one p-electron is missing from a complete shell, and the ground term is an inverted ^2P-term. The excited terms can be derived from the three lowest terms ^3P, ^1D and ^1S of F^+, but no term derived from ^1S and only two derived from ^1D are actually known. All the others arise from addition of an s-, p-,

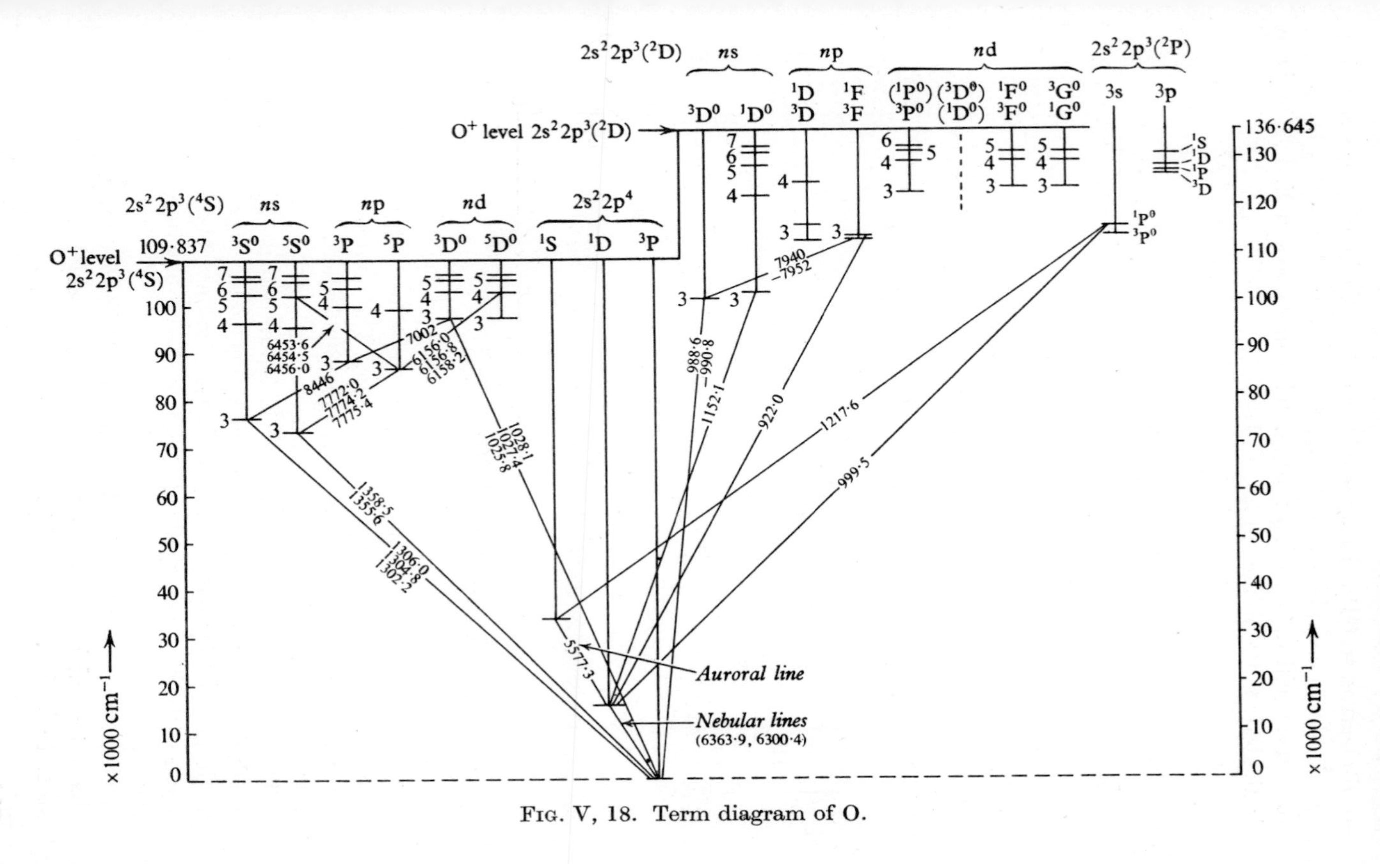

Fig. V, 18. Term diagram of O.

or d-electron to (^{3}P). In some of the (^{3}P)nd terms, the L, S coupling breaks down completely, so that two groups of close-lying levels can only be specified by their J values. They are marked X, Y, Z and are indicated as broad lines in fig. V, 19.

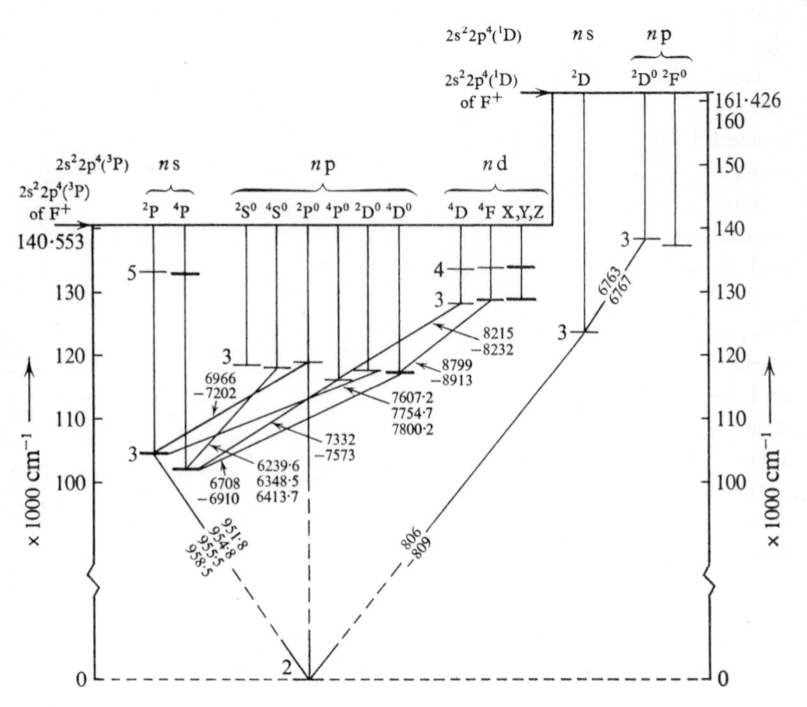

FIG. V, 19. Term diagram of F.

The wavelengths of the resonance lines are 955, 1380, 1576 and 2062 A for F, Cl, Br and I respectively. The doublet splitting of the ground term has the values 404, 881, 3685 and 7603 cm1 for these four elements.

The large magnetic interaction in iodine causes L, S-coupling to break down for many of the higher terms, so that the levels can only be designated by parity, J and the term value. The Landé g-values agree with the theoretical values for L, S-coupling only for the lowest terms.

The elements of the *eighth column* are the *noble gases* or *rare gases*, characterised by the closed electron shells which cause the ground state to be $^1S_0$.

A discharge in a rare gas emits most of the energy in the form of far ultra-violet radiation, since the lowest excited levels are about 135,000, 94,000, 80,000 and 68,000 cm^{-1} above the ground level for Ne, A, Kr and Xe. That these discharges look very bright and have allowed early and thorough spectroscopic study of the spectra is due to the fact that the rare gases are monatomic, so that most of the available energy is emitted as *atomic* radiation.

In all rare gases some of the lowest excited levels are *metastable*, having values of $J = 0$ or 2, and metastable atoms can exist in high concentration in glow discharges, giving rise to self absorption in some spectral lines.

Though many levels are known in all rare gas atoms, the coupling conditions are such that they make the description by term symbols and theoretical calculations of the energy values difficult.

The excited states of the atom are derived from the two levels $^2P_{\frac{3}{2},\frac{1}{2}}$ of the ground term of the ion. The doublet separation of these is fairly large, 782, 1432, 5371 and 10,537 cm^{-1} for Ne, A, Kr and Xe, so that the magnetic interaction within the core of the atom is comparable with, and partly larger than, the electrostatic interaction with the excited electron. For the latter, however, the magnetic interaction (assuming $l > 0$) tends to be smaller than the electrostatic interaction. The coupling conditions are therefore more aptly described by the j, l-scheme (see p. 286) than the j, j-scheme, and the former has come into use for the tabulation of the levels of the rare gases. Each level is then described by a symbol $(P_j)l[K]_J$.

Figure V, 20 shows some of the excited states of *neon*. According to their j, the levels are converging to the limits $^2P_{\frac{3}{2}}$ and $^2P_{\frac{1}{2}}$ of Ne^+, as indicated by the vertical arrows. For the term series $(^2P_{\frac{3}{2}})n$s and $(^2P_{\frac{1}{2}})n$s, the two schemes of j, j- and j, l-coupling become identical, since K is equal to one of the j. The doubling due to the coupling between the two j is only shown for $n = 3$; the levels $J = 0$ and 2 are metastable.

In the two term series $(^2P_{\frac{3}{2},\frac{1}{2}})n$p, the electrostatic and magnetic coupling of the outer electron are of similar magnitude and give rise to complex patterns, not all of whose components could be made visible in the figure. The fractions written next to the levels are values of K. In some of the $(^2P_{\frac{3}{2},\frac{1}{2}})n$d terms also, all the theoretically predicted levels have been observed, but they could not be shown in the figure. In these and the $(^2P_{\frac{3}{2},\frac{1}{2}})n$f levels, the K, s-coupling is weaker than the j, l-coupling, justifying the adoption of the j, l-scheme.

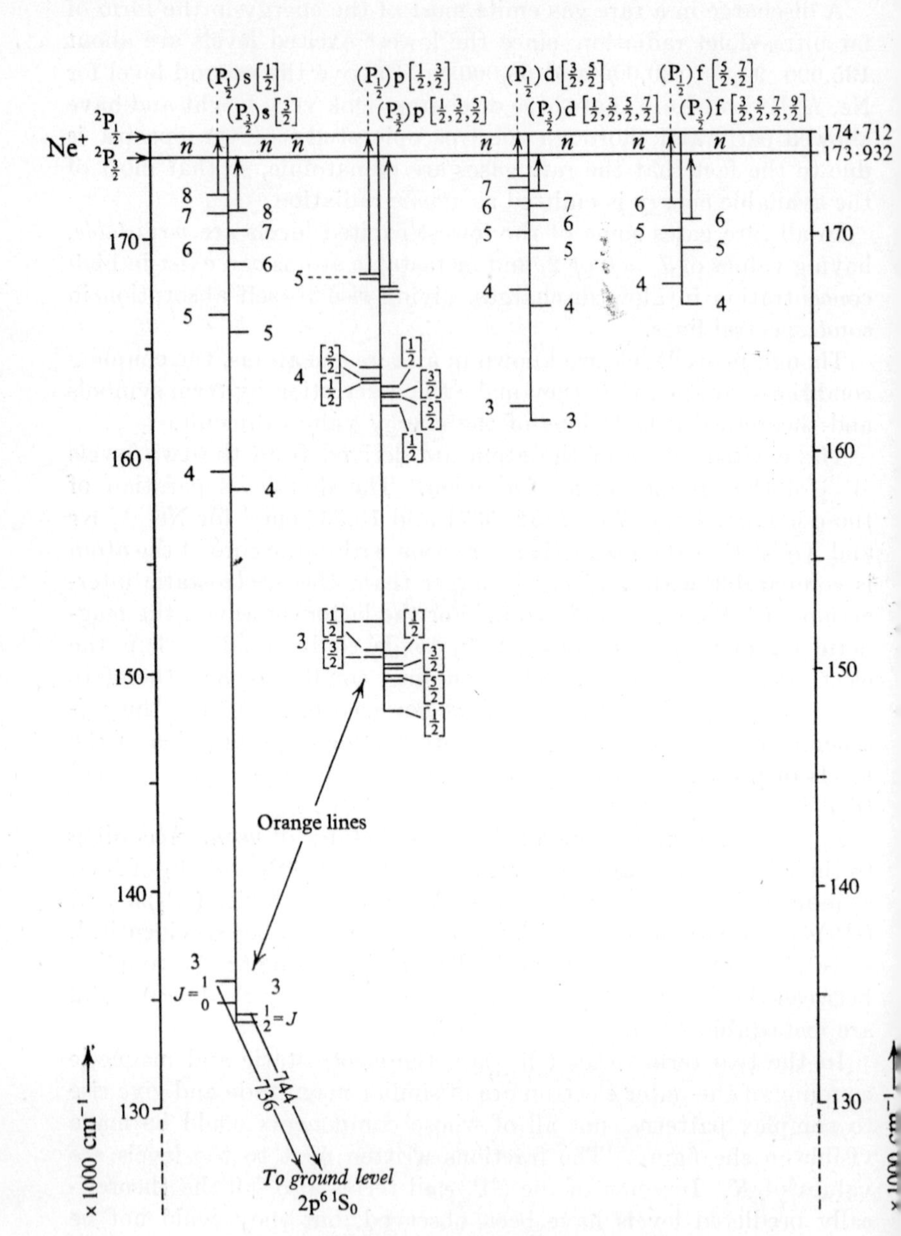

FIG. V, 20. Term diagram of Ne.

Transitions between the groups of levels $({}^2P_{\frac{3}{2},\frac{1}{2}})3p$ and $({}^2P_{\frac{3}{2},\frac{1}{2}})3s$ give rise to a large number of strong lines in the orange and near red; they cause the distinctive colour of neon discharges.

Conditions in the other rare gases are similar, but the width of the splitting of levels within a configuration is larger. Also in the iso-electronic spark spectra similar term structures have been found.

4. The elements of the d groups (long periods)

The sequences of elements in which the shells of 3d-, 4d-, and 5d-electrons are being filled in are known as the three long periods or the *iron-*, *palladium-* and *platinum*-groups. All these elements are also called *transition elements.*

The spectra of the long periods are generally more complex than those in the short periods; they are especially characterised by the great number and complexity of low terms. Any one configuration of d-electrons produces more terms and higher multiplicities than a configuration of p-electrons. The most striking features, however, are due to the small energy difference between electrons in nd- and $(n+1)$s-states. This causes all three configurations d^k, $d^{k-1}s$ and $d^{k-2}s^2$ to form low terms. Since these three configurations have the same parity, configuration interaction tends to be strong, even in low terms.

It follows from the Laporte rule that all the lower levels arising from the three mentioned configurations are *metastable.*

In spite of their complexity, many of the spectra have been well analysed, especially those in the first two periods. On the whole, the L, S-coupling scheme describes the structures well, and multiplets are mostly normal in the lower terms. Most of the lines arise from combinations of terms of the three *even* low configurations with the *odd* terms of configurations in which either a nd- or $(n+1)$s-electron has changed into a $(n+1)$p-electron.

In the last two elements of each group, the lowest configurations are $d^{10}s$ and $d^{10}s^2$ respectively, and the elements show many of the characteristics of alkali- or alkaline-earth elements, both in their spectra and their chemical properties.

A systematic description of individual spectra would appear un-rewarding in this context, and only some general features and regularities may be mentioned.

In the iso-electronic sequences, the spectra become more regular with increasing degree of ionisation, mainly because the nd states become increasingly more stable than the $(n+1)$s states. This is due to the same causes which have been discussed in IV.A.

The same tendency of increasing stability of d-electrons appears in the neutral atoms with increasing Z, but not in a simple form. In particular, a marked change always occurs in the middle of the period.

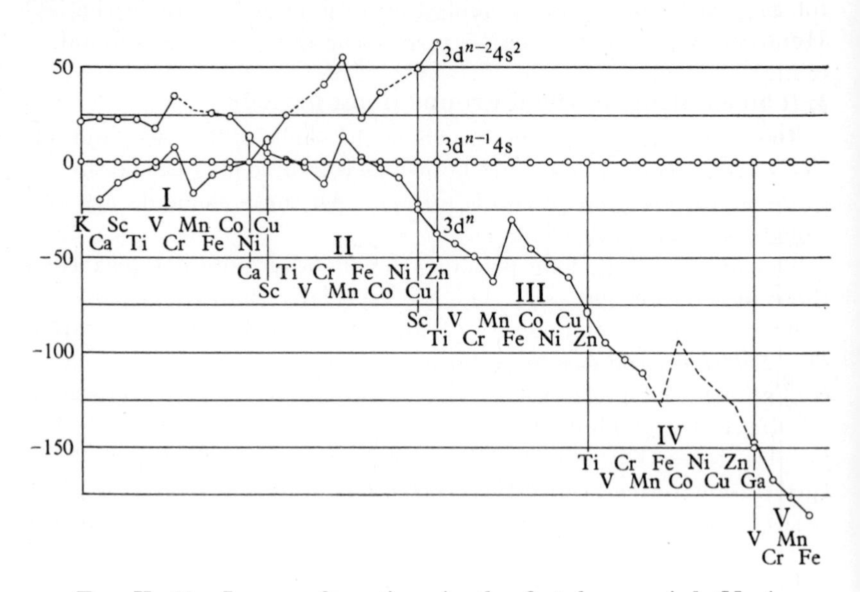

FIG. V, 21. Low configurations in the first long period (M. A. Catalan, F. Rohrlich & A. G. Shenstone, ref. 57).

Catalan, Rohrlich and Shenstone(57) have recently found most surprising regularities in the energy differences of the three low even configurations. The left parts of figs. V, 21 and V, 22 show the energies of $d^{k-2}s^2$ and d^k plotted, with the energy of $d^{k-1}s$ taken as zero level. For each configuration the term of highest multiplicity has been chosen. The authors find that not only the break in the middle of the period, but even minor characteristic features can be correlated with the theoretical expressions in terms of Slater integrals.

In addition, a striking but theoretically unexplained regularity was found: if the plot is continued to the iso-electronic sequences, so that the first spark spectrum of element Z is plotted on the same abscissa as the arc spectrum of $Z+8$, and so forth for the further spark spectra, a continuous curve arises in the first period, so that, e.g.

$$(3d^{10}-3d^9 4s) \text{ of Ni I} = (3d-4s) \text{ of Ca II},$$
$$(3d^9 4s^2-3d^{10}4s) \text{ of Cu I} = (4s^2-3d4s) \text{ of Sc II}.$$

In the second period the corresponding curve shows breaks, but they obey a very simple linear relation.

The ground configuration for each element of the first two long periods can be directly read off the graphs of figs. V, 21 and V, 22; for this configuration the lowest term is found from table V, 7 and Hund's rule, and the ground level as that of lowest J in the first, and of highest J in the second half of any period. The last two steps apply only to equivalent electrons, and tables such as V, 8 have to be consulted for other configurations.

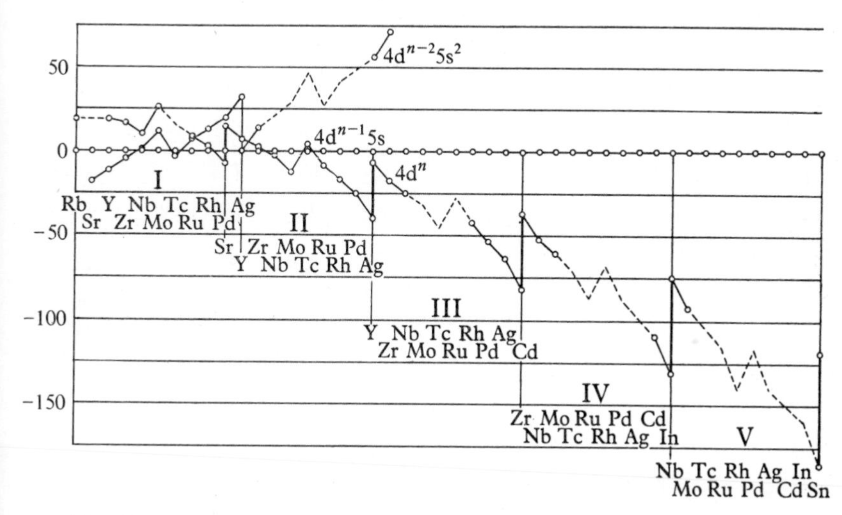

FIG. V, 22. Low configurations in the second long period (M. A. Catalan, F. Rohrlich & A. G. Shenstone, ref. 57).

The spectra of the third long period are much less well known, and coupling conditions do not generally follow the L, S-scheme.

The ground levels of the ions of the elements of the long periods are of interest to the theory of magnetic susceptibilities of solutions and solid salts of these elements.[58]

5. The rare earth elements and the actinium group

The spectra of the rare earth elements are extremely complex, with the exception of Eu, Gd and Lu where the half-filled and completely filled f-shell produces an S-term of the doubly or triply ionised atom, so that not more than two or three electrons contribute to the value of L.

The three single-electron states 4f, 5d and 6s have similar energies and all contribute to the low configurations. The known ground configurations and ground terms[59, 60] have been collected in table V, 15. The uncertain term symbols are given in brackets.

TABLE 15

Ground terms of rare earth elements
(uncertain terms in brackets)

57	La	$5d\ 6s^2$	$^2D_{3/2}$	64	Gd	$4f^7\ 5d\ 6s^2$	^9D
58	Ce	$(4f^2\ 6s^2)$	(^3H)	65	Tb	$(4f^9\ 6s^2)$	(^6H)
59	Pr	$(4f^3\ 6s^2)$	$^4I_{9/2}$	66	Dy	$(4f^{10}\ 6s^2)$	(^5I)
60	Nd	$4f^4\ 6s^2$	^5I	67	Ho	$(4f^{11}\ 6s^2)$	(^4I)
61	Pm	$(4f^5\ 6s^2)$	(^6H)	68	Er	$(4f^{12}\ 6s^2)$	(^3H)
62	Sm	$4f^6\ 6s^2$	^7F	69	Tm	$4f^{13}\ 6s^2$	^2F
63	Eu	$4f^7\ 6s^2$	^8S	70	Yb	$4f^{14}\ 6s^2$	^1S

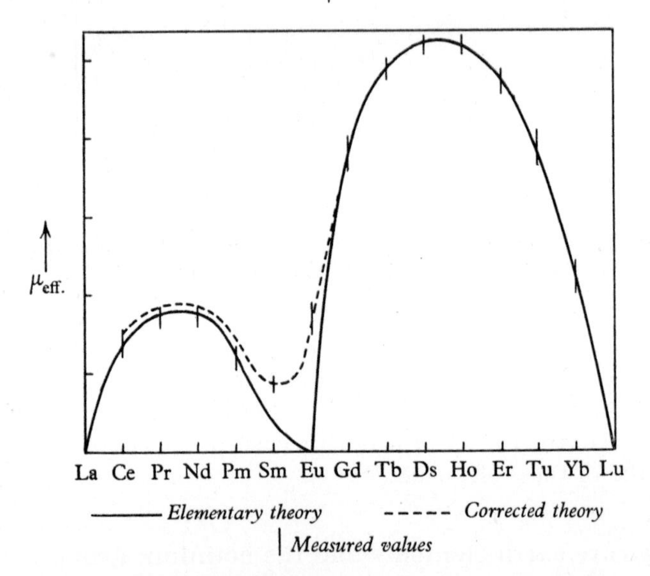

FIG. V, 23. Magnetic susceptibilities of triply ionised rare earth elements (J. H. Van Vleck, ref. 58).

The ground levels of the triply ionised rare earth atoms have the configurations f^k where k varies from 1 for Ce to 14 in Lu. These ions, as those of the d-groups, are paramagnetic. Hund[61] first showed that the Landé g-values derived theoretically for the ground levels accounted most satisfactorily for the susceptibilities measured for these ions in solutions and solid salts. Figure V, 23 shows the calculated susceptibility and the measured values.[58] The dotted line arises from a refinement of the theory in which higher multiplet levels have been taken into account.

TABLE 16

Ground terms of the actinide elements
(uncertain terms in brackets)

89	Ac	$6d\ 7s^2$	$^2D_{\frac{3}{2}}$
90	Th	$6d^2\ 7s^2$	$^3F_2$
91	Pa	$(5f^2\ 6d\ 7s^2)$	(^4K)
92	U	$5f^3\ 6d\ 7s^2$	^5L
93	Np	$(5f^4\ 6d\ 7s^2)$	(^6M)
94	Pu	$(5f^6\ 7s^2)$	—
95	Am	$5f^7\ 7s^2$	$^8S_{\frac{7}{2}}$
96	Cm	$5f^7\ 6d\ 7s^2$	^9D

In the elements from Ac upwards, 5f electrons appear, as well as 6d and 7s electrons. Only a very incomplete analysis of some of these very complex spectra has been possible so far, but work is actively in progress in several laboratories, with the use of modern spectroscopic equipment and automatic measurement and computing. The spectrum of Am has recently been partly analysed.[62]

Atomic beam resonance methods have provided a new means of determining the values of J and g of the ground levels, and studies of isotope shifts and h.f. structures can be expected to contribute to the identification of configurations. Table V, 16 shows the ground configurations and ground terms of elements in the actinide group.

CHAPTER

VI. Hyperfine Structure and Isotope Shift

A. INTRODUCTION

When spectra of heavy elements were photographed with large grating spectrographs, very fine structures of lines were often observed which could not be fitted into the scheme of interpretation described in the preceding chapters. With the use of interferometric methods and suitable light sources, a great number of such *hyperfine structures* (h.f.s.) have subsequently been resolved, also in the lighter elements. Following a suggestion by Pauli,[1] one found it possible to interpret these structures by assuming that many nuclei have an angular momentum, characterised by a quantum number I, and associated with a magnetic moment μ which is of the order of 1000 times smaller than a Bohr magneton. The interaction of this moment with the magnetic field due to the spins and orbital motions of the electrons causes the observed structures.

Anomalies in the spacings of these h.f. structures later led to the discovery of an electrostatic interaction due to deviations of the nuclear charge distribution from spherical symmetry which can be expressed by a quadrupole moment Q.

Different isotopes of an element have generally different values of I, μ and Q and show different h.f. structures. But also the centres of gravity of these patterns are often found to be displaced relative to one another. This *isotope shift* is most easily observed in isotopes having no spin and therefore no h.f. structure. The shifts are, generally speaking, of the same order of magnitude as h.f. structures and can lead to very complex patterns.

Isotope shifts are due to two entirely different causes. The difference in the nuclear mass for different isotopes gives rise to a shift which is appreciable for the lighter elements but decreases rapidly with increasing atomic weight. The large isotope shifts observed in heavy elements are due to differences in the charge distribution of the nucleus for different isotopes. The study of these *volume-* or *field-effects* has contributed to our knowledge of the size and shape of nuclei.

The smallness of most h.f. structures and isotope shifts requires special spectroscopic techniques for their study. The Fabry Perot interferometer is almost exclusively used now for achieving the high resolving power needed, up to 10^7 and more, and sometimes two etalons are used in series. One of the most serious difficulties is presented by the Doppler width of the lines; one can reduce it by cooling the discharge tube. This can most easily be done in the hollow cathode discharge which was introduced by Schüler and has become the most commonly used form of light source for high resolution spectroscopy. A far more powerful means of reducing Doppler widths is the atomic beam which can be used either for absorption or emission of spectral lines.[2–6] For elements composed of a large number of isotopes, i.e. elements of even Z, specimens enriched in any one isotope offer enormous advantages. Unfortunately the small quantities available generally preclude their use in the form of atomic beams.

Owing to these experimental difficulties, h.f. structures can often not be completely resolved but have to be derived from partly resolved or unresolved patterns. The accuracy and reliability of the result is generally the less, the less complete the resolution. Measurements of isotope shifts are less limited by the lack of resolution, provided that highly enriched isotopes are available. Their lines can then be photographed separately, and shifts much smaller than the line width can be measured. This method is laborious but capable of very high accuracy.

In the ground level and in very low metastable levels, h.f. structures can be investigated by atomic beam resonance methods which are capable of extremely high accuracy. They can even be applied to unstable nuclei whose radioactivity is used for the detection.

The so-called "double resonance" methods, developed by Brossel, Kastler, Bitter and others[7,8,9] also allow the study of h.f. structures of excited states, though these methods are somewhat restricted in their scope. In contrast to optical spectra, radio-frequency spectra are hardly affected by the Doppler width which, owing to the small frequencies involved, is usually negligible. The radiation width of the excited states, however, sets a limit to the accuracy of double resonance measurements. This method has proved of special value in the study of very narrow structures in the higher excited states of alkali atoms.

B. HYPERFINE STRUCTURE*

1. H.f. multiplets

The appearance of h.f.s. patterns is often very similar to that of ordinary fine structure multiplets. This is most frequently observed in heavier elements where the splitting tends to be wide enough to allow good resolution. Fig. VI, 1, shows an example of this from the spectrum of La. The regular decrease of spacings and intensities within the pattern points clearly to the interpretation as combination between two levels of which only one has an observable width of h.f. splitting, as happens very frequently. The appearance also suggests *interval-* and *intensity-rules* similar to those found in normal multiplets.

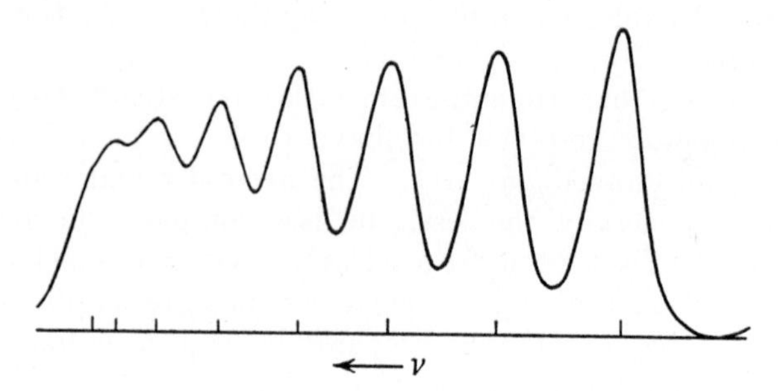

FIG. VI, 1. H.f.s. of La line 6250 A, traced from direct recording with Fabry–Perot etalon of spacing 0·75 cm (H. G. Kuhn & H. J. Lucas-Tooth, *J. Scient. Instr.* **35,** 413, 1958).

Plate 14 shows an example where the structure of both levels has been resolved. It represents an extreme case of a splitting which is so wide that it could be resolved by a grating and it is, in this respect, not to be regarded as typical. The interpretation as a transition between two levels with normal h.f. multiplet structure as shown in the figure is almost obvious.

In analogy to the quantum number s of the electron spin, we have to assume a new quantum number I which we associate with the angular momentum $\mathbf{I}$ of the nucleus and with its maximum value in any fixed direction by the relations

$$\begin{aligned} \mathbf{I}^2 &= I(I+1)\hbar^2, \\ I_z &= m\hbar. \quad (m=I, I-1, \cdots -I) \end{aligned} \tag{1}$$

* See also: H. Kopfermann, *Nuclear Moments*, Acad. Press. Inc. 1958.

I is a constant which is characteristic of any given isotope of an element. It has *half-integral* and *integral* values for nuclei of *odd* and *even mass* number respectively. No finite nuclear spin has ever been found for a nucleus containing an even number Z of protons and an even number N of neutrons, and it is very likely that all these *even–even* nuclei have the value $I = 0$. In a small number of cases this rule has been proved experimentally from band spectra.

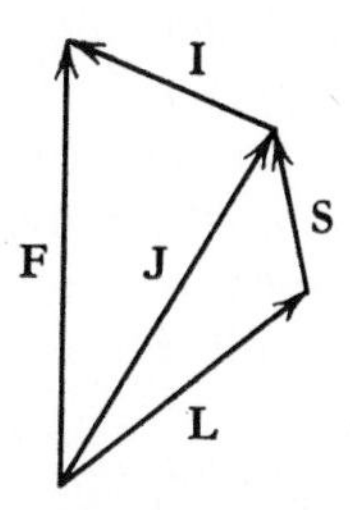

FIG. VI, 2. Vector diagram of hyperfine structure.

The vector addition of **I** and **J** (see fig. VI, 2) leads to the overall total angular momentum **F** of the atom including the nucleus, and to the h.f.s. quantum number F, according to

$$\mathbf{F}^2 = F(F+1)\hbar^2, \tag{2}$$

where F can have the values $F = J+I, J+I-1, \ldots -|J-I|$.

The quantum-theoretical basis of these rules of vector addition is exactly the same as for the multiplet structure, with the following substitutions:

$$\begin{matrix} L & S & J \\ \downarrow & \downarrow & \downarrow \\ J & I & F \end{matrix} \tag{3}$$

Each multiplet level, of given J, splits into $2I+1$ or $2J+1$ h.f. levels, for $I \leqq J$ and $I \geqq J$ respectively.

The validity of the coupling scheme on which these quantum numbers are based depends on the assumption that the interaction with **I** is small compared with the interaction between **S** and **L** (or, in the case of coupling other than Russell–Saunders coupling, compared with any of the interactions involved). In general, the smallness of the h.f. splittings compared with the multiplet splitting ensures that this assumption is justified to a greater extent than the corresponding assumptions on which Russell–Saunders coupling is based. Exceptions will be discussed in B.7.

2. Intensity ratio and determination of I

The mere assumption of the coupling scheme in which J is a good quantum number leads at once to statements on relative intensities within a h.f. multiplet, regardless of the nature of the coupling between **J** and **I**. In fact, formulae (III.88) and App. table 1 can be applied immediately, with the substitutions (3).

The physical basis of these relations can best be expressed in semi-classical language: the nuclear spin is so weakly coupled to the electronic system that it does not affect the total radiation of the atom with a given J. But by forcing **J** into a certain orientation with regard to **F** it affects the statistical weight of the level and causes a certain distribution of the radiation over the h.f.s. components. The situation is completely analogous to that governing the distribution of intensities within the components of a multiplet in Russell–Saunders coupling.

The formulae (III.88) imply the *sum rule*: within a h.f. multiplet, the ratio of the sums of the intensities of all transitions from two states with quantum numbers F and F' are in the ratio of their statistical weights $(2F+1) : (2F'+1)$.

If the splitting of one of the levels is negligibly small, the intensities of the lines are simply in the ratio of the values $2F+1$ of the level whose splitting causes the structure. This offers a means for determining the value of I, even if the structure consists of only two components, i.e. if $J = \frac{1}{2}$. This situation arises in the resonance lines of alkali or alkali-like spectra where the splitting of the ground term is of the order of ten times wider than that of the upper levels. The intensity ratios of such h.f. doublets, for different odd and even values of I, are listed in table VI 1.

TABLE 1

Intensity ratio of h.f. doublets

$I =$	$\frac{1}{2}$	$\frac{3}{2}$	$\frac{5}{2}$	$\frac{7}{2}$	$\frac{9}{2}$	$\frac{11}{2}$
	3	1·66	1·40	1·29	1·22	1·18
$I =$	1	2	3	4	5	6
	2	1·5	1·33	1·25	1·20	1·17

In determining the value of I from a measured intensity ratio, one only has to distinguish between either different half-integral or different integral values, according to whether the mass number is odd or even. It is evident that the accuracy needed for determining I is the greater the higher the value of I.

The measurement of intensity ratios in emission is especially difficult when the lower level is either the ground level or metastable; self absorption then tends to level out the intensities of the components, and careful measurements at different densities or currents are required. In absorption measurements, other difficulties arise, especially that of imperfect resolution, but with sufficient care, quite accurate values can be achieved by either method.

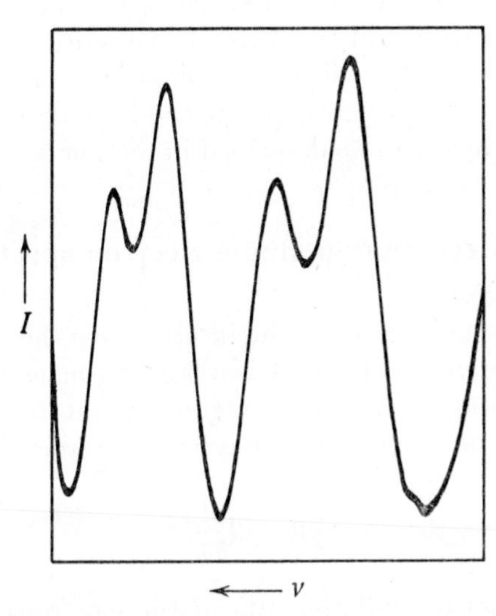

FIG. VI, 3. H.f.s. in the line $4\ ^2S_{1/2} - 4\ ^2P_{3/2}$ (7665 A) of K, from emission by atomic source ([1]K. W. Meissner & K. F. Luft, ref. 11).

Examples of h.f. doublets showing an intensity ratio of 1·66 ($I = \frac{3}{2}$) are shown in plate 15 and fig. VI, 3. Both spectra have been obtained by means of atomic beams, the first in absorption,[(10)] the latter in emission.[(11)] In the absorption experiment, light from an ordinary discharge tube containing potassium was sent through an atomic beam of potassium. Narrow absorption lines, whose Doppler width is caused only by the component of the velocities at right angles to the beam axis, appear on the background of the much wider emission lines having the full Doppler width of the source.

In the emission experiment, the atomic beam was bombarded by electrons.

If the value J of the level causing the splitting is sufficiently large, the nuclear spin I can be deduced from the number of components, provided the resolution is complete enough. Examples of this are the spectra of Protactinium[12] and Plutonium[13] whose term analysis is not known but where large numbers of h.f. quartets and doublets respectively were observed. Since J could be expected to be quite large in at least some of the levels, it could be concluded that the spins of these elements were $\frac{3}{2}$ and $\frac{1}{2}$.

If the h.f.s. pattern consists of at least three components but the value of J is not large enough to develop the full number $2I+1$ of components, the *interval rule* can sometimes be used for determining I, provided the validity of this rule can be relied on; the conditions for this to be permissible will be discussed in sections 4 and 7.

The most reliable spectroscopic method for determining I is the Zeeman effect which will be described in section 5.

3. Magnetic interaction in single electron spectra

a. The basic relations

The approximate validity of the interval rule shows that the main part of the interaction between **I** and **J** is of a magnetic nature. The simplest assumption is that of a magnetic point dipole of moment $\boldsymbol{\mu}$ in the centre of the nucleus, where $\boldsymbol{\mu}$ has the direction of **I**, i.e.

$$\boldsymbol{\mu} = \mathbf{I}\frac{\mu}{I\hbar}.\,* \tag{4}$$

The orbital motion and the spin of the electrons produce at the position of the nucleus a magnetic field whose time average in the direction of **J** can be written

$$\overline{\mathscr{H}_0} = \mathbf{J}\,\frac{\overline{\mathscr{H}_0}}{J\hbar} \tag{5}$$

where $\overline{\mathscr{H}_0}$ is the quantum-mechanical value of $\overline{\mathscr{H}_z}$ for $m = J$, i.e.

$$\langle n, J, m{=}J|\overline{\mathscr{H}_z}|n, J, m{=}J\rangle.$$

The magnetic energy is then

$$W = -\overline{\mathscr{H}_0}\cdot\boldsymbol{\mu} = -\frac{\overline{\mathscr{H}_0}\mu}{IJ\hbar^2}\mathbf{I}\cdot\mathbf{J} = -\frac{\overline{\mathscr{H}_0}\mu}{2IJ\hbar^2}(\mathbf{F}^2-\mathbf{I}^2-\mathbf{J}^2). \tag{6}$$

* This formula can be regarded as a definition of the scalar μ.

Considerations analogous to those in III.A.5 show that $\mathbf{J}^2$, $\mathbf{I}^2$, $\mathbf{F}^2$ and the energy form a commuting set of operators, so that the eigenvalues can be substituted, which leads to

$$\begin{aligned} W_F &= -\frac{\overline{\mathscr{H}_0}\mu}{2IJ}[F(F+1)-I(I+1)-J(J+1)] \\ &= A/2[F(F+1)-I(I+1)-J(J+1)]. \end{aligned} \tag{7}$$

The factor $F(F+1)$ in (7) implies the validity of the interval rule, in exact analogy to the factor $J(J+1)$ in L, S coupling.

For the *total width* of splitting between the levels $J+I$ and $|J-I|$ we find from (7)

$$\begin{aligned} \text{for } J \geqq I: \Delta W &= AI(2J+1), \\ \text{for } J \leqq I: \Delta W &= AJ(2I+1). \end{aligned} \tag{8}$$

The nuclear magnetic moment μ is usually expressed as a multiple μ' of the *nuclear magneton* μ_n. The latter is defined by the Bohr magneton μ_0 multiplied by the ratio of the masses of electron and proton,

$$\mu_n = \mu_0\frac{m}{m_p} = \mu_0/1836{\cdot}1. \tag{9}$$

Though the connection between magnetic and mechanical momentum in the nucleus is not as simple as it is for electrons in many atomic structures, it is often found convenient to define a *nuclear g*-value by the relations

$$\begin{aligned} \mu &= \mu_n g_I I = \mu_0 g_I I/1836{\cdot}1 \\ \mu' &= \mu/\mu_n = g_I I. \end{aligned} \tag{10}$$

If Dirac's theory of the electron applied also to protons, one would expect to find the value $g_I = 2$ for the latter. The actual value is $g_I = 5{\cdot}58$ ($\mu' = 2{\cdot}79267$) showing that the proton must be regarded as a complex particle.

The calculation of $\overline{\mathscr{H}_0}$ for any except hydrogen-like atoms is unfortunately difficult, and approximation methods, often of unknown and very limited accuracy, have to be used. For determinations of absolute values of μ the h.f.s. method is therefore not very accurate and has been superseded by methods of nuclear resonance wherever these can be applied. But the ratio of μ values of different isotopes of one element can be found accurately from h.f. structures because $\mathscr{H}_0$ is the same for different isotopes.

While the equations (4) to (7) apply quite generally, we shall now, in considering the absolute value of the splitting, confine ourselves to the effect of a *single electron.*

In quantum mechanics, both orbital motion and electron spin can be treated as current distributions of density $\mathbf{i}$, and the instantaneous magnetic field caused by them at the origin, which we can identify with the nucleus, is given by

$$\mathscr{H}_0 = \int \frac{\mathbf{r} \times \mathbf{i}}{r^3} \mathrm{d}\tau \tag{11}$$

where the integral is to be extended over the whole space.

For the orbital motion of a point electron the corresponding classical expression is

$$\mathscr{H}_0{}^L = -\frac{\mathbf{r} \times \mathbf{v}e}{cr^3} = \frac{-e}{mcr^3}\mathbf{L}. \tag{12}$$

This vector has a constant direction and its average value can be found if the time average $\overline{1/r^3}$ is known. For an actual electron, the presence of the spin causes the vector $\mathbf{L}$ to precess about $\mathbf{J}$ (fig. VI, 2) so that the time average in the direction of $\mathbf{J}$ is

$$\overline{\mathscr{H}_J{}^L} = \frac{-e}{mc}\overline{\frac{1}{r^3}}\mathbf{L}\cos(L, J) \tag{13}$$

where the angle (L, J) is constant.

In quantum mechanics, the expectation value, in the direction of $\mathbf{J}$, of $\mathscr{H}_0{}^L$ from (11) has to be calculated. This is again the product of the time average $\overline{1/r^3}$ and of constant factors expressing the angle dependence and containing only quantum numbers.

The effect of the electron *spin* can be ascribed to the field of a magnetic dipole of moment $-\boldsymbol{\mu}_0$. In averaging over the motion, one has to consider that the instantaneous field depends not only on $1/r^3$ and the angle between $\mathbf{S}$ and $\mathbf{J}$, but also on the angle between $\mathbf{S}$ and the radius vector. The calculation can be carried out semi-classically, with the Bohr-Sommerfeld model, and can again be expressed as the product of the time average $\overline{1/r^3}$ and the quantum numbers. For small quantum numbers, it has to be corrected in the usual way. This corrected result agrees with that obtained from the more rigorous calculation by means of Dirac's theory.[14,15]

The magnetic field, due to both orbital motion and spin, in the direction of $\mathbf{J}$ is found to be

$$\overline{\mathscr{H}_0} = -\frac{2\mu_0 L(L+1)}{J+1}\overline{\frac{1}{r^3}}, \tag{14}$$

and the splitting factor, from (7),

$$A = \frac{2\mu\mu_0 L(L+1)}{IJ(J+1)}\overline{\frac{1}{r^3}}. \tag{15}$$

For an s-electron, the field $\overline{\mathscr{H}_0}$ is produced by the action of the electron spin only, and the nucleus can be regarded as a magnetic point-dipole inside a spherically symmetric distribution of isotropic magnetisation $\mathbf{P}$. The calculation of the field in the centre of a uniformly magnetised sphere is a well-known problem in magnetostatics (or in electrostatics, with the corresponding substitutions), and the field strength turns out to depend only on the magnetisation, not on the radius, so that the result must hold for any spherically symmetric distribution. It is not obvious, however, if the force acting on the nuclear dipole is determined by $\mathscr{H}$ or by $\mathbf{B}$, though the latter appears to be a more likely choice.(16) A full theoretical analysis based on Dirac's theory of the electron spin, and experimental evidence shows that $\mathbf{B}$ is the relevant quantity, in accordance with the view that the spin is caused by circulating currents. The classical result $\mathbf{B} = 8\pi\mathbf{P}/3$ can then be applied directly, with $\mathbf{P} = -\psi_0^2\boldsymbol{\mu}_0$ giving the magnetic field

$$\mathscr{H}_0 (= B_0) = -\frac{8\pi}{3}\mu_0\psi_0^2 \tag{16}$$

and hence the magnetic energy and splitting factor

$$\begin{aligned} W_S = -\boldsymbol{\mu}\cdot\mathscr{H}_0 &= \boldsymbol{\mu}\cdot\boldsymbol{\mu}_0\frac{8\pi}{3}\psi_0^2 = \frac{16\pi}{3\hbar^2}\mu_0\mu_n g_I\psi_0^2\mathbf{I}\cdot\mathbf{S} \\ &= \frac{8\pi}{3}\mu_0\mu_n g_I\psi_0^2[F(F+1)-I(I+1)-S(S+1)], \end{aligned} \tag{17}$$

$$A_S = \frac{16\pi}{3}\mu_0\mu_n g_I\psi_0^2 = \frac{16\pi}{3I}\mu_0\mu\psi_0^2. \tag{18}$$

For a single s-electron, the expression in brackets assumes the values I and $-I-1$ for the two levels $F = I+\frac{1}{2}$ and $I-\frac{1}{2}$, giving the width of the splitting

$$\Delta W_S = \frac{8\pi\mu_0{}^2 g_I}{3\times 1836{\cdot}1}\psi_0{}^2(2I+1) = \frac{8\pi\mu_0{}^2\mu'}{3\times 1836{\cdot}1}\frac{2I+1}{I}\psi_0{}^2 \qquad (19)$$

Figure VI, 4 illustrates how the order of the h.f.s. levels can be found. For nuclei in which **μ** has the same sign as **I** (described as *positive* μ) the term order is normal: the levels of higher F have the higher energy. This also applies to $P_{\frac{3}{2}}$ levels; the fields of spin and orbital motion oppose one another, but the calculation shows that the latter predominates. This makes the width of splitting smaller than in the $P_{\frac{1}{2}}$ level; according to eq. (14): $\mathscr{H}_{\frac{1}{2}} : \mathscr{H}_{\frac{3}{2}} = 5 : 3$.

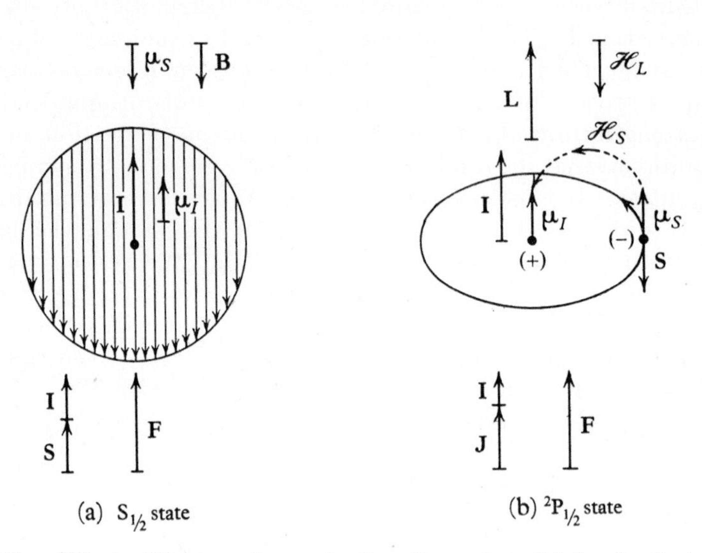

FIG. VI, 4. Diagram demonstrating the order of h.f.s. levels in $^2S_{1/2}$ and $^2P_{1/2}$.

*b. Hydrogen-like atoms**

In hydrogen-like atoms the eigenfunctions are known and the calculations can be carried out directly to a high degree of accuracy. The smallness of the splittings and the large Doppler width of lines emitted by light atoms, especially at high frequencies, makes it impossible to observe hyperfine structures in hydrogen-like atoms

* See also G. W. Series, *Spectrum of Atomic Hydrogen*, Oxf. Univ. Press 1957.

optically. Radiofrequency resonance methods have, however, in recent years led to an accurate study of the structures of the ground level 1s $^2S_{\frac{1}{2}}$ and even the metastable level 2s $^2S_{\frac{1}{2}}$ and have thus allowed a very accurate comparison with theory.

For S-levels, the value of ψ_0^2 (see table III, 4) is given by

$$\psi_0^2 = Z^3/\pi a_0^3 n^3. \tag{20}$$

Substitution of this value in (17), with $2\mu_0^2 = a_0^3 R\alpha^2 hc$, gives a relation known as the *Fermi formula*:

$$\Delta W_S = \frac{4hcR\alpha^2 Z^3}{3n^3 1836{\cdot}1} g_I(2I+1) \tag{21}$$

$$a_S = \frac{8hcR\alpha^2 Z^3}{3n^3 1836{\cdot}1} g_I, \tag{22}$$

$$a_S' = a_S/hc = g_I \times 0{\cdot}008487 Z^3/n^3 \text{ cm}^{-1}. \tag{23}$$

In this and further equations, splitting factors expressed in cm^{-1} are marked by a dash. For hydrogen-like levels with $L > 1$, the effect of the orbital motion has to be included. This gives the average value (see III.60).

$$\overline{\frac{1}{r^3}} = \frac{Z^3}{a_0^3 n^3 (L+\frac{1}{2}) L(L+1)} \tag{24}$$

and, as the result of perturbation calculation, the energy levels

$$W_F = \overline{\frac{1}{r^3}} \frac{\mu_0^2 g_I}{1836{\cdot}1} \frac{L(L+1)}{J(J+1)} [F(F+1) - I(I+1) - J(J+1)] \tag{25}$$

and the splitting factor

$$a_J' = \frac{R\alpha^2 Z^3}{n^3 (L+\frac{1}{2}) J(J+1) 1836{\cdot}1} g_I = \frac{0{\cdot}0031825 Z^3}{n^3 (L+\frac{1}{2}) J(J+1)} g_I \text{ cm}^{-1}. \tag{26}$$

The fact that the nucleus takes part in the motion increases r by a factor $(1+m/M)$ and causes a correction factor 0·9984, for hydrogen, in formulae 20–26, where R stands for R_∞. There is also a less important relativistic correction factor $1+\frac{3}{2} Z^2\alpha^2$.

In the earlier measurements of the h.f. splitting of the ground level of hydrogen(17) it was found that the observed value differed from that calculated with the known values of the constants by about 1 part in 1000. This and other spectroscopic discrepancies (see p. 216) led to the discovery of the *anomalous magnetic moment* of the electron spin. Owing to effects connected with the zero point energy of the radiation field, and thus with the Lamb shift, the magnetic moment of the electron spin is given by $1{\cdot}00116\,\mu_0$ (III.112). This brings the theoretical value of the splitting

$$\Delta\nu_{\mathrm{H}}{}^{(1s)} = 1418{\cdot}90 \pm 0{\cdot}03\ \mathrm{Mc/s}\ (= 0{\cdot}04733\ \mathrm{cm}^{-1})$$

in line with the most accurate experimental value

$$\Delta\nu_{\mathrm{H}}{}^{(1s)} = 1420{\cdot}4058\ \mathrm{Mc/s}\ (= 0{\cdot}04738\ \mathrm{cm}^{-1}).$$

More refined theories have introduced several other correction factors all of which are considerably smaller than the anomalous moment correction; for these the reader has to be referred to more specialised books (ref. p. 332).

c. Alkali-like atoms

Hyperfine structures in alkali-like spectra are of particular interest to spectroscopists. In contrast to hydrogen, the structures are generally wide enough to be accessible to optical methods, and the electronic structure of the atoms is simple enough to allow at least an approximate calculation of $\mathscr{H}_0$. Semi-empirical methods, in which $\overline{1/r^3}$ is derived either from the effective quantum number of the term or from the width of the ordinary doublet- (fine structure-)splitting have proved to be simpler and often more successful than purely theoretical calculations of $\mathscr{H}_0$. The first spectroscopic estimate of an absolute value of a nuclear magnetic moment was made by Jackson[18] by the latter method.

For S-states, the formulae (16), (17) and (18) are still valid, as far as the core electrons can be regarded as having definite spin and orbital quantum numbers, with vanishing resultants L and S. If exchange effects are included, this assumption is no longer valid; they lead to the spins of the core electrons having a small tendency to being orientated parallel to the spin of the valency electron. This magnetic *core polarisation* has been estimated[19] to increase $\mathscr{H}_0$ by 30 per cent in Li and by 5 per cent in Na. It can be neglected in most cases compared with the inaccuracies involved in the various approximations.

By considering the classical electron orbit as consisting of inner and outer portions, each of these elliptical with effective charge numbers Z_i and Z_a, Goudsmit[20] derived formulae which he corrected for small quantum numbers in the usual way and which turned out to apply even for S-states. For the latter, Fermi and

Segré[21] derived the same formulae by means of the Wenzel–Brillouin–Kramers method and modified them by the use of Dirac wave functions. They find *for an s-electron*:

$$\psi_S^2(0) = \frac{Z_i Z_a^2}{\pi a_0^3 n^{*3}}(1-\mathrm{d}\Delta/\mathrm{d}n), \tag{27}$$

$$\begin{aligned} a_S' &= \frac{8R\alpha^2 Z_i Z_a^2}{3n^{*3}1836{\cdot}1} g_I(1-\mathrm{d}\Delta/\mathrm{d}n) \\ &= 0{\cdot}00849 Z_i Z_a^2 g_I(1-\mathrm{d}\Delta/\mathrm{d}n)/n^{*3}\ \mathrm{cm}^{-1}, \end{aligned} \tag{28}$$

where n^* is the effective quantum number and Δ the quantum defect defined by $n^* = n-\Delta$. Z_a is to be taken as 1 for neutral atoms, 2 for singly ionised atoms, etc. Z_i is, for s-electrons, to be identified with the total charge number Z. The derivation of the formula assumes $Z_i \gg Z_a$. A more rigorous derivation of the formula has recently been given by Foldy,[22] but a theoretical estimate of its accuracy has not been possible. In fact, it has turned out to be remarkably accurate (see p. 340). The analogy of the relation with (22) is obvious.

The correction factor in brackets arises from the use of Dirac functions and is important for heavy elements only. It involves the change of Δ within a term series, i.e. the deviation from the Rydberg formula.

For alkali terms with $L > 0$, the relation (24) has to be replaced by (III.80), with the modifications mentioned on p. 167. From (III.80) and (III.82) the average of $1/r^3$ can be expresssed in terms of the doublet splitting constants a_{LS}':

$$\overline{1/r^3} = a_{LS}'/(a_0^3 R\alpha^2 Z_i).$$

Substitution in (25) and replacement of μ_0^2 by $\frac{1}{2}a_0^3 R\alpha^2 hc$ gives

$$W_F = \frac{hca_{LS}'g_I}{2Z_i 1836{\cdot}1}\,\frac{L(L+1)}{J(J+1)}[F(F+1)-I(I+1)-J(J+1)] \tag{29}$$

and the h.f. splitting factor

$$a_J' = \frac{a_{LS}'g_I}{1836{\cdot}1 Z_i}\,\frac{L(L+1)}{J(J+1)} = \frac{\Delta T g_I L(L+1)}{1836{\cdot}1 Z_i(L+\frac{1}{2})J(J+1)}\ \mathrm{cm}^{-1}, \tag{30}$$

where the fact has been used, that for a doublet ($S = \frac{1}{2}$, $J = L\pm\frac{1}{2}$) the term splitting is $\Delta T = a_{LS}'(L+\frac{1}{2})$. Z_i is to be taken as $Z-4$ for a p-electron and $Z-11$ for a d-electron.

d. Relativistic and volume corrections

The factor $1-\mathrm{d}\Delta/\mathrm{d}n$ in (27) was based on the use of only the large components of the Dirac functions. The full relativistic treatment based on all four components leads to further corrections[14,23] to be applied to (27), (28) and (30). The energy W_F of all terms has to be multiplied by a factor F_r which is a function of J and Z and is plotted in fig. VI, 5. For the terms with $L > 0$, the energy also has to be divided by a factor H_r, depending on L and Z and plotted in fig. VI, 6.

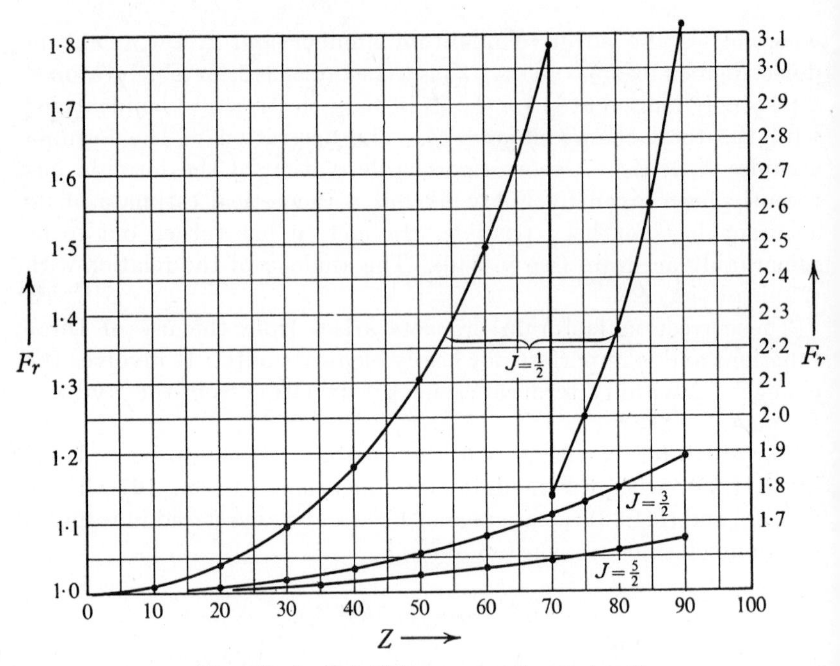

FIG. VI, 5. Relativistic correction factor F_r.

These relativistic factors depend on terms in $\alpha^2 Z^2$ of an expansion and are appreciable only for large Z where the velocity of the electrons near the nucleus becomes comparable with c.

The treatment of the nucleus as a point charge is well justified for hydrogen, whose nucleus has a radius of $1{\cdot}2 \times 10^{-13}$ cm, about 40,000 times smaller than Bohr's radius a_0 which gives the distance at which the ψ function of the 1s-electron drops to $1/e$ of its maximum value. The finite size of the nucleus causes only extremely small

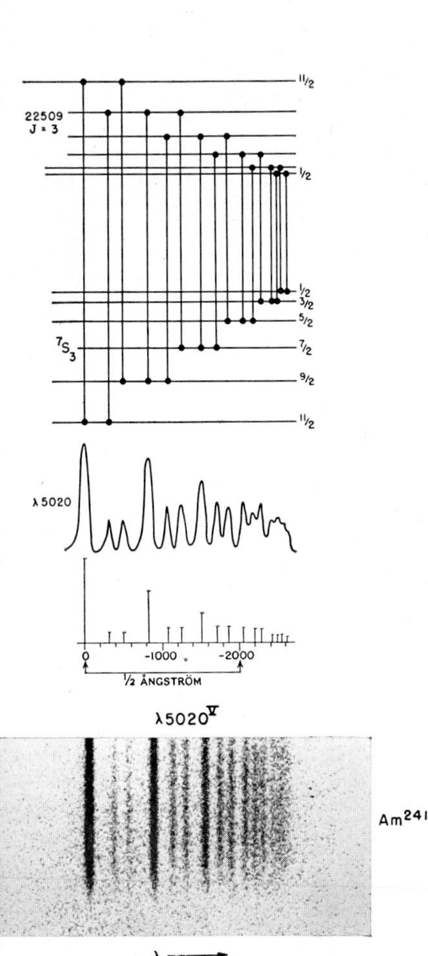

PLATE 14.
H.f.s. in the line 5020 A of Am^{241}
(M. Fred & F. S. Tomkins, ref. V.62).

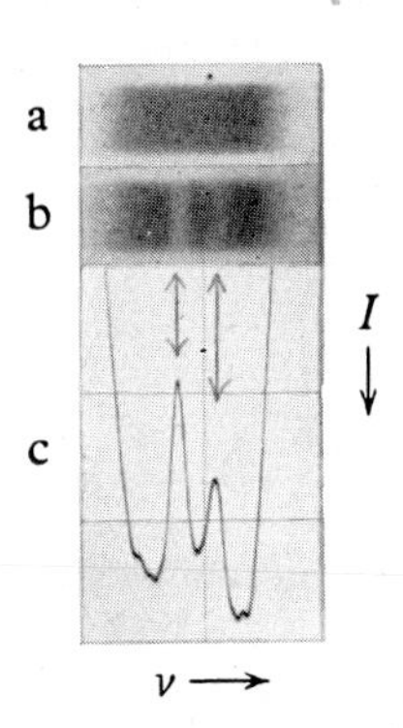

PLATE 15. H.f.s. in the line 4 $^2S_{1/2}$ – 4 $^2P_{3/2}$ (7665 A) of K. (a) light source alone; (b) with absorption by atomic beam; (c) photometer tracing of b. (D. A. Jackson & H. G. Kuhn, ref. VI. 10).

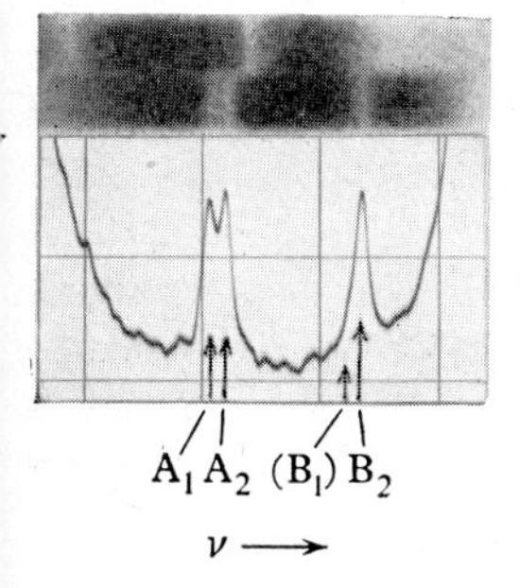

PLATE 16. H.f.s. in Na line 3 $^2P_{1/2}$ – 3 $^2P_{1/2}$ (5896 A) by absorption in atomic beam, with double etalon, showing the structure of the level 3 $^2P_{1/2}$ (F = 2 and 1). The upper spectrum shows the line 3 $^2S_{1/2}$ – 3 $^2P_{3/2}$. (D. A. Jackson & H. G. Kuhn, ref. VI. 29).

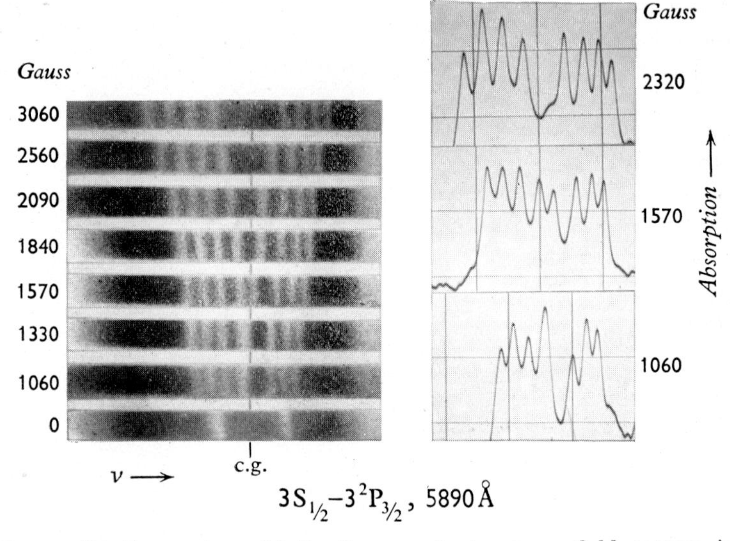

PLATE 17. Conversion of h.f.s. from weak- to strong-field pattern in Na line 3 $^2S_1/_2 - 3\ ^2P_3/_2$ (5890 A) in absorption by atomic beam, with double etalon of spacings 2 and 8 cm. (D. A. Jackson & H. G. Kuhn, ref. VI. 29).

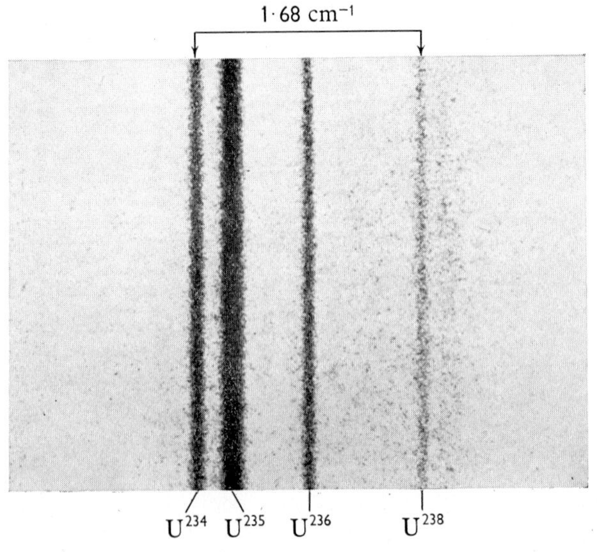

PLATE 18. Isotope shift in the line 4244·4 A of U, 6th order of 30 ft. grating, with mixture of 18, 55, 19 and 8 per cent of isotopes 234, 235, 236 and 238; the line due to 235 is slightly broadened by unresolved h.f.s. (by kind permission of Dr. M. Fred and Dr. F. S. Tomkins, of the Argonne Laboratory).

corrections. As we pass to heavier atoms, however, the nuclear radius increases and the ψ functions of the s-electrons contract. In Cs, e.g. ($Z = 55$), the corresponding effective radius of the 6s-electron, assuming it to be acted on by the full nuclear charge Ze, is $6a_0/55 \approx 5 \times 10^{-10}$ cm, as compared with the nuclear radius 6×10^{-13} cm, a ratio of about 800. This rough estimate suggests that for elements of Z above this value, the finite size of the nucleus can be expected to introduce correction factors which are by no means negligible; an s-electron will spend a small but not negligible part of its time inside the nucleus where the assumption of a point charge Ze and a magnetic point dipole μ are quite impermissible.

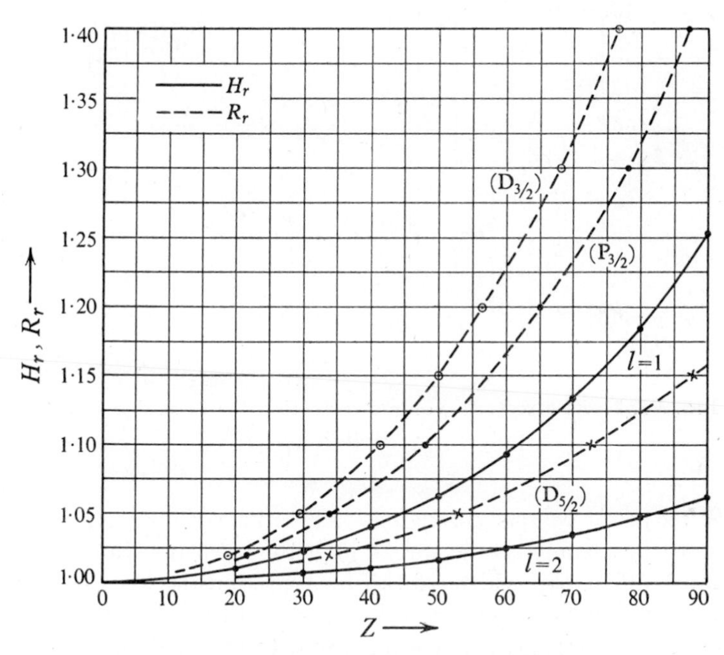

FIG. VI, 6. Relativistic correction factors H_r and R_r.

The electrostatic volume correction. Assuming the nucleus to be a uniformly charged sphere of radius $r_0 = 1{\cdot}2\sqrt[3]{(A)}10^{-13}$ cm, where A is the mass number, one can calculate the potential function; the result is shown qualitatively in fig. VI, 19. The change of the potential for $r < r_0$ will change the wave function, compared with that for a point charge, reducing its value slightly for small r. This will reduce the width of the h.f. splitting for s-electrons and, to a smaller degree, for $p_{\frac{1}{2}}$- electrons whose wave functions have a small peak at the origin. The calculation[24, 25] which has to be carried out with Dirac functions, leads to the following result. The right side of (27) and (28) has to be multiplied by a factor $(1-\delta)$ where the values of δ can be taken from fig. VI, 7. They depend quite critically on r_0 (approximately $\sim r_0^2$) but not very much on the details of the charge distribution: the assumption of a surface charge instead of a volume charge increases δ by about 25 per cent.

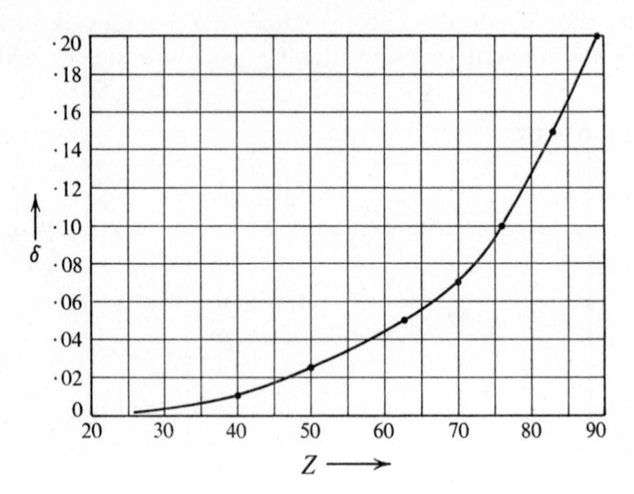

FIG. VI, 7. Electrostatic volume correction for h.f.s. due to s-electron.

The magnetic volume correction. The nuclear magnetic moment also must be assumed to be spread over the nuclear volume, and this too must affect the width of the h.f. splitting in an S-state; the value of $\mathscr{H}$ due to the electron decreases with r, so that any spread of the nuclear magnetisation will cause it to be acted on by a smaller average $\mathscr{H}$. The calculation[26, 27] leads to a further correction factor $(1-\epsilon)$ in the formula of a_J. The effect is proportional to the average value of r^2 of the distribution of the magnetic moment. It is small, except for very heavy elements; the value of ϵ has been estimated to be 0·004 for Rb, 0·005 for Cs and 0·03 for Tl.

While the electrostatic effect arises from the joint action of all the protons in the nucleus, the magnetic effect is due chiefly to one, unpaired, proton or neutron. This causes it to be much more specific: the effect of the particle spin is, e.g., relatively larger than that of the orbital motion of the proton inside the nucleus. In contrast to the correction term δ, the value of ϵ can thus be expected to differ considerably for two isotopes if their values of g_I differ appreciably. For nuclei of odd Z in which particle spin and orbital angular momentum are in nearly opposite direction, as indicated by an abnormally small value of g_I, the absolute value of ϵ can even be negative. An example is Au (see table VI, 2).

These views are now generally accepted as the explanation of the *h.f.s. anomaly* first found in Rb[28] and since measured in many elements. The ratio of the values of g_I or μ, for two isotopes, from nuclear resonance experiments was, in these cases, found to differ from the ratio from measurements of h.f. structures. The differences are always less than 1 per cent, and accurate radiofrequency resonance methods are required for their study.

With these corrections, the formulae (28) and (30) assume the form

$$L = 0 : a_J' = \frac{0{\cdot}00849 Z_i Z_a^2 g_I}{n^{*3}} F_r\left(1 - \frac{\mathrm{d}\Delta}{\mathrm{d}n}\right)(1-\delta)(1-\epsilon)\ \mathrm{cm}^{-1} \quad (31)$$

$$L > 0 : a_J' = \frac{\Delta T g_I L(L+1) F_r (1-\delta)(1-\epsilon)}{1836{\cdot}1 Z_i (L+\frac{1}{2}) J(J+1) H_r}\ \mathrm{cm}^{-1}. \quad (32)$$

TABLE 2

Hyperfine structure in $ns\ ^2S_{1/2}$-terms

Z		n	$\Delta\tilde{\nu}$ in 10^{-3} cm^{-1}	$1-\frac{\mathrm{d}\Delta}{\mathrm{d}n}$	$1-\delta$	$1-\epsilon$	$\mu'_{\mathrm{corr.}}$	$\mu'_{\mathrm{corr.}}/\mu'_{\mathrm{res.}}$
3	^{7}Li I	2	26·7	1·008	1·00	1·00	3·15	0·97
11	^{23}Na I	3	59·1	1·032	1·00	1·00	2·21	0·88
19	^{39}K I	4	15·4	1·062	1·00	1·00	0·391	0·92
20	^{43}Ca II	4	109	1·033	1·00	1·00	1·26	0·96
21	^{45}Sc III	4	670	1·031	1·00	1·00	4·59	0·96
37	^{85}Rb I	5	101	1·081	1·00	1·00	1·27	0·94
49	^{115}In III	5	3650	1·124	0·97	0·99	5·72	1·03
55	^{133}Cs I	6	307	1·101	0·96	0·99	2·58	1·00
79	^{197}Au I	6	210	1·424	0·89	1·11	0·136	0·95
80	^{199}Hg II	6	1358	1·248	0·88	0·97	0·518	1·03
81	^{205}Tl III	7	1348	1·100	0·88	0·97	1·62	1·00
83	^{209}Bi V	6	13200	1·14	0·86	0·98	4·26	1·04

Some numerical data for alkali-like S-terms are shown in table VI, 2.

Column 8 gives the values of μ in units of the nuclear magneton μ_n, as derived from (31) and column 9 the ratio of these to the values of μ from nuclear resonance. For elements of medium and high atomic weight, the ratios in column 9 are very close to 1, thus supporting the validity of the Goudsmit–Fermi–Segré formula and also of the volume corrections. For the lighter elements the formula gives systematically too low values of μ. This is not so surprising in view of the various approximations made, and the assumption $Z_i \gg Z_a$ contained in the derivation of the formula. The good agreement for Li is probably accidental.

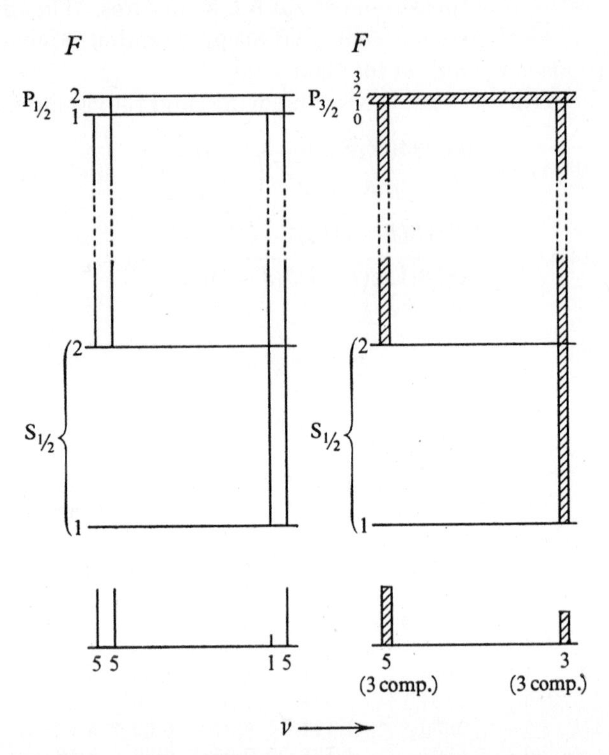

FIG. VI, 8. Term diagram of h.f.s. in transition $^2S - {^2P}$.

The values of $\mathscr{H}_0$ for the ground terms of the alkali atoms increase from $1{\cdot}3 \times 10^5$ gauss for Li to $2{\cdot}1 \times 10^6$ gauss for Cs.

In the lowest P-term of the alkali atoms, the width of splitting in the $P_{\frac{1}{2}}$ level is about ten times smaller than that of the $S_{\frac{1}{2}}$ ground term. This is illustrated by plate 16 showing the h.f.s. of the resonance lines of Na, in a spectrogram taken with a double etalon and using the absorption of light caused by an atomic beam of sodium.[29] The splitting of the $P_{\frac{1}{2}}$ level, of width $0{\cdot}006\ \mathrm{cm}^{-1}$, in the line $3s\ {}^2S_{\frac{1}{2}} - 3p\ {}^2P_{\frac{1}{2}}$ is clearly resolved in the components A_1 and A_2, though not in the other components, owing to the small intensity of B_1. In the other resonance line, the structure of the $P_{\frac{3}{2}}$ level is not resolved, owing to the smaller total width of splitting (smaller $\mathscr{H}_0$, see p. 332) and the larger number of components (see fig. VI, 8).

Even if the structure of one of the levels in a transition is unresolved, as in the case of the ${}^2P_{\frac{3}{2}}$ level in Na, the width of splitting of this level can often be deduced from the width of splitting of the blends measured in the two lines. Assuming the validity of the intensity rules, one can show that for any value of I the widths of splitting of the blends for the lines ${}^2S_{\frac{1}{2}} - {}^2P_{\frac{1}{2}}$ and ${}^2S_{\frac{1}{2}} - {}^2P_{\frac{3}{2}}$ are $\delta S_{\frac{1}{2}} + \frac{1}{3}\delta P_{\frac{1}{2}}$ and $\delta S_{\frac{1}{2}} - \frac{5}{9}\delta P_{\frac{3}{2}}$ where δ indicates the total splitting of the level. This gives the values of $\delta P_{\frac{1}{2}}$ and $\delta P_{\frac{3}{2}}$ if the theoretical ratio 5 : 3 of the latter is assumed.[30]*

The calculation of μ from different levels and different states of ionisation of one element provides convincing evidence for the validity of the theoretical formulae. The most thoroughly studied example is thallium as shown in table VI, 3.[25] The agreement with the nuclear resonance value of μ is greatly improved by the electrostatic volume correction in which uniform charge distribution has been assumed.

TABLE 3

Spectrum	Term	A	μ' (uncorrr.)	μ' (corr.)
Tl III	7s ${}^2S_{1/2}$	1·348	1·37	1·56
	8s ${}^2S_{1/2}$	0·565	1·38	1·58
	9s ${}^2S_{1/2}$	0·295	1·39	1·59
Tl II	6s 5g } 6s *n*s }	5·88	1·44	1·64
Tl I	6p ${}^2P_{1/2}$	0·710	1·49	1·55
	7s ${}^2S_{1/2}$	0·400	1·35	1·54
From nuclear resonance				1·628

* More general formulae for unresolved blends have been given by P. Brix, *Z. Phys.* **132,** 579, 1952.

4. Electric quadrupole interaction

In some h.f. structures in the spectrum of Europium, Schüler and Schmidt[31] discovered deviations from the interval rule which could not be explained by perturbations due to other levels. They ascribed them to an electrostatic interaction of the electrons with an electric quadrupole of the nucleus. On the basis of the theoretical treatment due to Casimir (ref. III. 101), this spectroscopic method has since supplied much valuable information on quadrupole moments and thus on shapes of nuclei.

Deviations of a charge distribution from spherical symmetry can be described in terms of multipole moments. The first moment of the charge distribution, in any given direction x, defined as

$$\int \rho x \, \mathrm{d}\tau,$$

is the x-component of the dipole moment where ρ is the charge density. This is zero for a nucleus, for any choice of the x-axis; no electric dipole moment has ever been found in any nucleus. The second moment, however,

$$\int \rho x^2 \, \mathrm{d}\tau,$$

has different values for different orientations of the x-axis if the nucleus has a non-spherical charge distribution, e.g. a non-spherical shape with uniform charge distribution.

If we assume axial symmetry about a z' axis, the electric quadrupole moment can be defined by the volume integral

$$eQ = \int \rho(3z'^2 - r^2) \, \mathrm{d}\tau = \int \rho r^2(3\cos^2\theta' - 1) \, \mathrm{d}\tau \tag{33}$$

where e is the elementary charge and ρ the proton charge density defined so that the integral

$$\int \rho \, \mathrm{d}\tau$$

over the whole space has the value Ze.

For uniform density, Q is positive for an elongated nucleus, and negative for a flattened nucleus.

The quadrupole moment is, by its transformation properties, a *tensor* whose components are, for a general choice of coordinate axes,

$$\int \rho x^2 \, \mathrm{d}\tau, \qquad \int \rho xy \, \mathrm{d}\tau, \qquad \int \rho xz \, \mathrm{d}\tau, \qquad ..$$

With the appropriate choice of a coordinate system (principal axes system), the components containing mixed products vanish, leaving only the diagonal elements containing x^2, y^2 and z^2. For a charge distribution of axial symmetry about the z' axis, the components with x'^2 and y'^2 are equal and in a definite ratio to the z' component which can then be used to define the quadrupole moment by (33).

For a nucleus of finite quadrupole moment, the electrostatic interaction will depend on the orientation, if the electronic charge distribution is not of spherical symmetry. We consider the electrostatic potential V due to an electron whose density distribution is determined by the Schrödinger function as $|\psi^2|$, or more rigorously by Dirac functions. Owing to the absence of an electronic dipole moment, $\partial V/\partial x, \partial V/\partial y$ and $\partial V/\partial z$ vanish at the origin. If ψ^2 also vanishes at the origin, and if the z axis is chosen as axis of symmetry of the electron distribution, it follows from Laplace's equation and from

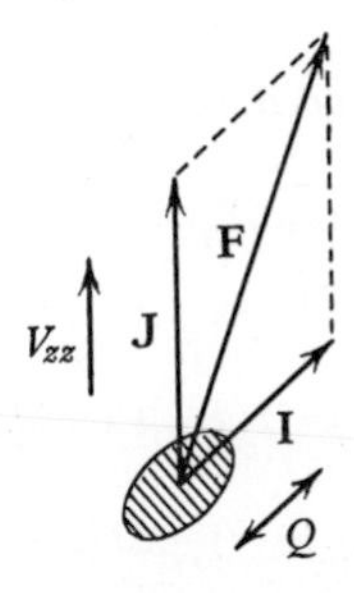

FIG. VI, 9. Nuclear quadrupole.

the axial symmetry that $\partial^2 V/\partial z^2 = -2\partial^2 V/\partial x^2 = -2\partial^2 V/\partial y^2$, so that $\partial^2 V/\partial z^2$, briefly written as V_{zz}, completely describes the second derivatives of the potential which form a *tensor* with the diagonal components V_{xx}, V_{yy}, V_{zz}. The energy of the nucleus of quadrupole moment Q in this potential field can be shown to be

$$W_Q = \tfrac{1}{4} V_{zz} e Q(\tfrac{3}{2}\cos^2\theta - \tfrac{1}{2}) \tag{34}$$

where θ is the angle between the axes of symmetry of nucleus and field (fig. VI, 9). The expression (34) can be regarded as the product of two symmetrical tensors, V'' and Q; and $Q(\frac{3}{2}\cos^2\theta - \frac{1}{2})$ as the component of the tensor Q in the z direction. As a tensor product, W_Q is a function of $\cos^2\theta$, with the period π.

Since z' and z are in the directions of $\mathbf{I}$ and $\mathbf{J}$, the angle θ can be expressed in terms of $\mathbf{I}^2$, $\mathbf{J}^2$ and $\mathbf{F}^2$ and a constant b:

$$W_Q(\text{class.}) = \frac{b}{4}(\tfrac{3}{2}\cos^2\theta - \tfrac{1}{2}) = \frac{b}{4}\,\frac{\tfrac{3}{2}(\mathbf{F}^2 - \mathbf{I}^2 - \mathbf{J}^2)^2 - 2\mathbf{I}^2\mathbf{J}^2}{4\mathbf{I}^2\mathbf{J}^2}. \tag{35}$$

According to the Bohr–Sommerfeld quantisation rules, the angular momenta in (35) have to be replaced by the corresponding quantum numbers F, I, J, and the resulting formula can be expected to hold in the limit of large quantum numbers. In this limit, it agrees in fact with the rigorous quantum-mechanical formula:

$$T_Q = \frac{W_Q}{hc} = \frac{B'}{4}\,\frac{\tfrac{3}{2}C(C+1) - 2I(I+1)J(J+1)}{I(2I-1)J(2J-1)} \tag{36}$$

where

$$C = F(F+1) - I(I+1) - J(J+1), \tag{37}$$

and

$$B' = \frac{1}{hc}eQ\overline{V_{JJ}(0)}, \tag{38}$$

where V_{JJ} is the second derivative of V taken in the direction of the vector $\mathbf{J}$.

The complete expression for the term value due to hyperfine structure is then from (7) and (36):

$$T_F = \frac{A'}{2}[F(F+1) - I(I+1) - J(J+1)] + \frac{B'}{4}\,\frac{\tfrac{3}{2}C(C+1) - 2I(I+1)J(J+1)}{I(2I-1)J(2J-1)}. \tag{39}$$

Whenever the quadrupole coupling constant B is comparable in magnitude with the magnetic h.f.s. constant A, the influence of the term (36) causes deviations from the interval rule. Qualitatively these deviations can be described by reference to the semi-classical formula which holds for large quantum numbers: owing to the factor $\cos^2(\mathrm{I}, \mathrm{J})$, the levels with the largest and smallest values of F, corresponding to $\theta = 0$ and π, are shifted in the same sense by similar amounts, while levels of intermediate F are shifted in the opposite sense.

Figure VI, 10 illustrates the quadrupole effect by two examples; they show the effect of different signs of the quadrupole moment for the arbitarily chosen case of $I = \frac{5}{2}$, $J = \frac{3}{2}$.

From the observed wave-number differences within a h.f. multiplet the values of A and B can be found by means of (39). For any pair of levels $F = a$ and b, the wave-number difference is found to be

$$\Delta\tilde{\nu}_{a\,b} = \tfrac{1}{2}A'(C_a - C_b) + \tfrac{3}{8}B'[C_a(C_a+1) - C_b(C_b+1)]/I(2I-1)J(2J-1). \tag{39a}$$

Since I, J and F are known, the two unknown constants A' and B' require two intervals $\Delta\tilde{\nu}$ for their determination. Further intervals can be used as checks.

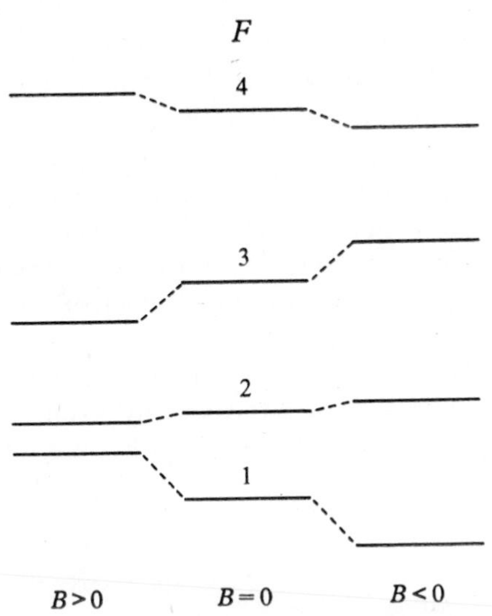

Fig. VI, 10. Influence of quadrupole interaction on h.f.s.

Other causes of deviations from the interval rule must be excluded for this method to be valid (see p. 358).

In order to find the value of Q from the observed B', it is necessary to calculate $V_{JJ} = \partial^2 V/\partial z^2$. For an electron cloud of charge density ρ the potential at the origin is given by the volume integral

$$V = \int \frac{\rho}{r}\,\mathrm{d}\tau.$$

From

$$r^2 = x^2+y^2+z^2 \quad \text{and} \quad \partial^2/\partial z^2(1/r) = (3z^2/r^2-1)/r^3 = (3\cos^2\vartheta - 1)/r^3$$

follows

$$V_{zz} = V_{JJ} = \int \rho \frac{3\cos^2\vartheta - 1}{r^3}\,\mathrm{d}\tau \tag{40}$$

if we assume again charge density and field to be zero at the origin.

In the quantum-mechanical calculation, ρ is to be replaced by $e\psi\psi^*$. Since ψ can, for the approximation of a single electron moving in a central force field, be written as a product $R(r)\,P(\vartheta)\,\Phi(\varphi)$ the integration can be carried out separately for the variables r, ϑ and φ, so that V_{JJ} is proportional to $\overline{1/r^3}$. For a single electron the calculation gives

$$B' = \frac{2J-1}{2J+2}\frac{e^2Q}{hc}\frac{\overline{1}}{r^3}R_r \text{ cm}^{-1}, \tag{41}$$

where R_r is a relativistic correction factor plotted in fig. VI, 6. For $J = \frac{1}{2}$, B' vanishes according to (41). This is due to the fact that $\psi\psi^*$ is spherically symmetrical for all wave functions with $J = \frac{1}{2}$, e.g. for the levels $^2S_{\frac{1}{2}}$ and $^2P_{\frac{1}{2}}$: quadrupole interactions do not exist for s-electrons. It can be shown quite generally from symmetry considerations that a quantum-mechanical system with an angular momentum quantum number $\frac{1}{2}$ (or 0) cannot have a quadrupole moment. This applies to a system of electrons with the resultant $J = \frac{1}{2}$ and to nuclei with $I = \frac{1}{2}$.

The value of $\overline{1/r^3}$ can be found by means of (III.80) from the ordinary doublet splitting, or alternatively from the magnetic h.f.s. factor A by means of (15), provided μ is known.

TABLE 4

Quadrupole moments from $^2P_{3/2}$ levels

	B/A	$Q \times 10^{24}$ cm^2	level	reference
^{23}Na	0·14	0·10	$3P_{3/2}$	(32)
^{39}K	0·90	0·11	$5P_{3/2}$	(33)
^{85}Rb	1·1	0·29	$6P_{3/2}$	(34)
^{87}Rb	0·15	0·14	$6P_{3/2}$	(34)
^{133}Cs	0·007	−0·003	$7P_{3/2}$	(35)
^{27}Al	0·199	0·15	$3P_{3/2}$	see (36)
^{69}Ga	0·328	0·19	$4P_{3/2}$	see (36)
^{71}Ga	0·163	0·12	$4P_{3/2}$	see (36)
^{113}In	1·83	0·75	$5P_{3/2}$	see (36)
^{115}In	1·86	0·76	$5P_{3/2}$	see (36)
^{35}Cl	0·268	−0·079	$3P_{3/2}$	see (36)
^{37}Cl	0·253	−0·062	$3P_{3/2}$	see (36)
^{79}Br	0·435	0·33	$4P_{3/2}$	see (36)
^{81}Br	0·337	0·28	$4P_{3/2}$	see (36)
^{127}I	1·385	−0·82	$5P_{3/2}$	see (36)

Since quadrupole interaction requires a value of J of more than $\frac{1}{2}$, but $1/r^3$ decreases rapidly with l, the $P_{\frac{3}{2}}$ levels of single-electron spectra are the main sources of information on quadrupole moments of nuclei. This includes the alkali-like spectra, and also the earth-like spectra in which the ground term is $ns^2\, np\; ^2P_{\frac{1}{2},\frac{3}{2}}$. The level $J = \frac{3}{2}$ is usually only a little above the ground level $J = \frac{1}{2}$ and is thus accessible to atomic beam resonance methods. The B values of these levels in the earth-like elements are therefore known with considerable accuracy. The same applies to halogens where the ground configuration $ns^2\, np^5$ is equivalent to a single p-electron (see p. 312).

Table VI, 4 shows values of B/A and Q for elements of the types mentioned. In some cases the A value of the $P_{\frac{3}{2}}$ level has been used for eliminating perturbing influences due to other terms and thus obtaining a more accurate value of Q.

5. Zeeman effect in hyperfine structures

a. Weak fields

In very weak magnetic fields, each individual h.f.s. line splits into several components, each having a definite state of polarisation; the width of splitting is proportional to $\mathscr{H}$. The Zeeman splitting of the lines can be reduced to the magnetic splitting of the h.f.s. levels. In optical spectra, the h.f.s. splitting is usually not much wider than the resolution limit, set by either Doppler width or instrumental limitations. Optical measurements of magnetic splitting in *weak fields*, i.e. of splittings small compared with the h.f.s., have therefore been possible in very few cases only. In the atomic beam resonance techniques, however, the weak field effect can often be measured and used for determining I.

The phenomena and the theory of the Zeeman effect of h.f. structures in weak fields and of the transition to the effects in strong fields, are in many respects analogous to the ordinary anomalous Zeeman effect in multiplets and its transition into the Paschen–Back effect in strong fields. The terms "weak" and "strong" have to be taken on different scales for h.f. and multiplet structures, so that a field of a few thousand gauss which is *strong* in its action on a h.f.s. is *weak* in its action on a multiplet structure.

The vector diagram fig. VI,11(a) shows the coupling conditions in a weak field. The vectors **J** and **I** precess jointly about the resultant **F** which in its turn precesses more slowly about the external field $\mathscr{H}$ assumed to be in the direction of the z-axis. With the substitution of **F**, **J** and **I** for **J**, **L** and **S**, the magnetic energy can be calculated exactly as in III.F.2 and 4 by taking components of the magnetic moments of nuclear spin and electron moment in the direction of **F** and then taking components in the direction of $\mathscr{H}$. In eq. (III.93), $\mu^{(L)}$ and $\mu^{(S)}$ have to be replaced by $\mu^{(J)} = -g_J J \mu_0$ and $\mu^{(I)} = g_I I \mu_0 m_0/M$ and in (III.101), γ_0 and $2\gamma_0$ by $g_J\gamma_0$ and $-g_I\gamma_0/1836$.

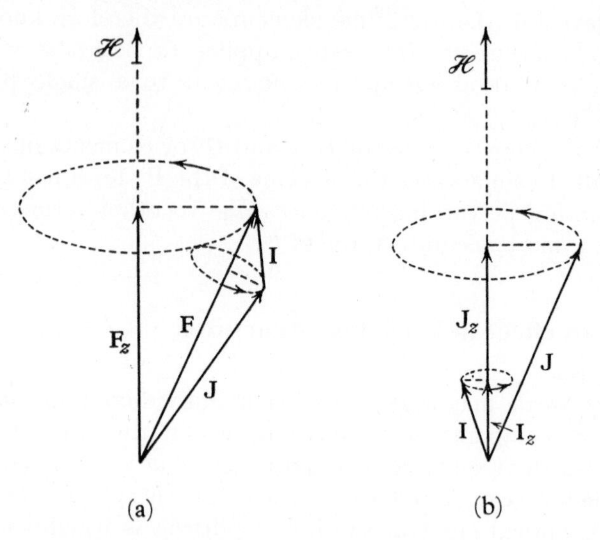

FIG. VI, 11. Vector diagram for (a) weak, (b) strong magnetic field.

The nuclear magnetic moment is, however, so much smaller than the electron moment that its interaction with the external field can be entirely neglected for most purposes. Substituting $g_J\gamma_0$ for γ_0 in the first term of (III.101), F, J, I for J, L, S and omitting the second term we obtain for the energy of interaction with the external field

$$W_m = -\mathscr{H}\mu_z = \mathscr{H} m g_F \mu_0/1836 \tag{42}$$

where

$$g_F = g_J \frac{F(F+1)+J(J+1)-I(I+1)}{2F(F+1)} \tag{43}$$

and m can have the values

$$m = F, F-1, \ldots -F.$$

The Zeeman levels are uniformly spaced, with energy differences of $\mathscr{H} g_F \mu_0/1836$. This spacing, and thus g_F, can be measured in atomic beam resonance in weak field transitions. I can be derived from g_F by means of (43) since g_J and J are known. One finds, e.g. for the h.f.s. level with $F = J+I$ that $g_F = g_J J/(J+I)$ from which I can be found. It is to be noted that the value of μ does not affect the splitting in weak fields, in the approximation considered.

b. Strong fields

As the magnetic effect, with increasing field strength, becomes comparable with the h.f.s. splitting and finally larger, the Zeeman effect undergoes a transformation. The resulting strong field effect is described as Paschen–Back effect of the h.f.s. or as *Back–Goudsmit effect.*[37] The spectral line in question appears split into the π- and σ-components of the anomalous Zeeman effect, and each of these components is again split into several equidistant h.f.s. components, as the result of the nuclear spin; one might say that the Zeeman effect of the h.f.s. has been transformed into a h.f.s. of the Zeeman effect.

The strong field quantisation is illustrated by fig. VI, 11b. The strongest interaction—apart from the interactions forming **J**—is now the magnetic interaction of **J** with the field, causing a rapid precession of **J** about $\mathscr{H}$, and the formation of equidistant anomalous Zeeman levels with quantum numbers $m_J = J, J-1, .. -J$. The next largest interaction is that of the nuclear magnetic moment $\boldsymbol{\mu}$ with the magnetic field $\mathscr{H}(0)$ due to the electrons; it will be remembered that this is of the order of 10^6 gauss and thus much larger than the external field $\mathscr{H}$. Owing to the precession of **J**, $\mathscr{H}(0)$ has a constant component $\mathscr{H}(0)_z$ and oscillating components $\mathscr{H}(0)_x$ and $\mathscr{H}(0)_y$. In calculating the interaction of **I** with $\mathscr{H}(0)$ which is the weaker one and thus causes the slower precession, we can average over the more rapid $\mathscr{H}$, J precession, so that $\overline{\mathscr{H}(0)_x} = \overline{\mathscr{H}(0)_y} = 0$. The strong coupling between **J** and $\mathscr{H}$ thus produces an effective field $\mathscr{H}(0)_z$ much stronger than $\mathscr{H}$. **I** precesses about this field, forming the quantum numbers

$$m_I = I, I-1, .. -1.$$

Relative to the Zeeman level without nuclear spin, the energy is given classically by

$$W_{\text{class.}} = -\mathscr{H}(0)_z\mu\cos(\mathscr{H},\mathrm{I}) = -\mathscr{H}(0)_J\mu\cos(\mathscr{H},\mathrm{J})\cos(\mathscr{H},\mathrm{I}) \tag{44}$$

and quantum-theoretically by the corresponding relation

$$W_m = -\frac{\mathscr{H}(0)\mu m_J m_I}{IJ} = Am_J m_I, \tag{45}$$

where the relation between $\mathscr{H}(0)$ and A has been taken from (7), and where the interaction between the external field and the nuclear magnetic moment has again been neglected.

Equation (45) shows that the h.f.s. of the strong field components consists of equidistant levels whose spacing is independent of $\mathscr{H}$. The total width of this structure is of the same order as the h.f.s. in the absence of a field.

The selection rule for m_I can be derived similarly to that for m_s in the Paschen–Back effect. In the approximation of electric dipole radiation, the orientation of the nuclear spin coupled firmly to the external field cannot make any contribution. To this approximation the electromagnetic radiation does not act on the nuclear spin, giving the selection rule

$$\Delta m_I = 0 \text{ (for el. dipole radiation).} \qquad (46)$$

Owing to (45) the h.f.s. of each anomalous Zeeman component consists of $2I+1$ equidistant lines; also their intensities are equal. Whenever this structure can be resolved it provides a very reliable method of determining I. It is entirely independent of the value of J, and it has therefore been possible to apply it to the ground levels of alkali atoms, with $J = \frac{1}{2}$.

Plate 17 and the diagram fig. VI, 12 show the transition from weak to strong fields for the h.f.s. of one of the resonance lines of sodium, 3s $^2S_{\frac{1}{2}}$ – 3p $^2P_{\frac{3}{2}}$.(29) The Doppler width was reduced by the use of an atomic beam of Na causing absorption of light from a sodium discharge tube. The high resolving power, about 5×10^6, was achieved by means of two Fabry–Perot etalons in series, of spacings 2 and 8 cm.

In the weakest fields, one of the h.f.s. components is seen to split into 3, the other into 5 Zeeman components. With increasing field strength, however, one of the latter gradually joins the other group, so that finally two groups of 4 equidistant components are formed showing that for sodium $I = \frac{3}{2}$. The non-linear dispersion of the etalon spectrum does not make the equality of the spacings immediately obvious. The measured positions of the components were found to be in close agreement with the theoretical calculation which will be discussed below.

In contrast to optical spectra, transitions in *atomic beam resonance*, and other types of paramagnetic resonance, are caused by magnetic dipole interaction with the radiation field, but it is the electron spin, or the vector **J** on which the magnetic field of the radiation acts, causing transitions $\Delta m_J = \pm 1$. The nuclear magnetic moment is so much weaker that its direct interaction with the alternating magnetic field can be neglected, so that the selection rule $\Delta m_I = 0$ holds, for the same reasons as those given for electric dipole transitions. Only in pure nuclear resonance experiments, of the Bloch–Purcell type or

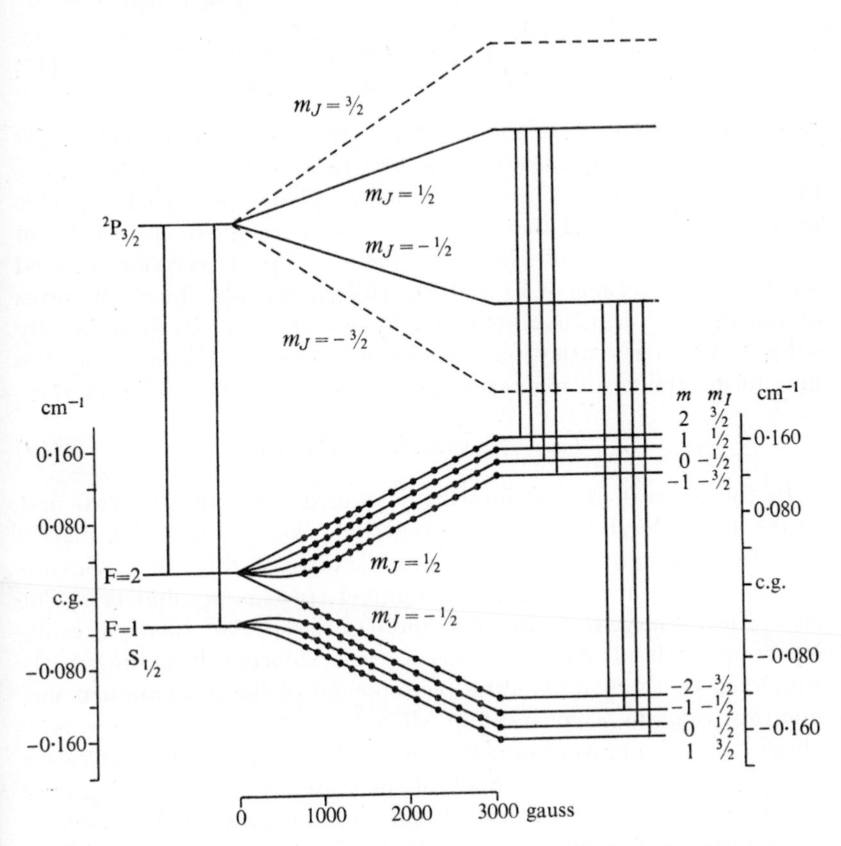

FIG. VI, 12. Conversion from weak- to strong-field pattern, quantitative term diagram of 3 $^2S_{1/2}$ – 3 $^2P_{3/2}$ of Na (I = 3/2); curves are calculated, dots measured values (D. A. Jackson & H. G. Kuhn, ref. 29).

in molecular beam resonance, are the transitions $\Delta m_I = \pm 1$ caused by direct interaction of the radio-frequency magnetic field with the nuclear magnetic moment. Radio-frequency resonance in atomic beams now forms the most accurate method of studying atomic energy levels in magnetic fields of different strength.

c. General theory for arbitrary fields

The quantum-mechanical treatment for the general case of intermediate fields follows so closely the lines of the theory described in III.F.4 that only a brief summary need be given here.

For any level defined by a fixed value of J, the perturbation

$$H' = -\frac{\mathscr{H}_0\mu}{IJ\hbar^2}\mathbf{I}\cdot\mathbf{J}+\frac{\mathscr{H}\mu_0 g_J}{\hbar}J_z\left[-\frac{\mathscr{H}\mu}{\hbar}I_z\right] \tag{47}$$

has to be applied where the last term can usually be neglected. In weak fields the eigenvalues of the first term of (47), of given F, are taken separately and the second term is applied as perturbation. This leads to the weak field Zeeman levels, each characterised by a value of $m = m_F$, with the results discussed before. This description is based on the eigenfunctions of $\mathbf{F}^2$ and F_z. In strong fields, the eigen-states of the second term, characterised by m_J and m_I, are individually subjected to a perturbation due to the interaction described by the first term; the resulting term values are, from (45) and (III.97):

$$T' = \mathscr{H}\mu_0 m_J g_J/hc + A' m_J m_I. \tag{48}$$

In the general case of intermediate field strengths the two first terms in (47) have to be treated together. The perturbation matrix is no longer diagonal in F but connects states of different F which is now no longer a good quantum number, just as m_J and m_I. The energy levels have to be found as roots of the secular equation resulting from both perturbation terms. The solution is again greatly simplified by the fact that the z-component of the total angular momentum remains a constant in all fields, so that $m_J+m_I = m$ is always a good quantum number. A secular equation can then be set up separately for each m. Each of the states $m = J+I$ and $-J-I$ occurs only once; the weak field and strong field wave functions become identical for these, and the energy is a linear function of $\mathscr{H}$.

For levels with $J = \frac{1}{2}$, such as the ground state of Na, there are two states for each value of m, except for the highest and lowest m, and the term values are roots of quadratic equations. The results can conveniently be expressed in terms of the parameter $x = \mathscr{H}\mu_0 g_J/(hc\Delta T)$ where ΔT is the width of the h.f. splitting without field, in cm^{-1}. The displacement of any level from the c.g. of the pattern is then given by the "Breit–Rabi formula"(38):

$$T_m = -\frac{\Delta T}{2(2I+1)}\pm\frac{\Delta T}{2}\sqrt{\left(1+\frac{4m}{2I+1}x+x^2\right)}. \tag{49a}$$

For $m = \pm(I+\frac{1}{2})$:

$$T_m = \frac{I}{2I+1}\Delta T \pm \tfrac{1}{2} g_J \mu_0 \mathscr{H}/hc. \tag{49b}$$

The interaction of the nuclear moment with the external field has again been neglected.

The value of (49b) becomes identical with (48), with the substitution from (8), for $J = \frac{1}{2}$.

Figure VI, 12 gives a quantitative diagram of the levels in intermediate fields for the ground state of Na. The dots mark the observed levels. The splitting of the $^2P_{\frac{3}{2}}$ level is not shown though it has been included in the calculation. For this level the field could be regarded as strong, and formula (48) could be used.

6. Structures due to several electrons

a. Introductory remarks

In those spectra in which more than one electron contribute to the h.f.s., the quantum numbers J, I and F retain their validity, with the exception of special cases to be discussed in section 7. The general structure of the h.f. multiplet is the same as that described in B.1–3 and by formula (39). The intensity rules still hold, and also the interval rule, provided that quadrupole interaction is negligible. The value of I can be deduced in the same way as for one-electron spectra.

Absolute values of μ and Q, however, can be determined from spectra with several electrons to only a small accuracy and by somewhat laborious methods. Ratios of these values, for different isotopes may be derived directly from spectra of any kind.

In the term analysis of complex spectra and especially for the assignment of configurations, the study of h.f. structures—and of isotope shifts—can be a valuable help.

The theory proceeds in two steps, the introduction of magnetic dipole and electric quadrupole interaction constants a and b for the individual electrons, and the calculation of the resulting factors A and B for the whole atom. The first of these steps tends to be a crude, semi-empirical estimate, and the second leads to rather complex mathematics, except in a few simple cases.

b. Ideal coupling conditions[39]

We assume that the coupling between the electrons can be described by a simple scheme, in particular the Russell–Saunders scheme where a level is characterised by L and S as well as J, though this assumption is not often well justified. Mainly magnetic interactions will be discussed.

For each electron outside closed shells or sub-shells a coupling constant a is defined as in (7). Its value can be estimated by means of the Goudsmit–Segré formulae, if the necessary data are available. The h.f.s. constant A_J for the whole atom can be derived most simply by using the vector diagram, taking components and introducing the usual modifications for small quantum numbers. In view of the small accuracy required, one generally neglects the direct influence of all except s-electrons. The procedure may be illustrated by the example of a two-electron configuration where the first is an s-electron and the second a p-, d-, f- . . electron.

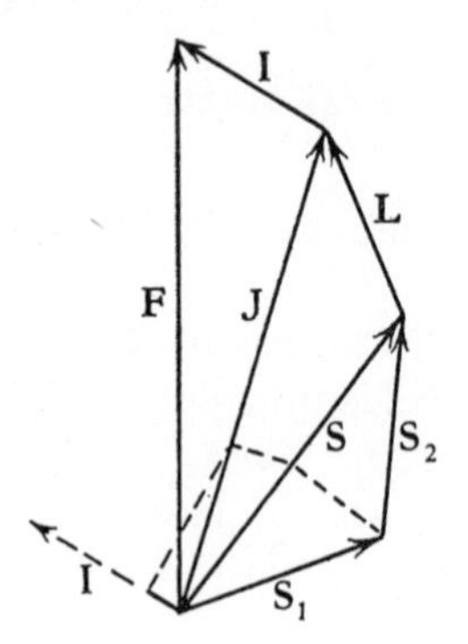

Fig. VI, 13. Vector diagram of h.f.s. coupling in two-electron spectrum.

Figure VI, 13 shows the vector diagram in which $\mathbf{S}_1$ and $\mathbf{S}_2$ have to be imagined as precessing very rapidly about $\mathbf{S}$, the latter and $\mathbf{L}$ less rapidly about $\mathbf{J}$, and $\mathbf{J}$ and $\mathbf{I}$ still less rapidly about $\mathbf{F}$. For finding the magnetic energy of the nuclear spin due to the field of the s-electron it is necessary to calculate the average value of $\cos(S_1, I)$. By successively averaging over precessions in order of decreasing speed, we find

$$W_F = a_S{}^{(1)}(|\mathbf{I}||\mathbf{S}_1|(|\mathbf{I}||\mathbf{S}_1|/\hbar^2) \cos(\mathrm{S}_1, \mathrm{S}) \cos(\mathrm{S,J}) \cos(\mathrm{I, J}). \quad (50)$$

When the cosines are expressed by the sides of triangles, and with the usual substitutions of $F(F+1)$ for F^2, etc., the term value can be written, for $l_1 = 0$, $l_2 = L$,

$$W_F = \tfrac{1}{2}A_J[F(F+1)-I(I+1)-J(J+1)],$$

$$A_J = a_s\frac{J(J+1)+S(S+1)-L(L+1)}{2J(J+1)} \times \frac{S(S+1)+s_1(s_1+1)-s_2(s_2+1)}{2S(S+1)}$$

For $s_1 = s_2 = \frac{1}{2}$, this formula gives the following result for the three triplet levels ($S = 1, J = L+1, L, L-1$) and the singlet level ($S = 0$):

	$^3X_{L+1}$	^3X_L	$^3X_{L-1}$	^1X_L
$A_J =$	$a_s/2(L+1)$	$a_s/2L(L+1)$	$-a_s/2L$	0

where X stands for the symbols S, P, D, . . for $L = 0, 1, 2$ . . and a_s is the splitting factor for the s-electron. For $L = 1$, we find $A_J = a_s/4$ for both levels $^3P_2$ and $^3P_1$.

In alkaline earth spectra, the ground state $s^2\ ^1S_0$ cannot have any structure since $J = 0$. According to the above relations, also all other singlet levels should have no h.f.s. In fact they show only very narrow structures, due either to the effect of the electron with $l > 0$ or to deviations from L, S-coupling.

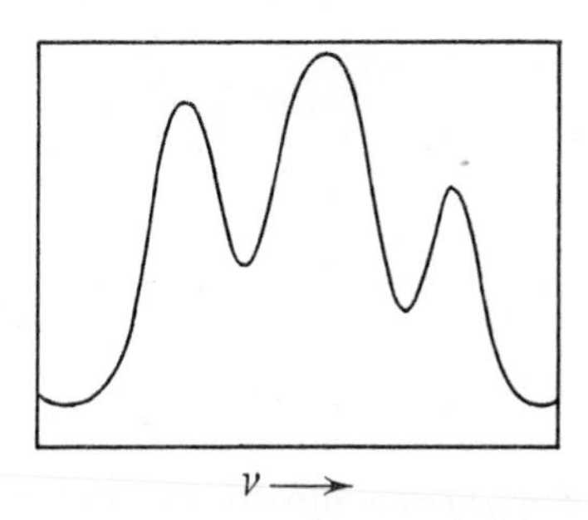

FIG. VI, 14. H.f.s. of line 6103 A in Ca^{43}, enriched isotope sample (F. M. Kelly, H. G. Kuhn & Anne Pery, ref. 40).

All triplet terms of the simple two-electron spectra have the configuration sl where the s-electron is in its lowest possible state and l stands for s, p, d, . . . Their h.f.s. is mainly due to the low s-electron and is therefore of the same order of magnitude for all terms.

In measurements of h.f. structures in the spectra of the alkaline earth elements, a difficulty arises which is common to all elements of even atomic number. All these elements have several abundant isotopes of even mass number; they have no nuclear spin and give rise to a strong central peak in the h.f.s. pattern which often prevents the complete resolution of the structures in the odd isotopes, even in enriched samples. Figure VI, 14 shows an example of this in the structure of the Ca line λ6103 (4s4p $^3P_0 - $ 4s5s $^3S_1$). It is a photometer tracing[40] of a Fabry-Perot pattern. The source was a hollow cathode discharge, cooled with liquid hydrogen and containing $\frac{1}{2}$ mg of Ca in which the concentration of the naturally rare isotope 43 had been increased to 75 per cent. The central component of the h.f. triplet appears much too strong and consists in fact of a blend of the h.f. component of Ca43 with the line due to residual Ca40.

For perfect j, j-coupling, the splitting factor A_J can be calculated from the a factors of the individual electrons by methods similar to those used for L, S-coupling. Assuming again the first to be an s-electron, the second a p-, d- . . . . electron, and neglecting the direct interaction of the latter with the nucleus, one finds

$$A_J = a_s \frac{J(J+1)+s_1(s_1+1)-j_2(j_2+1)}{2J(J+1)}. \tag{52}$$

For the configuration sp, the values of A_J for the two groups of levels are

for $j_2 = l+\frac{1}{2}$

$$J = l+1; \quad A_J = a_s/2(l+1); \qquad J = l: \quad A_J = -a/2(l+1)$$

for $j_2 = l-\frac{1}{2}$:

$$J = l-1: \quad A_J = -a_s/2l; \qquad J = l: \quad A_J = a_s/2l.$$

c. Sum rules and quantum-mechanical methods

In many practical cases the coupling conditions conform neither with the L, S- nor the j, j-scheme to any degree of accuracy, and more general methods have to be employed for deriving A_J from the a-values of the individual electrons.

A simple procedure often leading to partial solutions is the use of *sum rules*[41] the principle of which was explained in V.A. Though first developed by Pauli before the advent of quantum mechanics, it can be most easily understood in terms of matrix methods: the diagonal sum of a matrix is always equal to the sum of the eigenvalues, independently of the scheme of representation. Again, as shown in V.A.8, the sum need only be extended over the sub-matrix defined by one value of any quantum number which is a good quantum number in both schemes used.

For h.f. structures, we consider the following coupling schemes: (a) that in which the energy is diagonal in a magnetic field strong enough to produce Back–Goudsmit effect, but not strong enough to cause Paschen–Back effect. It is defined by the quantum numbers J, m_J, m_I; (b) the scheme in which states are described by m_I and the quantum numbers j_1, m_{j_1}, j_2 m_{j_2} . . of the individual electrons.

This would make the energy diagonal only for j, j-coupling in a strong magnetic field; that this condition is rarely met in practice does not affect the validity of the method.

In both cases I is uncoupled from J or the j_k, so that in both schemes m_I and $m_J = \Sigma\, m_{j_k}$ are good quantum numbers corresponding to classically constant angular momenta. For any fixed pair of values m_I, m_J the sum of the diagonal elements of the energy will be the same. Using (45) we find

$$m_J \sum A_J = \sum m_{j_1} a_{j_1} + m_{j_2} a_{j_2} + \cdots, \tag{53}$$

where the common factor m_I has been omitted. The sum on the left has to be taken over all A_J of the same J, and that on the right for all combinations for which $m_{j_1} + m_{j_2} + \ldots = m_J$.

TABLE 5

Sum rule for configuration sp

$L = 1,\ S = 0,\ 1$	$l_1 = 0$	$l_2 = 1$
$J = 2,\ 1,\ 1,\ 0$	$j_1 = \frac{1}{2}$	$j_2 = \frac{1}{2},\ \frac{3}{2}$
m_J	m_{j_1}	m_{j_2}
2	$\frac{1}{2}$	$\frac{3}{2}$
1	$\frac{1}{2}$	$\frac{1}{2}$
1	$\frac{1}{2}$	$\frac{1}{2}$
1	$-\frac{1}{2}$	$\frac{3}{2}$

The configuration sp may be chosen as a simple example. Table VI, 5 gives the possible positive values of m_J. For $m_J = 2$, the sum has only one term:

$$m_J \sum A_J = 2A_2 = \tfrac{1}{2}a_s + \tfrac{3}{2}a_{p_{3/2}}; \qquad A_2 \approx \tfrac{1}{4}a_s.$$

For $m_J = 1$, there are three possible pairs m_{j_1}, m_{j_2}:

$$m_J \sum A_J = 1(A_2 + A_1 + A_1') = \tfrac{1}{2}a_s + \tfrac{1}{2}a_{p_{1/2}} + \tfrac{1}{2}a_s + \tfrac{1}{2}a_{p_{3/2}} - \tfrac{1}{2}a_s + \tfrac{3}{2}a_{p_{3/2}};$$

$$A_2 + A_1 + A_1' \approx \tfrac{1}{2}a_s.$$

Since the negative values of m_j give the same results, and the relation for $m_j = 0$ vanishes identically, all the information that can be obtained from the sum rule, when the direct effect of the p-electron is omitted, is:

$$A_{J=2} = \tfrac{1}{4}a_s.$$

$$A_{J=1} + A'_{J=1} = \tfrac{1}{4}a_s,$$

together with the trivial fact that the level $J = 0$ has no structure.

The value of $A_{J=2}$ agrees with that found for L, S-coupling and for j, j-coupling, and the sum of the A values for $J = 1$ agrees with the above result for both types of coupling.

If the splitting factor is measured for $J = 2$ and one of the two possible levels $J = 1$, the other can be found by means of the sum rule.

Various other methods have been used for calculating splitting factors from those of the individual electrons. Breit and Wills[42] and Schmidt[43] use specific wave functions, while Trees[44] has applied Racah's more general techniques of tensor operators to the calculation of h.f. structures. The same methods have been used for calculating quadrupole interaction constants B. The most general treatment using Racah's method and including interactions with higher nuclear moments is due to Schwartz.[45]

7. Perturbations, forbidden lines and anomalous coupling

In chapter V.A.12 certain irregularities have been discussed which occur in the arrangements and strengths of lines within a series or within a multiplet. Such *perturbations* also occur in h.f. structures. The regular structures of series and multiplets are interpreted by the theory in terms of certain interactions expressed for every level by a definite set of quantum numbers; the latter are associated with definite constants of the classical motion, such as angular momenta of electrons or groups of electrons and are based on certain approximations. Whenever two energy levels happen to be close together, these approximations may break down completely, and irregularities described as perturbations occur. No clear distinction can be made, however, between obvious perturbation phenomena and mere inaccuracies in the energies of levels or transition probabilities derived from first-order approximations. Such inaccuracies can often be accounted for by higher orders of approximation involving other energy levels of the zero approximation system in the form of non-diagonal elements.

Two levels can perturb one another if the following conditions are fulfilled:

(i) the levels have the same parity;
(ii) they have the same value of F;
(iii) their values of J differ by 0 or ± 1.

The perturbation shows itself in a mutual "repulsion" of the energy levels and in changes of the transition probabilities; in particular, transitions $\Delta J = \pm 2$ can occur as dipole radiation.

Mathematically, the perturbation is described by the influence of non-diagonal elements of the matrix of the magnetic interaction with the nucleus: the operator $W = -\overline{\mathscr{H}_0} \cdot \boldsymbol{\mu}$ in (6) has, apart from the diagonal elements, given in (7) also non-vanishing elements

$$\langle n, J, F \| \overline{\mathscr{H}_0} \cdot \boldsymbol{\mu} \| n', J', F \rangle$$

where $J' = J$ or $J \pm 1$. The solution of the secular equation leads to eigenvalues differing from the diagonal elements $W_{J,F}$. If this difference is δ and the total h.f.s. contribution to the energy is W_F, the ratio δ / W_F is of the same order of magnitude as the ratio of the h.f.s. splitting to the difference between the perturbing terms.

Deviations from the interval rule are the most striking manifestations of perturbations. They occur when multiplet levels of J differing by 0 or 1 lie close together and have the same parity.

An example of this is found in two levels of Hg,[46] both arising from the configuration 6s 6d, with $J = 2$ and $J = 1$. The coupling is intermediate between L, S- and j, j-coupling, so that $^1D_2$ and $^3D_1$ would not be an accurate description. The h.f.s. levels of the isotope Hg^{201} ($I = \frac{3}{2}$) show a distinct deviation from the interval rule in the level $J = 2$ (see fig. VI, 15): the distance $\frac{7}{2} - \frac{5}{2}$ is evidently much too small. This can be understood from the assumption that the levels $F = \frac{1}{2}, \frac{3}{2}, \frac{5}{2}$ with $J = 2$ are depressed by the influence of the levels of the same F with $J = 1$, while the level $F = \frac{7}{2}$ has no counterpart with $J = 1$ and remains unperturbed.

The order of magnitude of the effect can be estimated by the rule given, from the width of the h.f.s. splitting and the difference of the terms $J = 2$ and $J = 1$ and agrees with the observed facts. The ratio of the displacements of the levels with different F can be derived from the known dependence of the matrix elements on F (ref. III, 101), but the measurements are hardly accurate enough to allow a comparison with the theory.

The perturbation by the level $J = 1$ also causes transitions to become allowed which would be forbidden for a pure state with $J = 2$. The line due to the transition from 6s6d ($J = 2$) to 6s6p $^3P_0$ was observed[47] to appear only in the odd isotopes of Hg. It showed, for the isotope 201, the three h.f.s. components from the levels $F = \frac{1}{2}, \frac{3}{2}, \frac{5}{2}$.

The transition probability of a line caused by perturbation, as a fraction of the transition probability from the perturbing level, is of the same order of magnitude as the ratio of the level shift due to perturbation to the distance from the perturbing term.[48]

The appearance of similar forbidden lines as the result of h.f.s. perturbation has also been studied in Cd.[49]

Perturbations by levels of the same J need not cause deviations from the interval rule but can affect the width of splitting, i.e. the value of A_J. The perturbing level may belong to the same configuration, but more striking effects occur when it belongs to a different configuration, i.e. by *configuration interaction.*

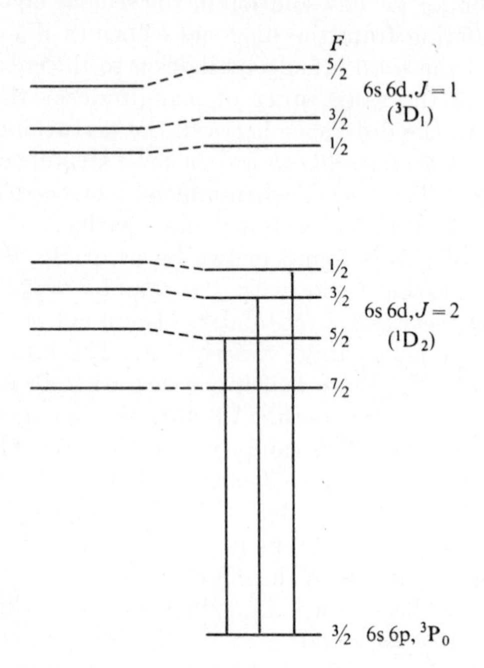

FIG. VI, 15. Perturbation of h.f.s. levels in Hg.

A very small admixture of a configuration having a strong h.f.s. interaction can radically change the h.f.s. of a term whose principal configuration would produce only a very narrow structure. This fact presents one of the most serious difficulties in investigations of h.f. structures of the higher excited levels in any but the simplest spectra.

Even ground states can be strongly affected by configuration interaction. This is shown in the elements of the earth type. The most thoroughly studied example is the ground term of Ga, $4s^2\ 4p\ ^2P_{\frac{1}{2},\frac{3}{2}}$. According to (15) the splitting factors of the levels $J = \frac{1}{2}$ and $J = \frac{3}{2}$ should be in the ratio 5 : 1. Relativistic corrections increase this value to 5·41, while measurements have shown the ratio to be 7·02. By the use of approximate wave functions Koster[(50)] showed that the discrepancy was due to admixtures of the two terms ^{2}P of the configuration 4s 4p 5s, contributing together $2\frac{1}{2}$ per cent of the probability density of the ground term. The contribution of this admixture raised the ratio of the splitting factor to 6·67, in sufficiently good agreement with the experimental value. The corrected eigenfunction could be used for calculating the nuclear quadrupole moment from the constant B derived from the observed structure of $P_{\frac{3}{2}}$.

Another interesting example of a ground state in which the simple configuration symbol cannot account for the h.f.s. is that of Mn, $3d^5 4s^2\ {}^6S_{\frac{5}{2}}$. In the lowest term of a half-filled shell such as d^5, the h.f.s. constant A_J must be expected to vanish completely, owing to the symmetry in the relative orientation of the orbital angular momenta. The spherically symmetrical probability density of a half-filled shell (see p. 96) and the fact that the spins of all electrons are parallel causes the effect of a uniformly magnetised shell in the centre of which the magnetic field strength vanishes.* In fact, the ground term of Mn shows a small h.f.s. splitting which obeys the interval rule to a high degree of accuracy.[(51)] Also the ground term of Mn^{++}, $3d^5\ {}^6S_{\frac{5}{2}}$, has a h.f.s.

The explanation[(52)] appears to lie in exchange effects between the 3d electrons whose spins are not paired and the paired 2s-, 3s- and (for Mn) 4s-electrons. This results in a slight difference between the wave functions of the s-electrons whose spin is parallel and those whose spin is antiparallel to the spins of the d-electrons. It is the same effect which accounts for part of the doublet splitting of the $2\ {}^2P-$ term of Li (see p. 168). The effect was found to be of about the right magnitude for both Mn and Mn^{++}. The lowest term which could cause perturbations is 17,000 cm^{-1} above the ground level, and the effect would be too small to account for the observed splitting.

The assumption that the h.f.s. interaction is small compared with other interactions breaks down entirely in some anomalous coupling conditions.

If in a spectrum of the alkaline earth type one electron is excited to levels of increasing n and l, the magnetic interaction causing the triplet structure becomes very small while the h.f.s. interaction, caused by the low s-electron, remains approximately the same. The condition can then be reached where the h.f.s. interaction is

* The formula $\mathbf{B} = 8\pi\mathbf{P}/3$ (p. 331) holds again, but $\mathbf{P}_0 = 0$ except for s-electrons.

comparable with the L, S-interaction and finally larger than the latter and even larger than the electrostatic interaction causing the singlet–triplet term difference.

Such anomalous coupling conditions have been treated theoretically by Güttinger and Pauli[(53)] and have been studied in the spectrum of Al II by Paschen and his co-workers and more completely by Suwa.[(54)]

The anomalies become apparent in the lines $4\,^3P - 4\,^3D$ and $4\,^3P - 5\,^3D$ and are extreme in the series $4\,^3F - nG$ where the coupling between the nuclear spin and the spin of the s-electron is predominant. The observed patterns agree well with the results of the theory.

Similar anomalies have been observed in the triplet terms $1s\,nd\,^3D$ of the isotope of helium of mass number 3.[(55)] The multiplet structure is very narrow while the h.f.s. due to the 1s-electron is comparatively large. The large Doppler width unfortunately only allowed partial resolution of the patterns, but the agreement with the theory was very good.

Table App. 4 gives a list of values of I, μ' and Q as derived from hyperfine structure work and by means of other methods. Some doubtful values are included, and many have been rounded off. For details and literature the reader may consult the article ref. (36).

C. ISOTOPE SHIFT*

1. Displacement of lines and of levels

In studying the facts and causes of isotope shifts, one has to reduce the observed displacements of spectral lines to displacements of levels or terms, and this raises a point of definition. In comparing the energies of two different mechanical systems, namely two atoms with different nuclei, we must define a state for which we count the energies as equal. This can, e.g. be the state where all electrons are at rest at infinity; with a reference level thus defined, the isotope shift of a level of an atom with several electrons is the difference, for the two isotopes, in the energies released in bringing up the electrons from infinity. This will include the contribution of all electrons, those in closed shells as well as those outside, to the isotope shift. More conveniently one defines as reference level of equal energy of two isotopes the state in which only one of the outer electrons has been removed to infinity. In order to derive the isotope shift of a term, on

* See also: H. Kopfermann, *Nuclear Moments*, Acad. Press Inc. 1958.

this definition, one has to measure the isotope shift of several lines in a series and extrapolate to the series limit. This can be omitted if there is reason to believe that the isotope shift of one of the levels of a transition is negligibly small, so that removal of the outer electron from this level requires the same energy for both isotopes. An example will be discussed in the section on mass shifts, though the problem of a reference level arises quite independently of the causes of the shift.

It has to be remembered that the isotope shift observed in a spectral line can be regarded as the difference of the displacements of the lines of each isotope from some fictitious position it would have if the cause of the effect were made to vanish, and that this displacement of the line of any one isotope arises from the difference of the displacement of the two levels.

2. Mass shift

The simplest example of an isotope shift, caused by the motion of the nucleus owing to its finite mass, has been described in III.A; in atoms or ions containing only one electron, the term value is reduced in the ratio $1/(1+m/M) \approx 1-m/M$. For two isotopes, whose masses differ by ΔM, the relative term difference is given by

$$\Delta T/T = \Delta[1/(1+m/M)] \approx -\Delta(m/M) \tag{54}$$

and the same relation applies to the relative wave-number difference of the lines $\Delta\tilde{\nu}/\tilde{\nu}$.

Since the displacement due to the mass effect reduces the term value (raises the energy), and since this displacement decreases with increasing M, the line due to the *heavier* isotope has the *greater wave-number*.

Also for the lighter atoms with several electrons, isotope shifts of similar order of magnitude are observed, and they too are ascribed to the fact that the nucleus takes part in the orbital motion. Their magnitude is, however, often noticeably different from the *normal mass shift* defined by (54), so that the view that only the motion of *one* electron determines the isotope shift must be regarded as too crude.

If we assume the centre of gravity to be at rest and the electrons in an atom to have the momenta $\mathbf{p}_1$, $\mathbf{p}_2$ . . , the momentum of the nucleus is equal to their negative sum, so that its kinetic energy is

$$T_N = \frac{(\mathbf{p}_1+\mathbf{p}_2+\ \ldots)^2}{2M} = \mathbf{p}_1{}^2/2M+\mathbf{p}_2{}^2/2M+\ \ldots$$
$$+\mathbf{p}_1\cdot\mathbf{p}_2/M+\mathbf{p}_1\cdot\mathbf{p}_3/M+\ \ldots$$

If, in quantum mechanics, T_N and the $\mathbf{p_k}$ are taken as operators, the Schrödinger equation can be solved easily if the mixed terms are omitted. The square terms, together with the corresponding terms of the kinetic energy of the electrons, form a term $\mathbf{p_k}^2/2\mu$ for each electron, where μ is the *reduced mass*, and the energy eigenvalues are rigorously equal to those for infinite M, multiplied by $M/(M+m)$; this gives the normal mass shift (54).

The mixed terms, which depend on the relative direction of the motion of the individual electrons, are more difficult to evaluate. They can be treated as perturbation terms

$$-(\hbar^2/M)\sum_{k,j}\nabla_k\nabla_j$$

in the Hamiltonian, and the first-order perturbation energy is found to be

$$\Delta W = \frac{\hbar^2}{M}\sum_{k,j}\int \psi^*\nabla_k\nabla_j\psi\,\mathrm{d}\tau. \tag{55}$$

This is called the *specific mass shift.*

Hughes and Eckart[56] have calculated the specific shift for some simple atoms with the assumption that ψ is a linear combination of products of single-electron eigenfunctions. For two electrons, this approximation gives

$$\Delta W = \pm(\hbar^2/M)\left|\int u^*\nabla v\,\mathrm{d}\tau\right|^2. \tag{56}$$

ΔW vanishes except for electrons whose eigenfunctions u and v have values of l differing by ± 1. In the spectra of He and Li^+, with one 1s-electron, only the terms 1s np ^{1,3}P have a specific mass shift in the approximation of single-electron eigenfunctions. The + sign in (56) applies for the eigenfunction which is symmetric in the space coordinates and antisymmetric in the spins, i.e. for singlet terms. The specific effect then increases the normal effect (raises the energy level); the electrons have to be imagined as moving in phase with one another. In the triplet terms, for which the − sign applies, the electrons tend to move in opposite directions, and the mass shift is reduced.

If hydrogen-like wave functions with effective nuclear charge numbers Z_i are used for u and v, Z_1 for the inner (1s-) and Z_2 for the outer (np-) electron, the specific term shift becomes

$$-\Delta W/hc = \Delta T = \mp\frac{128}{3}\frac{m}{M}R_\infty Z_1^5 Z_2^5 n^3(n^2-1)\frac{(Z_1n-Z_2)^{2n-4}}{(Z_1n+Z_2)^{2n+4}} \tag{57}$$

where it is to be noted that T is opposite in sign to W.

The isotope shift in Li^+ was first measured by Schüler[57] in the line 5485 A, 1s2s $^3S_1 - $ 1s2p $^3P_0$ and found to be 1·14 cm^{-1}. For the normal mass effect we find

$$\Delta\tilde{\nu} = \tilde{\nu}\Delta(m/M) = 18230(1/6-1/7)/1836 = 0{\cdot}24 \text{ cm}^{-1},$$

in disagreement with the observed value. The specific effect acts only on the P-term and, though causing a shift of the term of less than the normal effect, it makes a comparatively large contribution to the shift of the line. Determining Z_1 and Z_2 by a variation method minimising the energy of the state, Hughes and Eckart find $Z_1 = 2{\cdot}98$, $Z_2 = 2{\cdot}16$ and a specific term shift of 0·85 cm^{-1}. The total calculated shift of 1·09 cm^{-1} agrees fairly well with the experimental value. Figure VI, 16 shows the relative positions of the levels, with the two stages of displacement, by the normal and the specific effect.

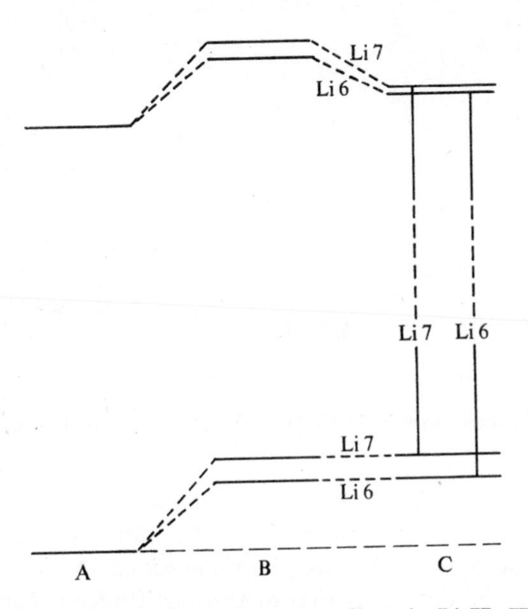

FIG. VI, 16. Isotope shift due to mass effect, in Li II (H. Kopfermann, *Nuclear Moments*, Acad. Press Inc. 1958).

For the three-electron configuration $1s^2np$ the formula (57) applies without modification, with the + sign. On account of the Pauli principle, only one of the 1s-electrons contributes to the specific effect. For the term $1s^22p$ ^2P of Li the specific shift according to (57) is 0·08 cm^{-1}. For the resonance lines 2 $^2S - 2$ ^2P this has to be added to the normal shift of 0·20 cm^{-1}. The resultant theoretical shift of 0·28 cm^{-1} is to be compared with the experimental value of 0·35 cm^{-1}.[58] The addition of the specific shift thus improves the agreement but the remaining discrepancy is well outside the experimental limits of error.

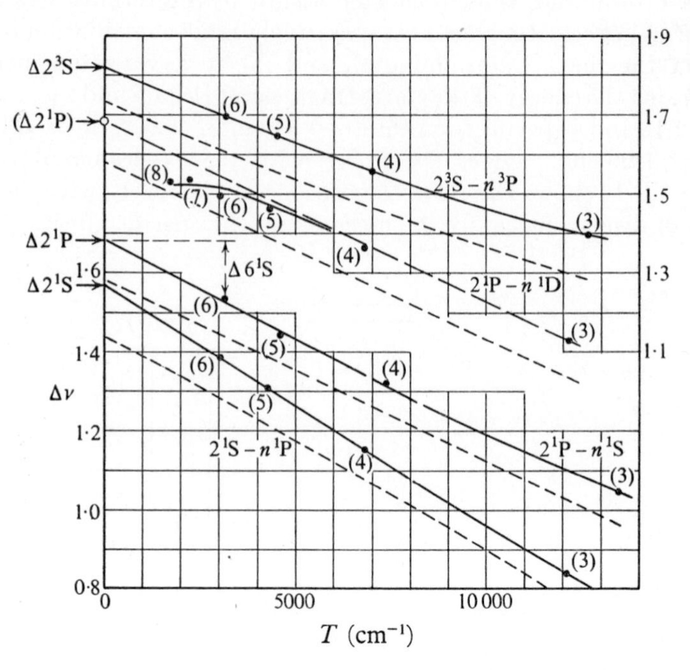

FIG. VI, 17. Isotope shift in He (L. C. Bradley & H. G. Kuhn, ref. 59).

More extensive studies of isotopic mass shifts in two-electron spectra have been carried out in He, for the isotopes of masses 3 and 4.[59,55] Measurement of shifts in successions of lines in several series allowed the determination of term shifts by extrapolation to the series limit, essentially independently of any theory. Figure VI, 17[59] shows the isotope shifts of the lines plotted as function of the term value T_n of the upper state. The extrapolation for $T_n = 0$, i.e. $n = \infty$, gives the shifts of the lower terms; the shifts of the serial terms T_n can be read on the graph as vertical distances from the limit as indicated for Δ6 ¹S. The n ¹D terms are perturbed by the n ³D terms so that no reliable extrapolation is possible for the series 2 ¹P − n ¹D.

The dotted lines are calculated from normal mass shift with addition, for P-terms, of the specific shifts from (57). The results show distinct deviations from the Hughes–Eckart theory. Even the S- and D-terms which should only have the normal shift show an additional specific shift.

For some of the terms of helium, isotope shifts have now been calculated more rigorously with the use of variational wave functions;[60, 61] in table VI, 6, the results are compared with the observed values.[59, 55] The improvement on the theory of Hughes and Eckart (last column) is very striking.

TABLE 6

Isotope shift in He (in cm^{-1})

	obs. shift	Specific shift		
		observed −normal mass shift	refined theory	Hughes-Eckart
$2\,^1S$	1·57	0·13	0·15	0
$2\,^3S$	1·82	0·10	0·076	0
$2\,^1P$	1·68	0·46	0·454	0·36
$2\,^3P$	1·08	−0·64	−0·636	−0·56

Isotope shifts have been measured in a number of the lighter elements, and quite generally it has been found that the theory of Hughes and Eckart in its original form can be regarded as not more than a useful first approximation. Calculations with more accurate wave functions could make a valuable contribution to the subject.

The specific mass effect undoubtedly decreases rapidly with increasing atomic number, and for the elements above the middle of the periodic table it is probably safe to assume that it is of the order of magnitude of the normal mass effect. There are, however, exceptional cases where the mass effect is unexpectedly large, e.g. in some lines involving transitions of d-electrons in Zn II and Cu.[62] Unfortunately, no simple rules appear to exist which would allow even the crudest estimates of the magnitude of the specific shift in elements with many electrons.

Whatever the method of calculation, the resulting term shift due to mass effects must be proportional to $1/M$. In isotopic sequences A, $A+1$, $A+2$, .. or A, $A+2$, $A+4$, .., the following relations must hold:

$$\Delta(A, A+1)/\Delta(A+1, A+2) = 1+2/A$$

and

$$\Delta(A, A+2)/\Delta(A+2, A+4) = 1+4/A; \tag{58}$$

isotope shifts due to mass effects thus change linearly and, for $A \gg 1$, very slowly in a sequence of isotopes.

3. Volume shift

a. The principal facts

While isotope shifts decrease with atomic number A, for small values of A, and become very small in the range of $A \approx 30$ to 50, the heavier elements show large shifts in some of their spectral lines. It became soon evident that these shifts could not be ascribed to mass effects. As Pauli and Peierls(63) first suggested they are caused by the finite size of the nuclear charge distribution which differs for different isotopes and causes a slightly different electrostatic field to act on the electrons in the immediate vicinity of the centre of attraction. The quantitative formulation of these views is due to Breit and Rosenthal(64) and forms the basis of the understanding of this so-called *volume effect* or *field effect*. Since the influence of the *shape* of the nucleus on the effect was recognised by Brix and Kopfermann(65) it has become a valuable method for studying deformations of nuclei as function of the neutron number.

The most striking features of the volume shifts are the following:

(i) The magnitude of the shift depends mainly on the *configuration* of the two levels, so that it is practically the same for different lines of a multiplet. Appreciable shifts are observed only when the number of s-electrons differs in the two terms. This leads to the assumption of a term shift caused by s-electrons.

(ii) The sign of the shift is obtained correctly on the assumption that the s-electron raises the level of the heavier isotope relative to the lighter one.

(iii) The lines of the even isotopes are always arranged in the order of their mass numbers, and the separations tend to be of the same order of magnitude for different isotopes in one line. They are not, however, strictly equal but can show trends and even abrupt changes within the sequence of isotopes. The ratios of the shifts between different isotopes are generally the same or nearly the same in different lines of the same element.

(iv) The lines of the odd isotopes—or the centroids of their h.f.s. patterns—do not lie midway between the adjacent even isotopes, but are shifted towards the isotope of lower A. This effect which occurs to a varying degree in different elements is known as *odd–even staggering*.

Isotope shifts are generally small and are difficult to measure accurately. The earlier work was mainly confined to the very heavy elements where the shifts tend to be large and the Doppler width is small. Samples of elements enriched in any one particular isotope have become available only in recent years, but even with the use of such samples the accurate measurement of isotope shifts is laborious. Not many measurements exist in which the ratio of shifts, for pairs of isotopes, has been determined to an accuracy better than 10 or 20 per cent.

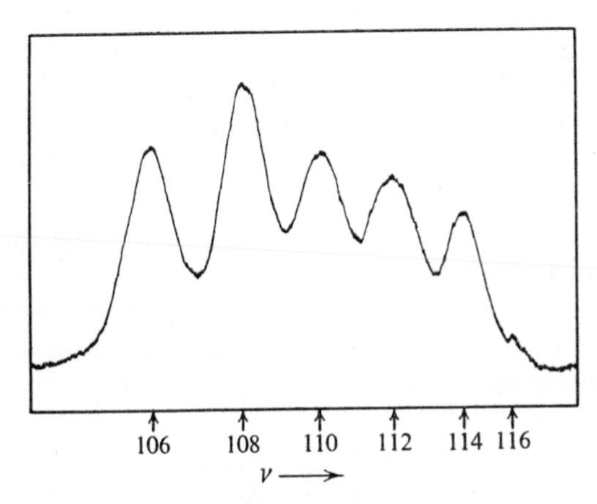

FIG. VI, 18. Isotope shift in the line 4416 A of Cd II, with enriched isotopes (H. G. Kuhn & S. A. Ramsden, ref. 66).

Plate 18 shows an example of the very large shifts found in the heaviest elements. The narrowness of the h.f.s. in the line chosen makes the odd-even staggering directly visible.

As an example of a comparatively large isotope shift, fig. VI, 18[(66)] shows the photometer tracing of a Fabry–Perot spectrogram of the Cd II line λ 4416 ($4d^{10}\ 5p\ ^2P_{\frac{3}{2}} - 4d^9\ 5s^2\ ^2D_{\frac{5}{2}}$). In this transition *two* s-electrons are converted into p- and d-electrons respectively, thus causing a large shift, as first observed by Schüler and Westmeyer.[67] The spectrogram shown here was taken with a liquid air-cooled hollow-cathode discharge tube containing 1 mg of cadmium enriched in the rarer isotopes 106 and 108. The sample contained enough of the abundant isotopes 110, 112 and 114 to give strong fringes, and even the isotope 116, present in only small quantity, is just visible.

In the study of volume shifts it is important to establish whether any appreciable part of the shift is due to mass effects, other than the normal effect which can easily be allowed for. This is especially necessary for elements of only moderately large mass numbers ($A < 100$) and for lines showing a small total shift. If lines of different multiplicity but of the same configurations show the same isotope shift, the specific mass effect is not likely to be significant. Also if the ratio of shifts for two pairs of isotopes differs appreciably from (58) and is found to be the same in different lines, the contribution of mass effects can be assumed to be small.

b. Theory for spherical nuclei

The original theory[64] assumed that the nucleus has a spherically symmetrical charge distribution. Different radial distributions have been discussed, but we confine ourselves to the assumption of uniform charge distribution within a sphere of radius r_0. The electrostatic potential acting on the electron is then given by the full line in fig. VI, 19 which, for $r < r_0$, runs above the dotted line representing the potential of a point charge. If the difference of these two potentials is $\delta V(r)$, the corresponding difference in the energy of an electron charge cloud of spherically symmetrical charge density $\rho_e(r)$ is

$$\delta W = \int_0^{r_0} \rho_e(r) \delta V(r) 4\pi r^2 \, dr. \tag{59}$$

This assumes $\rho_e(r)$ to be the same for both potential curves so that δW represents the first-order perturbation energy. For two different isotopes of mass numbers A and $A + \Delta A$ the potential differs by $\Delta\delta V$ and the first-order energy difference is

$$\Delta\delta W = \int_0^{r_0} \rho_e(r) \Delta\delta V(r) 4\pi r^2 \, dr. \tag{60}$$

The value of $\rho_e(0)$ and therefore the isotope shift vanishes for all except s- and $p_{\frac{1}{2}}$-electrons, and it is usually negligibly small for the latter.

An estimate of orders of magnitude shows that δW can have values of as much as 100 cm^{-1} in typical cases, but since atoms with point nuclei do not exist it is not an observable quantity. The observable isotope shift $\Delta\delta W$ is of the order of magnitude of 0·005 to 1 cm^{-1}.

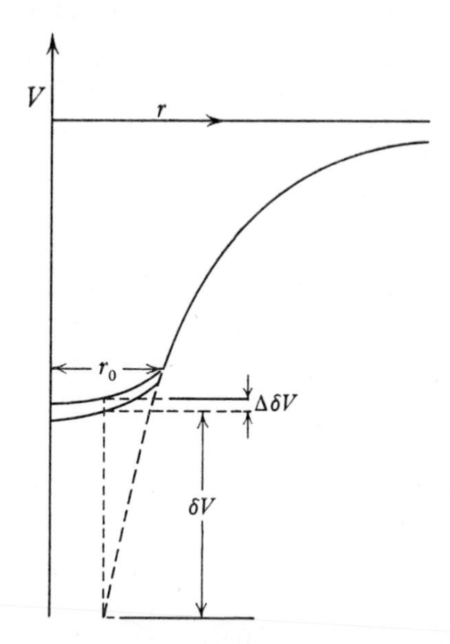

FIG. VI, 19. Potential curve of nucleus of radius r_0, with uniform charge density.

If $\rho_e(r)$ is derived from Schrödinger functions as $e|\psi_0|^2$ it is practically constant within the nucleus and can be taken out of the integral as a factor. The isotope shift of the term can then be written as the product of the factor $|\psi_0|^2$ depending only on the electron state, and a nuclear factor C, strictly speaking a factor depending on the properties of two isotopic nuclei. If ΔT is the term difference, one defines C by the relation

$$\Delta T = |\psi_0|^2 \frac{\pi a_0{}^3}{Z} \times C(Z, r_0, \Delta r_0/r_0). \tag{61}$$

For heavy atoms it is necessary to replace the Schrödinger function by Dirac functions and to correct the latter for the finite nuclear charge distribution. The correction can, however, be included in C, so that (61) still holds; with the use of the Rydberg constant R_∞:

$$C(Z, r_0, \Delta r_0/r_0) = \frac{12R_\infty(\rho+1)}{(2\rho+1)(2\rho+3)\Gamma^2(2\rho+1)}\left(\frac{2Zr_0}{a_0}\right)^{2\rho}\frac{\Delta r_0}{r_0} \tag{62}$$

where $\rho = \sqrt{(1-\alpha^2 Z^2)}$ and Γ indicates the "Gamma function". For connecting r_0 with the mass number, the usual assumptions are

$$\begin{aligned} r_0 &= R_0\sqrt[3]{A}, \qquad R_0 = 1{\cdot}2\times 10^{-13}\text{ cm}, \\ \Delta r_0/r_0 &= \tfrac{1}{3}\Delta A/A. \end{aligned} \tag{63}$$

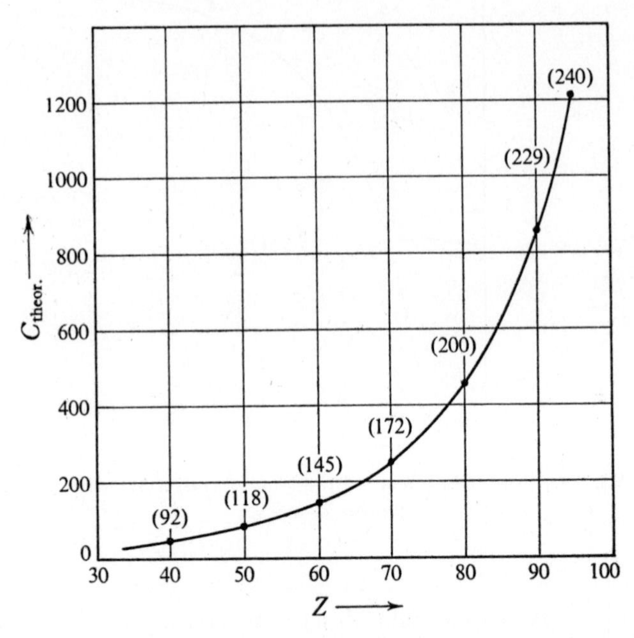

Fig. VI, 20. Calculated values of isotope shift constant C for spherical nuclei.

Putting $\Delta A = 1$ we find from (62):

$$C_{\text{theor.}} = F(Z)(2ZR_0/a_0)^{2\rho}A^{2\rho/3-1} \tag{64}$$

where $F(Z)$ has been written for the first factor in (62). Since ρ is of the order of 1 (for $A = 50, 70, 90$, $\rho = 0{\cdot}93, 0{\cdot}86, 0{\cdot}75$ respectively), eq. (64) shows that $C_{\text{theor.}}$ decreases very slowly with A. It depends on Z to a rather higher power than 2. Refinements of the theory[68–71, 25] have led to modifications affecting mainly the absolute value. Figure VI, 20 gives a plot of $C_{\text{theor.}}$ against Z. Results by

Humbach[69] were used, but corrected for the now generally accepted value $R_0 = 1{\cdot}2 \times 10^{-13}$ cm (instead of 1·4). For each Z the value of A of the mainly occurring isotope has been assumed and is given in brackets.

c. Comparison with experiment[72,73]

Experimental values of C derived from (61) require the knowledge of $\psi_{(0)}{}^2$, a problem which has been discussed in B.3.c. It can either be derived with the use of the Fermi–Segré formula as

$$C_{\text{exper.}} = \Delta T n^{*3}/Z_a{}^2(1 - \mathrm{d}\Delta/\mathrm{d}n) \tag{65}$$

or from the experimentally determined h.f.s. splitting constant A_s by means of (18). This gives, with the inclusion of the correction factors:

$$C_{\text{exper.}} = \frac{\Delta T}{A_s} \times \frac{8\pi R_\infty \alpha^2}{3 \times 1836{\cdot}1} Z g_I F_r(1-\delta)(1-\epsilon). \tag{66}$$

The latter method is only possible if the h.f.s. of an odd isotope with known g_I has been measured and if both the h.f.s. and the isotope shift are due to an unpaired s-electron. If, e.g. a configuration n s n'd is perturbed by n s^2, the perturbation will contribute to the isotope shift but not to the h.f.s.; the calculation of $\psi_{(0)}{}^2$ from the h.f.s. would then lead to wrong results.

The proportionality of the isotope shift to the electron density has been tested in a few cases in which the isotope shift and the hyperfine structure were measured for two odd isotopes in several alkali-like terms. This was especially done for Tl III where the terms 6s ^{2}S, 7s ^{2}S and 8s ^{2}S gave values of the isotope shift in nearly the same ratio as the hyperfine splitting constants, thus confirming the validity of (66) to within about 20 per cent.[25]

The determination of $C_{\text{exper.}}$ is restricted to spectra in which the isotope shift has been measured for levels whose configuration is both simple and fairly pure. A further difficulty arises from *screening effects*. Though p- and d-electrons have no direct effect on the isotope shift, it has been found, e.g. that the configuration 5d^{10} 6s 6p in Hg I shows a smaller isotope effect than 5d^9 6s 6p in Hg II (261 against 350 cm^{-1}). This must be ascribed to the screening effect of the additional 5d-electron which causes the density distribution of the 6s-electron to be shifted away from the nucleus.

It is also the screening effect which causes a configuration ns^2 to show usually less than twice the isotope shift of ns.

In a transition $ns - np$ the isotope shift will arise primarily from the presence of the ns-electron in the lower and its absence in the upper state. But the presence of the ns-electron has the secondary effect of screening part of the effective nuclear charge acting on the s-electrons of the closed shells, causing them to expand and thus to reduce the isotope effect.

Theoretical estimates of the magnitude of screening effects are difficult,[69, 72] but appear to indicate that they are generally not very large and sometimes compensated by third-order effects.

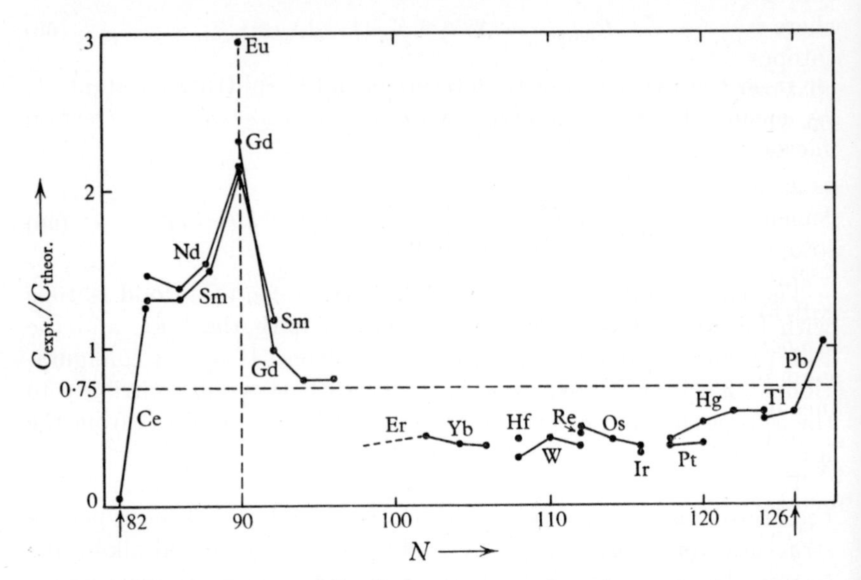

FIG. VI, 21. Isotope shifts for neutron numbers from 82 upwards.

For the various reasons mentioned the derivation of $C_{\text{exper.}}$ is often impossible and at best subject to uncertainties of about 10 to 20 per cent. Nevertheless, enough experimental results have become available to allow a comparison of the values of $C_{\text{exper.}}$ and $C_{\text{theor.}}$, with the aim of testing the underlying theory. Figure VI, 21 shows a plot of $C_{\text{exper.}}/C_{\text{theor.}}$ as function of the neutron number N, for values of $N > 82$.[74]

If the theory based on spherical nuclei applied strictly, all points would lie on a horizontal line of ordinate 1. It has to be remembered that, owing to the uncertainties of the calculation, the comparison of C-values for different elements is less accurate than the comparison for different isotopes of the same element. The plot thus consists of many short curves, one for each element, but the height of each curve relative to the adjacent ones is somewhat uncertain.

The following main features are evident:

(1) the absolute value of $C_{\text{theor.}}$ is, on the average, too high;

(2) the points do not lie on a horizontal line but form a curve with a sharp peak at $N = 90$, and minima near 80 and 120.

Especially interesting is the comparison of C values for two different elements having several isotopes of the same neutron number. This overlap of curves occurs frequently for elements of even Z, where it is found that two elements Z and $Z+2$ have often several isotopes of the same N. If at least three such iso-neutronic pairs exist one can compare the ratio of at least two C values, a comparison which is unaffected by the ψ values used in the calculation. It was first found for Nd and Sm[(75)] that the ratio was very nearly the same, and this has since been observed for several other pairs of elements. This shows directly that the specific features of the isotope shift are primarily determined by the neutron number.

This view is strengthened by the marked increase in the isotope shift found when the number of neutrons exceeds one of the *magic numbers* 82 or 126 which indicate the closure of a shell and mark a point of special stability of the nucleus. In Ce, the shift between lines due to ^{138}Ce and ^{140}Ce ($N = 80$ and 82) was found to be about an order of magnitude smaller than that between ^{140}Ce and ^{142}Ce ($N = 82$ and 84).[(76)] In Pb, a much smaller but fairly distinct increase was found for the neutron pairs 124, 126 and 126, 128.[(77)]

If we take the extreme view that the influence of the neutron configuration can be treated separately and the ratio of the values C of two different isotope pairs can be assumed to be entirely independent of Z, we can express this by writing it as a product:

$$C = f(Z)g(N, N') \tag{67}$$

where $f(Z)$ describes a dependence of Z probably similar to that contained in the first two factors of (62) while $g(N, N')$ describes the deviation of $C_{\text{exper.}}$ from $C_{\text{theor.}}$ apart from a mere scale constant. It is the latter function which contains the more specific and more interesting effects of the neutron number on the properties of the nucleus causing the isotope shifts.

If we accept (67) we can derive the ratio of two successive g values directly from the ratio of isotope shifts Δ of two pairs of isotopes:

$$_Z\Delta(N, N')/_Z\Delta(N', N'') = {_Z}C(N, N')/_ZC(N', N'') = g(N, N')/g(N', N'') \tag{68}$$

which is independent of Z. The first step in (68) has eliminated $\psi_{(0)}^2$ by application of (61), and the second the charge dependence by means of assumption (67). A direct plot of isotope shifts within one element as function of N represents C on an arbitrary scale, assuming that mass shifts can be neglected. If we accept the validity of (68) we can continue the plot from element Z to $Z+2$, $Z+4$, .. by scaling the isotope shift of each element so that the shifts of iso-neutronic pairs of adjacent elements agree. This provides a plot of $g(N, N')$ over an extended range of N-values. The method requires very accurate measurements of isotope shifts if any significant information is to be gained. It has been applied systematically to the sequence $_{44}$Ru, $_{46}$Pd, $_{48}$Cd, $_{50}$Sn, $_{52}$Te(66, 78, 79, 80, 81, 82) each of which contains a large number of even isotopes. The results showed that, in fact, ratios of shifts of iso-neutronic isotope pairs are very similar, but certainly not exactly equal, so that (67) does not hold rigorously (fig. VI, 22). The shifts, after deduction of the normal mass shift, were

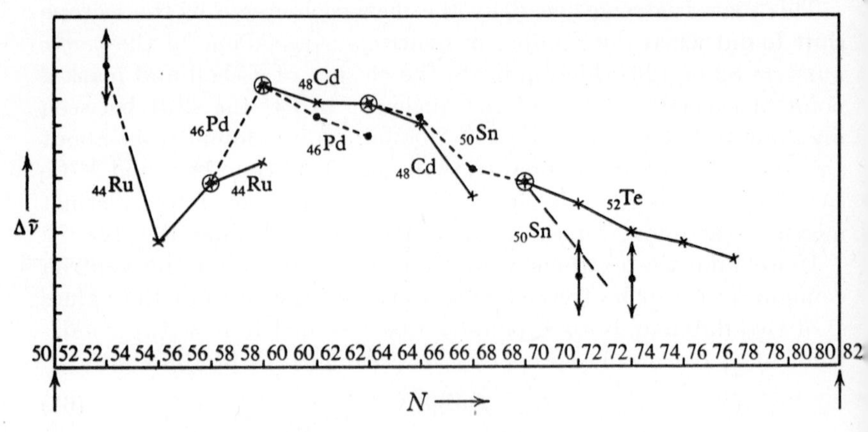

FIG. VI, 22. Relative isotope shifts from $N = 50$ to 82.

normalised by equating the shift of the lightest pair of Z with that of the iso-neutronic pair of $Z-2$. This procedure is consistent though somewhat arbitrary, since shifts of other iso-neutronic pairs are not quite equal. Nevertheless, the curve shows a definite trend which continues from element to element, and it should give an approximate picture of the dependence of the nuclear electrostatic potential field on the neutron number. The irregular changes of slope are genuine and not caused by experimental errors.

The mentioned slight change of slope from element to element appears to be systematical; one could express it formally by a further factor $g'(Z, N, N')$ in (67).

d. Nuclear deformation

A number of observations had suggested that the shape of many nuclei differs from that of a sphere: the anomalous variations of isotope shifts with N, described in the last section, the large values of some electric quadrupole moments, regularities in the spacings of low-lying nuclear energy levels and finally the Coulomb excitation of these levels. A basic theory of deformation of nuclei was developed by Bohr, Mottelsohn and others [83] and a large number of data were provided by measurements of Coulomb excitation. Our knowledge of nuclear shapes is still somewhat crude, but it has been possible to correlate it with at least some of the observed facts of isotope shift. [84]

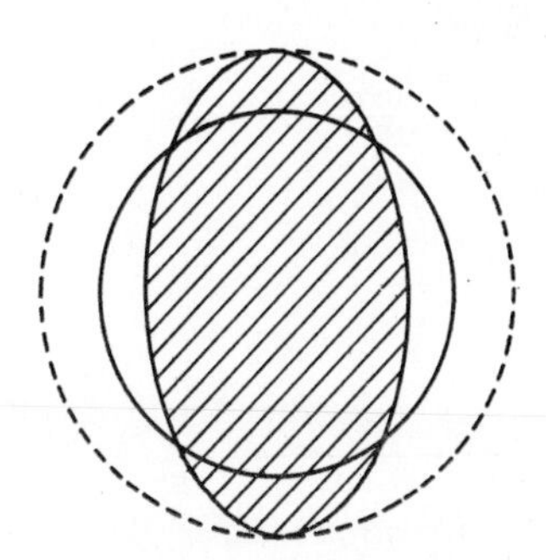

FIG. VI, 23. Nuclear deformation.

Nuclei of even Z and even N have no nuclear spin and must therefore be assumed to have spherical symmetry. They can nevertheless be deformed in the sense that one has to imagine a spherical charge distribution to be elongated (positive deformation) or flattened (negative deformation), and the resulting ellipsoid of revolution to be averaged over all possible orientations in space. This deformation causes the charge to be spread over a larger range of r (see fig. VI, 23), thus increasing ΔT.

If the addition of one or two neutrons does not change the shape but only increases the volume, the isotope shift will be somewhat increased as the result of the deformation. If the addition of the neutrons decreases the deformation, the isotope shift will be smaller than in the first case. If, however, the additional neutrons increase the deformation, the shift will be increased above that due to the mere volume change. Anomalously large isotope shifts such as occur at $N = 90$ have to be explained by the assumption that, for this neutron number, the deformation increases rapidly with N.

If β is the parameter describing the deformation (it is proportional to the quadrupole moment), δW is proportional to β^2, and the isotope shift $\Delta\delta W$ proportional to $\Delta N \times d(\beta^2)/dN$.

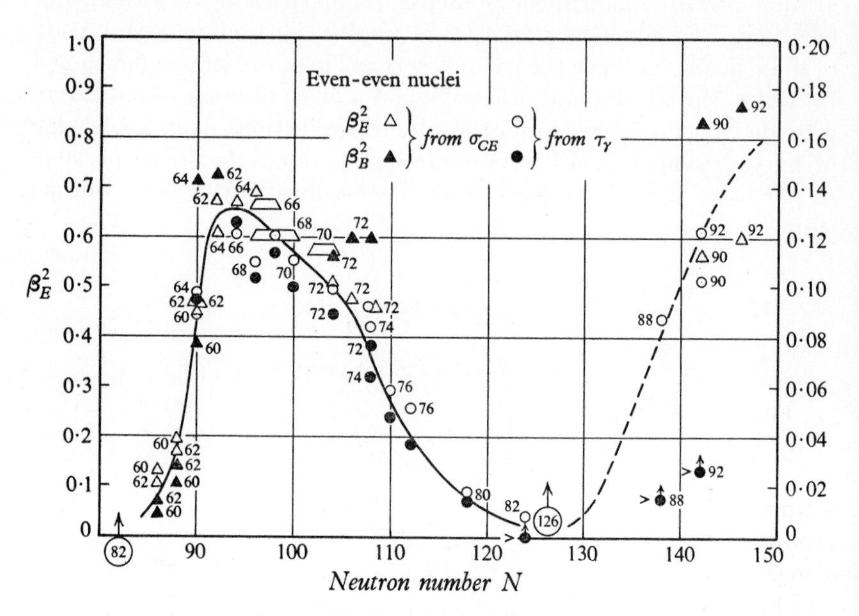

FIG. VI, 24. Deformation parameters β^2 for $N > 82$, from Coulomb excitation (G. M. Temmer, ref. 83).

In the range between the magic numbers $N = 82$ and $N = 126$, the deformations of nuclei, or at least their relative values, are fairly well known from Coulomb excitation and level structure. Figure VI, 24[83] shows a plot of these data from the two sources mentioned, whereby a discrepancy in the absolute values has been removed by the choice of different scales. The derivative of this curve matches the curve of fig. VI, 21 satisfactorily again with the appropriate choice of scale, especially the pronounced maximum in the latter curve at $N = 90$.

For the elements between the magic numbers $N = 50$ and 82, conditions are not quite so simple, and the influence of Z appears to be more marked. The deformation data from Coulomb excitation shown in fig. VI, 25[85] do not fall on a smooth curve, but if one considers only the slopes, there appears to be a certain continuity.

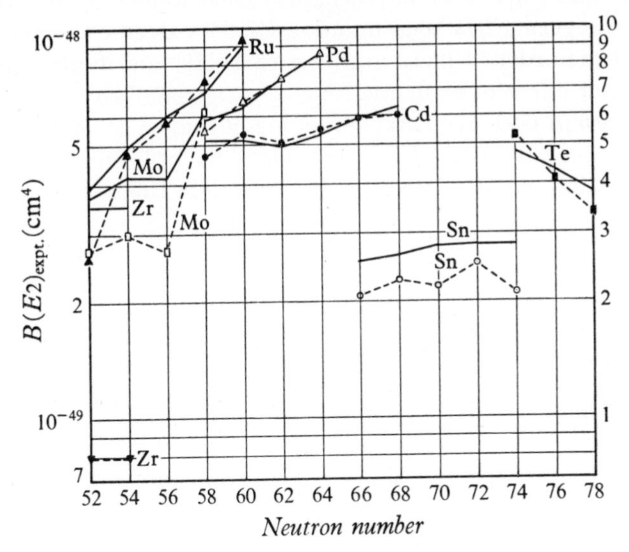

FIG. VI, 25. Deformation parameter β^2 for N from 52 to 78, from Coulomb excitation (P. H. Stelson & F. K. McGowan, ref. 85).

Adjustment of the ordinate scale individually for each element would produce a moderately continuous curve, with a sharp rise at 56–58, a maximum between 62 and 70 and a decrease for larger N. The strikingly low absolute values for Sn must be ascribed to the magic number 50 of protons in this element. The negative slope of the isotope shift curve in fig. VI, 22 from $N = 58$ upwards would indicate a negative second derivative $d^2(\beta^2)/dN^2$, in agreement with the convex shape of the curve fig. VI, 25 in this range of N.

The limited accuracy of the Coulomb excitation data does not allow the comparison with isotope shifts to be carried very far, but there appears to be a general consistency between them. The slight change of slope from element to element in fig. VI, 22, which was mentioned before, indicates some dependence of $d(\beta^2)/dN$ on the nuclear charge; the cause of this is uncertain.

The nuclear model as described here is certainly an over-simplification. Such factors as compressibility of nuclear matter must have an influence, especially on the absolute value of the isotope shift, but too little is known about them.

The *odd–even staggering* mentioned before has been found to exist in practically all cases where lines of both even and odd isotopes have been observed. But this has been done in only a small number of elements, and only very few accurate measurements are available. Marked staggering has been found in Cd,[66] extreme staggering just reversing the order of mass numbers in Te,[80] and indications of an even greater effect exist in Ba.[86]

The cause of odd–even staggering is not known, in spite of several attempts to explain it. Until the facts have been studied more systematically, one can hardly hope to explain them in a convincing way.

CHAPTER

VII. Width and Shape of Spectral Lines

A. THE DIFFERENT CAUSES OF LINE WIDTH

The term "spectral line" implies that the light emitted by atoms consists of a number of perfectly monochromatic radiations. The width of a spectral line observed in an ordinary spectrograph is, in fact, found to decrease as the spectrograph is improved; it is a property of the instrument rather than of the radiation. The half-value width on the wavelength scale, $\Delta\lambda$, or on the wave-number scale, $\Delta\tilde{\nu}$, is used for defining the instrumental *resolving power* $R = \lambda/\Delta\lambda = \tilde{\nu}/\Delta\tilde{\nu}$.

When the resolving power is sufficiently increased, by the use of large gratings or even interferometers, one finds that the width of the lines depends on the conditions in the source itself. It must then be assumed that the radiation has a genuine spread of wave-numbers for each "line", where this word is now used in a somewhat extended instead of the strictly mathematical sense.

This chapter deals with the intensity distribution $I(\tilde{\nu})$ which a spectrograph of infinite resolving power would show for each spectral line. Experimental work in this field requires very high resolving power, and our knowledge is very largely based on theory supported by comparatively crude experiments.

The lines emitted by a gas discharge are found to become narrower as gas pressure and current density are reduced: a line emitted by a glow discharge is narrower than that emitted by an arc operated at atmospheric pressure. This difference is due to the greater density of atoms, ions and electrons colliding with the radiating atom or disturbing it by their approach. These effects are described as *pressure broadening*. Connected with it is usually a *pressure shift* of the line.

When gas pressure and current density are progressively reduced, the line width approaches asymptotically a finite value which is found to depend on the temperature. Most of this residual width is due to *Doppler effect* caused by the random motion of the radiating atoms. It can be reduced only with some difficulty, either by cooling the gas or by the use of atomic beams in place of ordinary gases.

When the Doppler width is allowed for or is reduced sufficiently to become negligible, the remaining width is known as *natural width* or *radiation width*. It is due to the finite length of the wave train emitted even by an entirely undisturbed atom at rest. In optical spectra, the natural width is usually much smaller than the Doppler width and is therefore difficult to measure directly. The width of most spectral lines emitted by glow discharges is predominantly caused by Doppler effect.

An additional and often very important cause of broadening arises when spectral lines, whose lower level is either the ground state or a metastable state, are observed in emission. It is the *self absorption* whose influence on intensity measurements was discussed in I.D.3.

In the following paragraphs, we shall proceed in the reverse order, by first considering a single, undisturbed atom at rest, then allowing for the effect of random motion and finally of the various disturbing influences by other particles.

B. THE NATURAL OR RADIATION WIDTH

The natural line width is due to the radiation process itself which has been treated in II.C. It is closely connected with the life time of excited states which can sometimes be measured independently.

The simplest model of an emitter is the classical electron oscillator. It shows only one spectral line, of frequency ν_0, which can be observed in absorption, spontaneous emission or fluorescence. In the first two modes of observation, the shape of the line is the same and is given by (II.100) or (II.97). The half value width is $\gamma_0/2\pi c$ cm^{-1} and the life time $1/\gamma_0$ sec. They can be calculated from the frequency alone, with the use of the accepted values of m and e for the electron. There are a few cases in which one can expect this simple, classical picture to form a good approximation. In sodium vapour, e.g., more than 95 per cent of the absorption in a spectral range extending from the far infra-red to soft X-rays, is concentrated in one line in the yellow; the fact that this yellow line is a close doublet can be neglected for the present purpose and can subsequently be included without affecting the principal argument. The life time measured for the spontaneous emission of these lines and their natural widths are, in fact, found to agree with the classical values within the limits of error (see under D). Similarly good agreement is found for the resonance lines of other alkali atoms.

If we interest ourselves in the other, much weaker lines of an alkali atom or in lines of more complex spectra, the classical theory can no longer be expected to apply since it does not even give the correct frequencies.

A quantum-mechanical description of the radiation process was first given by Dirac.[1] An atom which is slowly losing energy to the radiation field cannot strictly be considered as a conservative system, and the discrete energy levels are replaced by maxima of probability in a continuous energy "spectrum" (see fig. VII, 1). The half value width ΔE_j of this energy distribution is connected with the life time $\Delta\tau$ of the state according to the principle of indeterminacy. The value of ΔE_j of a level is determined by the sum of the transition probabilities of all the transitions to lower levels:[2, 3]

$$2\pi\Delta E_j/h = 1/\tau_j = 2\pi\Delta\nu_j = \gamma_j = \sum_i A_{ji}. \tag{1}$$

The width of a *line* due to radiation damping, or *natural width*, is equal to the sum of the widths of the two levels,

$$\begin{aligned} \gamma_{kj} &= \gamma_k + \gamma_j, \\ \Delta\nu_{kj} &= \Delta\nu_k + \Delta\nu_j. \end{aligned} \tag{2}$$

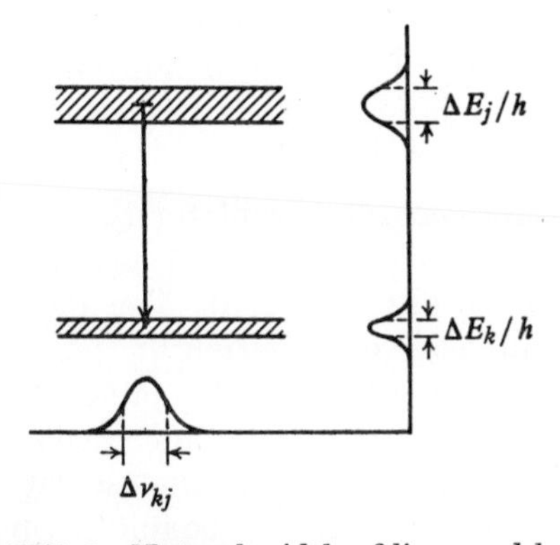

FIG. VII, 1. Natural width of lines and levels.

The probability distribution of each of the states, as indicated in fig. VII, 1 on the right, is given by a dispersion function (II.100). More complete treatments of the radiation field have shown that the interaction of the electron with the radiation field not only causes broadening of the energy levels but also accounts for small displacements which were discovered in hydrogen-like spectra (see p. 121).

For combinations with the ground state, the natural line width is equal to the width of the excited state. The resonance doublet of sodium may again be chosen for an illustration of the theory. According to the f-sum rule, the sum of the f-values of all lines of the absorption series must be equal to 1. The strength of all the higher lines, however, and of the continuous absorption is so small, that approximately

$$f(3S_{\frac{1}{2}}-3P_{\frac{1}{2}})+f(3S_{\frac{1}{2}}-3P_{\frac{3}{2}}) = 1.$$

The Burger–Dorgelo intensity rule for the emission of the two resonance lines shows for the transition probabilities that

$$A(3S_{\frac{1}{2}}-3P_{\frac{1}{2}}) = A(3S_{\frac{1}{2}}-3P_{\frac{3}{2}}),$$

so that according to eq. (II.113) the ratio of the f_{ij} values is 1: 2, or

$$f(3S_{\frac{1}{2}}-3P_{\frac{1}{2}}) \approx \tfrac{1}{3}, \qquad f(3S_{\frac{1}{2}}-3P_{\frac{3}{2}}) \approx \tfrac{2}{3}.$$

The width of the two lines which is merely given by the A values, is equal. The greater strength of one line in emission is due to the greater statistical weight of the level $P_{\frac{3}{2}}$, causing a greater number of atoms to exist in this state. The greater strength of the line in absorption can be said to arise from the greater number of states of the level $P_{\frac{3}{2}}$, as compared with $P_{\frac{1}{2}}$, to which transitions from $3S_{\frac{1}{2}}$ are possible.

Apart from the values of f and the definition of the damping factor γ, given by (2), the fundamental facts of the width and shape of spectral lines are in agreement with the classical model:

(a) *Spontaneous emission.* The oscillator is, as the result of a very rapid process such as an electron impact, given an initial amplitude which decays exponentially (fig. II, 8). The Fourier analysis of the motion gives the intensity distribution of the emission line (II.97).
(b, c) *Fluorescence* and *absorption.* The action of an external, oscillating field of frequency ν on a harmonic oscillator of frequency ν_0 is shown in fig. VII, 2, a and b. In a, the oscillator is assumed to be initially at rest, in b it is assumed to be initially oscillating. In both cases the oscillator, after an initial period of change, settles down to an *undamped* oscillation whose amplitude depends on the values of ν and ν_0, with a maximum fo $\nu = \nu_0$. The oscillation continues as long as the primary wave motion persists. If the latter is perfectly monochromatic, and thus continuing for an infinitely long time, the scattered light is equally monochromatic. This applies to both ordinary Rayleigh scattering and resonance fluorescence which differ only in amplitude from one another.

The view that the absorption of light leads to an excited state with subsequent emission of a light quantum is, in this respect, misleading;

the width of the fluorescence line is not affected by the radiation width of the excited state. The classical view, however, leads to the correct result. In the rigorous quantum-theoretical treatment,[2, 3] the absorption of monochromatic light and its re-emission are described by a single quantum process in which the energy of the excited state remains undetermined.

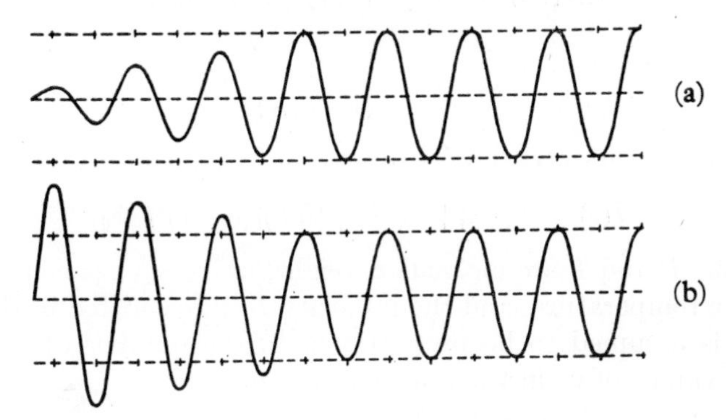

FIG. VII, 2. Forced oscillations of damped oscillator, (a) starting at rest, (b) starting at large amplitude.

The absorption by a classical, damped oscillator has exactly the same spectral distribution as the spontaneous emission. This can be shown quite generally for any kind of damping force.[4] The presence of the radiation field causes a slight increase in the damping and thus in the width, an effect which is proportional to the radiation density and has been observed in radio-frequency spectra. Dirac's theory gives essentially the same result.[3] The slight, additional width is, in this theory, ascribed to the fact that the radiation field causes up- and down-transitions between excited and ground state, thus shortening the life time of both by an amount proportional to the radiation density. These effects correspond to the initial increase and decrease in the amplitude of the wave motions in fig. VII, 2a and b, though the quantitative description is outside the scope of classical theory.

C. DOPPLER WIDTH

If a monochromatic light source is moving with a velocity whose component in the line of sight is v_x, the frequency appears shifted by an amount $\nu_0 v_x/c$ compared with the frequency ν_0 of the source at rest. Owing to this *Doppler effect*, the random motion of the atoms of a gas causes the spectral lines emitted or absorbed by these atoms to be broadened. If the lines emitted by atoms at rest are considered as infinitely narrow, i.e. if natural line width and broadening by external influences are neglected, the intensity distribution follows immediately from Maxwell's distribution law. The relations

$$\mathrm{d}N(v_x) \sim \exp[-(Mv_x^2/2RT)]\,\mathrm{d}v_x$$

and

$$\nu = \nu_0(1 - v_x/c)$$

give directly the intensity as function of frequency:

$$I(\nu) = I_0 \exp\{-(Mc^2/2RT)[(\nu_0 - \nu)^2/\nu_0^2]\} \tag{3}$$

where M, R and T are molecular weight, universal gas constant and absolute temperature, and the intensity $I(\nu)\,\mathrm{d}\nu$ emitted in the interval $\mathrm{d}\nu$ is assumed to be proportional to the number of atoms $\mathrm{d}N$ having values of v_x in the appropriate range.

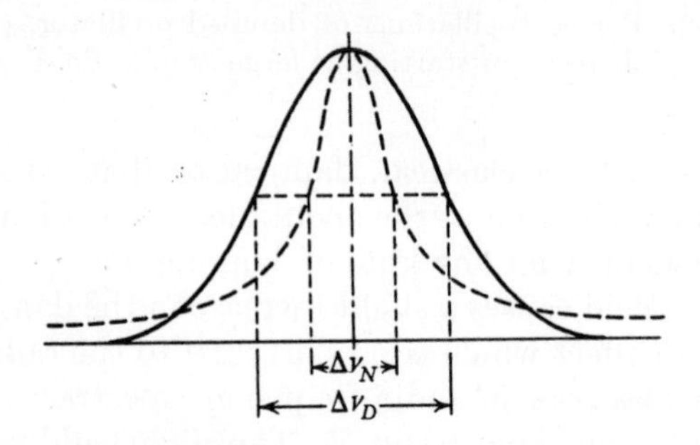

FIG. VII, 3. Line profiles due to Doppler effect (———) and to radiation damping (– – – –).

The half-value width in wave-numbers (fig. VII, 3) is found from (3) to be

$$\Delta\tilde{\nu}_D = \frac{2\tilde{\nu}_0}{c}\sqrt{\left(\frac{2RT\ln 2}{M}\right)} = 7{\cdot}16 \times 10^{-7}\tilde{\nu}_0\sqrt{(T/M)}. \tag{4}$$

Figure VII, 4 gives a set of curves showing the Doppler width as function of $\lambda = 1/\tilde{\nu}_0$ and T/M.

For the example of the resonance lines of sodium emitted at room temperature one finds $\Delta\tilde{\nu} = 0{\cdot}043\ \mathrm{cm}^{-1}$. A comparison with the natural width shows that the Doppler effect is the most important cause of line broadening in most practical cases, in the optical range of the spectrum. This applies in particular to the lighter elements; the great Doppler width of the Balmer lines, e.g. has retarded several fundamental discoveries more than any other experimental difficulty.

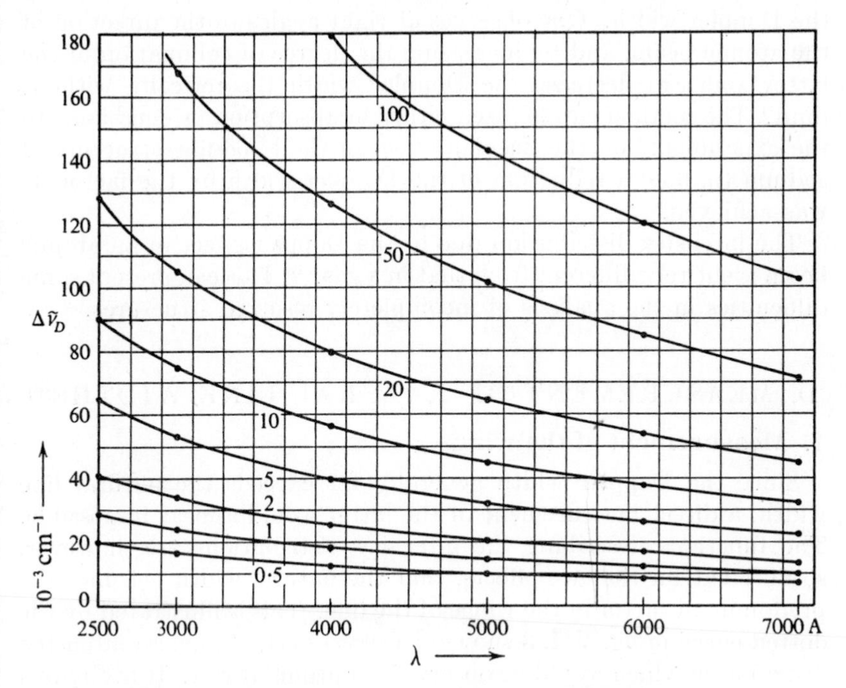

FIG. VII, 4. Doppler half-value width as function of wavelength. The number next to each curve is the parameter T/M.

Formula (4) shows further:

(1) $\Delta\tilde{\nu}_D \sim \tilde{\nu}$. For the investigation of any given term structure, such as a hyperfine structure, the Doppler width becomes the less important the smaller the frequency. A red spectral line is, from this point of view, more favourable than one in the violet or ultra-violet, and in radio-frequency spectroscopy the Doppler width is generally quite negligible.

(2) $\Delta\tilde{\nu}_D \sim \sqrt{T}$. The Doppler width can be reduced by cooling of the source. Especially in the form of hollow cathode discharges, first introduced by Schüler, light sources are often cooled with water or liquid air in order to allow higher resolution. Even liquid hydrogen and liquid helium have been used as coolants. The temperature of

the radiating atoms is, however, always higher, and usually very much higher than that of the walls of the tube, and Doppler temperatures below 100°K can only be achieved in favourable cases and in very weak discharges.

Atomic beam sources are a much more powerful means of reducing the Doppler width. One observes at right angles to the direction of the atomic beam, and by increasing the degree of collimation of the latter, one can decrease the Doppler width theoretically without limit. The method can be used either in absorption or emission. In the experiments on the Zeeman effect of the hyperfine structure of sodium (p. 350) a reduction of the Doppler width by the factor 30 was achieved.

The intensity distribution due to the Doppler effect in an atomic beam is entirely different from that in a gas.[5] This can present some difficulties in the analysis of incompletely resolved structures.

D. MEASUREMENT OF NATURAL LINE WIDTHS[6]

1. Measurement of the wings

Since the Doppler width generally far exceeds the natural line width, a direct measurement of the latter would appear impossible. The functions describing the intensity distributions are, however, so different for the two effects, that the natural width becomes the predominant factor in the wings of the line. This is illustrated by the dotted curve in fig. VII, 3 showing a plot of (II.97), i.e. an intensity distribution which would be observed for atoms at rest. It is obvious that the Doppler effect can be neglected at sufficiently large distance from the maximum. In practice, measurements in this range can readily be carried out in absorption where the smallness of the atomic absorption coefficient can easily be offset by a greater gas density or length of path. For the resonance lines of sodium, such measurements[7] have led to a value of the radiation half-value width of each of the lines of $0{\cdot}33 \cdot 10^{-3}$ cm^{-1} which agrees closely with the value of $0{\cdot}34 \cdot 10^{-3}$ cm^{-1} calculated for a classical electron oscillator of the same frequency.

2. Anomalous dispersion

The change of refractive index is a direct consequence of the absorption and offers a convenient and accurate method of studying the contour of an absorption line at distances from the maximum at which the influence of the Doppler effect can be neglected. In the vicinity of an absorption line the refractive index is almost entirely due to the anomalous dispersion by this line and is described by the formula (II.103). Measurements by means of a *Jamin* refractometer, in the form of *Roschdestwensky's "hook method"*[8, 9] have provided the most reliable radiation widths of absorption lines. The results are usually expressed in terms of f-values whose connection with transition probabilities has been discussed in II.D. For resonance lines, the value of A is, according to (1) and (2) identical with that of γ.

Of the numerous results of such measurements, only two may be mentioned. For the resonance lines of Na, Ladenburg and Thiele[10] find the total value of f to be 1·05, very close to the value 1 of the classical oscillator. For the lines 2537A($6\,^1S_0 - 6\,^3P_1$) and 1849A ($6\,^1S_0 - 6\,^1P_0$) of Hg, Ladenburg and Wolfsohn[11] find the f-values 0·025 and 1·19 respectively, giving life times of the states $6\,^3P_1$ and $6\,^1P_1$ of $1{\cdot}14 \times 10^{-7}$ and $1{\cdot}30 \times 10^{-9}$ sec., according to the relation $1/\tau = 2\pi\Delta\nu$ between life time and radiation width of resonance lines.

3. Width due to several transitions

If transitions to more than one level are possible from the excited state, or if the line in question is a combination between two excited states, the relation between radiation width, transition probabilities and life times is given by (1) and (2).

It is found, e.g. that the second member of the principal series of Cs, $6\,^2S - 7\,^2P$, has a natural width to which not only the transition itself but also the other possible transitions from $7\,^2P$ to $7\,^2S$ and $5\,^2D$ contribute.[12]

Very striking examples of the influence of the final state on the radiation width of a line have been found in Ne and He. Among the visible lines of Ne, those whose lower levels form strong combinations of extremely high frequency with the ground state show an anomalously large width.[13] Similarly, the line 6678A($2\,^1P - 3\,^1D$) of He was found to have an anomalous width[14] due to the transition from $2\,^1P$ to $1\,^1S$ (584A) whose great strength and high frequency cause the state $2\,^1P$ to have a very short life time.

4. Life time

In direct measurements of the life time of excited states, it has proved difficult to achieve an accuracy of better than $0{\cdot}5 \times 10^{-8}$ sec.; the method is therefore restricted to fairly long life times. From fluorescence experiments in which primary and fluorescence light were passed through Kerr cells acting as light shutters, the life time of the term 3 ^2P of Na has been determined, in good agreement with its value from dispersion measurements.[15]

Measurements of the decay time of resonance fluorescence lead to the correct value of the life time of the excited atomic state only in the limit of very low densities. Owing to the extremely high absorption coefficient at resonance, a considerable amount of multiple scattering can occur and can cause a longer decay time. This process is known as *imprisonment* of resonance radiation.

Pulsed excitation by slow electrons with observation of delayed coincidences by means of a photo-multiplier has recently been used successfully[16] for the measurement of life times of triplet states in helium. This method is somewhat similar to that used in nuclear Physics. It represents a refinement of an older method[17] by which the comparatively long life time of the level 6 $^3P_1$ of Hg was measured, in good agreement with the known f-value.

One of the earliest methods of measuring life times, due to Wien, used the decay of the intensity along the path of positive rays (canal rays), but the experimental conditions were not very well defined in these experiments which are now of merely historical interest.

For very long life times, the decay of the intensity of fluorescence along the path of an atomic beam has been used[18, 19]; the life time of the level 5 $^3P_1$ of Cd has been found, in this way, to be $2{\cdot}5 \times 10^{-6}$ sec. This value has been confirmed by other methods.[20]

5. Line contour

The complete contour of a line due to the combined effect of radiation and Doppler broadening can be immediately written down as an integral since both effects are entirely independent and can be simply superimposed. Introducing the natural half-value width $\Delta\nu_n = \gamma/2\pi$ and the Doppler half-value width $\Delta\nu_D = 2(\nu_0/c)\sqrt{(2RT \ln 2)/M}$, we can write for the intensity which would exist at ν in the absence of Doppler effect:

$$I(\nu) = \frac{C}{1+\left[\dfrac{2(\nu-\nu_0)}{\Delta\nu_N}\right]^2}. \tag{5}$$

Owing to Doppler effect, the intensity at any point $\nu-\delta$ on the frequency scale of the line without Doppler broadening contributes according to its value from (5) multiplied by the Doppler function (3) with the frequency difference δ. The resulting intensity function is then

$$I(\nu) = C\int_{-\infty}^{+\infty} \frac{\exp\left\{-\left[\frac{2\delta\sqrt{(\ln 2)}}{\Delta\nu_D}\right]^2\right\} d\delta}{1+[(2/\Delta\nu_N)(\nu-\nu_0-\delta)]^2}. \qquad (6)$$

This *folding* integral cannot be solved analytically, but values have been tabulated by Voigt and others.[21, 22] The distribution (6) is often called "Voigt profile".

E. PRESSURE BROADENING*

1. The two different approaches

The interactions of a radiating atom with other atoms, ions or electrons have effects on the spectral lines which are of a very complex nature. In order to make the problem mathematically manageable, one has to introduce simplifying assumptions whose justification is often not obvious and whose validity is limited to certain aspects of the phenomena. Experimental data are frequently not very accurate or refer to conditions outside the range of existing theories.

The interest in the phenomena of pressure broadening and pressure shift is two-fold. On the one hand their study is able to provide information on the nature of the interactions between atoms, on the other hand it can be used as a means of exploring the conditions in the source, e.g. in stellar atmospheres or discharge plasmas.

The theory has developed from two opposite viewpoints. Following Lorentz[23] one can consider the effect of other atoms as due to collisions breaking the otherwise unperturbed wave train. The line width is then determined not by the natural life time τ but by the smaller time τ_c between two collisions. The Fourier analysis leads to a width proportional to $1/\tau_c$; the effect is a *kinetic* one, depending on the relative velocity of the atoms.

The opposite, *statistical* approach was first applied by Holtsmark[24] to the effect of ions, and by other authors to that of neutral atoms. It neglects the motion of the atoms which are regarded as being at rest but distributed in space in a random fashion. The characteristic frequency of the radiating atom is changed by the presence of another atom or ion according to a certain function of the mutual

* For fuller treatment the reader is especially referred to: A. Unsöld, *Physik d. Sternatmosphären*; 2nd ed. Springer, 1955, and to the review articles by H. Margenau and W. W. Watson, *Rev. Mod. Phys.*, **8**, 22, 1936; S. Y. Ch'en and M. Takco, *Rev. Mod. Phys.*, **29**, 20, 1957; and R. G. Breene, *Rev. Mod. Phys.*, **29**, 94, 1957.

distance. Each atom can be regarded as emitting one definite frequency at any given time, but the random effect of many radiating atoms leads to a continuous intensity distribution whose calculation is a purely *statistical* problem.

It will be shown in the following paragraphs how the collision theory has been modified by a more refined treatment of the collision process in terms of intermolecular forces, and how the statistical theory has been modified to take molecular motion into account. Even these refined theories are imperfect in various respects, and few definite comparisons with experimental facts have been possible so far.

2. The collision theory

If conditions in a gas are such that the average time $\overline{\tau_c}$ between collisions is much shorter than the natural life time $1/\gamma$ of the excited state of the atom considered, each wave train can be regarded as undamped but abruptly cut off by the collisions. The Fourier analysis of a periodic function whose amplitude is constant in a certain range of the variable and zero outside this range is well known from the treatment of Fraunhofer diffraction by a slit. It consists of a central peak followed by zero points at $\Delta\nu = \pm 1/\tau_c$ and approximately equidistant, weaker maxima. If the time τ_c between two collisions were the same for all atoms of the gas, the collision broadening would show distinct satellite bands whose spacing would allow the collision time to be measured most conveniently. In actual fact, free paths and collision times are distributed according to a Gaussian error curve. The averaging over this distribution leads to an intensity profile without subsidiary maxima, and of the same type (II.97) as that due to radiation broadening. The life time $1/\gamma$ is simply replaced by $\frac{1}{2}\tau_c$. This theory due to Lorentz leads to the correct result of a proportionality of line width with gas density. If $1/\gamma$ and τ_c are of similar magnitude, γ in (II.97) has to be replaced by $\gamma + 2/\tau_c$.

The values of the collision radius derived from observed pressure broadening are noticeably larger than the gas-kinetic collision radii. This is qualitatively plausible if we consider that in contrast to a gas-kinetic collision involving a pronounced change in the direction of motion, an *optical collision* in the sense of Lorentz's theory only requires a change in phase, sufficient to cause incoherence of the light waves emitted before and after the collision.[25] A quantitative treatment of optical collisions requires a knowledge of the laws of interaction between atoms, especially at comparatively large distances. We assume that the characteristic frequency ν_0 of the atom, the frequency of an emission- or absorption-line, is altered as the result of the presence of another atom at distance r by an amount $\Delta\nu_0(r)$ which is usually a function of the type

$$\Delta\nu_0(r) = C/r^n \tag{7}$$

which vanishes for $r = \infty$. This is a purely classical picture, though the values of $\Delta\nu_0(r)$ may be calculated by quantum-mechanical methods for two atoms assumed to be at rest.

The passage of another atom near the radiating atom causes a temporary, small change in the frequency of the "atomic oscillator", resulting in a phase change by the amount

$$\epsilon = 2\pi \int_{-\infty}^{+\infty} \Delta\nu_0(t)\,\mathrm{d}t. \tag{8}$$

ϵ can be calculated if the function $\Delta\nu_0(r)$ is given, and the path and therefore the function $r(t)$ is known. For sufficiently distant encounters the mechanical forces are so small that the path can be regarded as a straight line and the only parameter, apart from the speed, is the collision parameter ρ defined by the normal distance of one particle from the path of the other (fig. VII, 5).

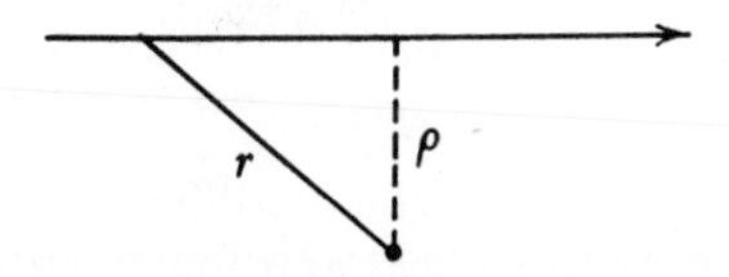

FIG. VII, 5. Definition of collision radius.

An encounter with parameter ρ can be regarded as an optical collision if ϵ is of the order of 1 or more. Weisskopf[25] defines the optical collision radius ρ_0 by the condition

$$\epsilon = 1. \tag{9}$$

The value of ρ_0 is easily found from (8) to be

$$\rho_0 = \left(\frac{2\pi C c_n}{v}\right)^{1/(n-1)}, \tag{10}$$

giving the half-value width of the line

$$\Delta\nu_{\frac{1}{2}} = \left(\frac{2\pi C c_n}{v}\right)^{2/(n-1)} vN \text{ sec}^{-1} \tag{11}$$

where N is the number of atoms per cc. The constant c_n has the values π, 2, $\pi/2$, 4/3, $3\pi/8$ for $n = 2, 3, 4, 5, 6$ respectively. For definitions of a collision radius with a value of ϵ other than 1, the constant c_n has to be divided by ϵ.

Quantum mechanics shows that the frequency change due to a neutral atom is described by $n = 6$ for the relevant range of radii, with a negative value of C (decrease of frequency) which can be calculated in simple cases (see section 3).

Table VII, 1 shows that the theoretical collision radii are in reasonably good agreement with experimental values, but neither theory nor experiments can claim great accuracy.

TABLE 1

Optical collision radii in 10^{-8} cm

	A	Ne	He
Na(5890, 5896A)			
exper.	11·2	6·2	5·6
theor.	8–10	6–8	5–6
Hg(2537A)			
exper.	7–9	6	4–5
theor.	5·4–8·3	4·4–6·7	3·2–5·0

In Lorentz's theory, even in the modified form due to Weisskopf, the broadened line is symmetrical and undisplaced. Lenz,[26] Lindholm[27] and others[28,29] have carried the collision theory a step further by a proper Fourier analysis of the passages of atoms for different values of ρ. For $n = 6$, the result can be expressed by replacing (9) by the assumption $\epsilon = 0{\cdot}61$ in the calculation of the line width.

In addition, however, these refined theories give a shift of the maximum which has for $n = 6$, the value

$$\delta\nu = \Delta\nu_{\frac{1}{2}}/2{\cdot}8. \tag{12}$$

The origin of the shift is easy to understand qualitatively. Taking the example of a negative C, a small phase change will always arise from a very gradual, temporary, increase of the wavelength (fig. VII, 6a), and the Fourier analysis must lead to a smaller average wave-number. A large, fairly sudden increase of wavelength due to a closer approach will not have such an effect but will merely cause an almost instantaneous change of the phase by any value (fig. VII, 6b); this mere chopping action widens the line but does not shift it. This leads to the following result: encounters outside the Weisskopf radius ρ_0 are mainly responsible for the shift of the line and its asymmetry, while those inside ρ_0 cause most of the broadening effect.

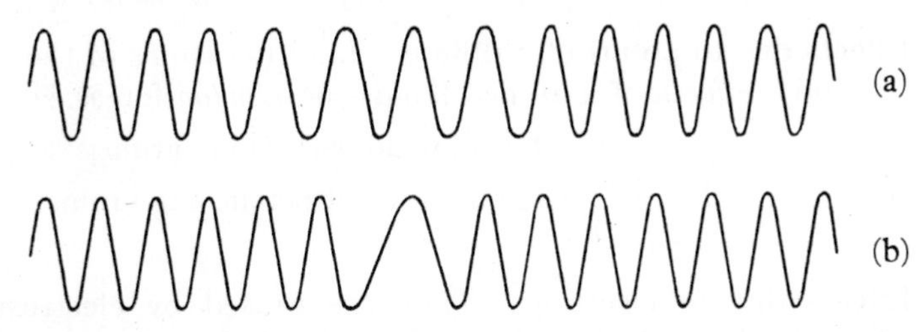

FIG. VII, 6. Wave trains for (a) distant, (b) close optical collision.

The relation (12) can be tested experimentally. The ratio of width to shift was found to have widely varying values, ranging 0·5 to about 5, and even more in a few cases. A predominance of values between 2 and 4 is sometimes regarded as a confirmation of the theory.

3. Statistical theories

The statistical theory treats the radiating atom and the perturbing particles as if they were at rest while the emission or absorption of light takes place, so that the frequency ν_0 is a function of the mutual distances alone. This assumption implies firstly that the atoms are moving sufficiently slowly, and secondly that the interaction with the light wave does not cause any appreciable change in position and momentum of the atom. The first is a purely classical assumption which is the more justified the lower the temperature and the heavier the atoms. The second is equivalent to assuming the validity of the *Franck–Condon principle* normally applied to electron transitions in vibrating molecules.[30] How well it holds is a question of quantum theory. We shall, at first, take both assumptions for granted and subsequently discuss the limits of their validity.

The following long-range interactions have to be considered.

1. Effects due to ions (Stark effects).

   (a) hydrogen or hydrogen-like ions: linear Stark effect.

   $$\Delta\nu_0 \sim F \sim 1/r^2 \text{ for a single ion;}$$

   for several ions, the fields have to be added vectorially. Each line splits symmetrically into several components.

(b) Other atoms: quadratic Stark effect.

$$\Delta\nu_0 \sim F^2 \sim 1/r^4 \text{ for a single ion.}$$

2. Effects due to atoms of the same kind, involving a level forming a strong combination with the ground state: resonance effects.

$$\Delta\nu_0 \sim 1/r^3 \text{ for a single perturbing atom.}$$

3. Effects due to atoms of a different kind (or atoms of the same kind): effects of *Van der Waals*- or *London*-forces.[31]

$$\Delta\nu_0 \sim 1/r^6 \text{ for a single perturbing atom;}$$

$$\Delta\nu_0 \sim \sum_k 1/r_k{}^6 \text{ for several perturbing atoms.}$$

4. Effects due to electrons. They are caused by electrostatic fields, but the basic assumptions of the statistical theory do not apply, owing to the small mass of electrons.

The probability of finding another atom (or ion) at a distance r much smaller than the average distance between atoms is proportional to the density, that of finding *two* at this distance proportional to the *square* of the density, and the latter probability is very much smaller than the former. If the exponent n is sufficiently large, a given value $\Delta\nu_0$ can be either due to the influence of one atom at a given distance or of two atoms at only slightly greater distance. The probability of the double encounter can be neglected compared with that of the single one for sufficiently small r, i.e. large $\Delta\nu_0$. The intensity in the wing of the pressure-broadened line can therefore be ascribed to *single encounters*. Its value at any given $\Delta\nu_0$ is simply proportional to the probability of finding one atom at the distance r at which the interaction causes the frequency change $\Delta\nu_0$:

$$I(\Delta\nu)\,\mathrm{d}\nu = A4\pi r^2\,\mathrm{d}r. \tag{13}$$

With the assumption $\Delta\nu = C/r^n$, differentiation and substitution leads to the intensity distribution[32]

$$I(\Delta\nu) = \text{const. } \Delta\nu^{-(n+3)/n}. \tag{14}$$

Closer to the maximum, however, the intensity is due to interactions at large distances; it will be proportional to the probability of the perturbation $\Delta\nu_0$ being caused by either one or by two, three . . atoms, in each of which cases a number of combinations of distances have to be considered. Moreover it turns out that the basic assumption of the statistical theory is a good approximation in the wings but not near the maximum of the line (see below). The *single-encounter* approximation is the more useful the higher the value of n.

In the following, we shall briefly deal with the cases of different values of the exponent n.

(1) The Stark effect broadening ($n = 2$ and 4) has been treated by Holtsmark[24] and others.[33] The problem is mathematically complicated by the fact that the fields due to different ions have to be added vectorially, and by the low value of n, especially for the linear effect, which increases the importance of multiple interactions. Figure VII, 7[34] shows the calculated intensity distribution for linear and quadratic Stark effect. The abscissa is the relative frequency shift $\Delta\nu_0/\overline{\Delta\nu_0}$ where $\overline{\Delta\nu_0}$ is the Stark shift which would be caused by one ion at the average distance a defined by

$$\frac{4\pi}{3}a^3N = 1. \tag{15}$$

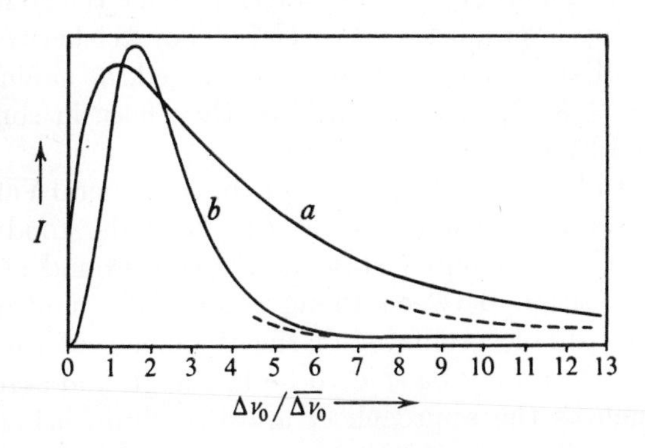

FIG. VII, 7. Holtsmark distribution for (a) linear, (b) quadratic Stark effect. The dotted curves show the effect of single encounters only. (A. Unsold, ref. 34.)

The value of $\overline{\Delta\nu_0}$ is different for each Stark component, and the total intensity distribution arises from a superposition of several curves of the kind shown in fig. VII, 7. For the linear effect, each curve has a branch for negative $\Delta\nu_0$ exactly similar to that for positive $\Delta\nu_0$. The superposition of all components can show a minimum at the undisturbed frequency, and this has been observed in hydrogen lines.

(2) If an atom in its ground state is brought near an excited atom of the same kind, the electrostatic interactions cause the same effect which has been discussed in III.B.4 for two electrons in the same atom: the identity of the electrons leads to two states in both of which the energy of excitation can be regarded as oscillating rapidly from

one atom to the other. The potential energy shows the same dependence on the internuclear distance as for two electric dipoles, i.e. proportionality to $1/r^3$. The Franck–Condon principle, postulating r and the kinetic energy to remain constant during the transition, causes $\Delta\nu_0$ to be proportional to $1/r^3$.

On account of their dipole character these forces combine, for large numbers of atoms, in a peculiar way which is known from the classical theory of disperson.[35] The resulting effect is very much smaller than that from a mere addition of the effects of individual atoms. As a consequence of this, one can assume that these resonance forces play only a small part in the wings of the line, though at very low densities they can cause collision broadening with a large optical collision radius.

(3) The most general type of interaction is due to the forces whose potential energy is proportional to $1/r^6$. They act between any pair of like and unlike atoms, excited or unexcited, at sufficiently large distances ($\approx > 3$ A). For like atoms, they may in some cases be overshadowed by the resonance forces.

According to F. London[31], the origin of the general Van der Waals forces of attraction between atoms can be understood as follows. The existence of an excited state above the ground state, with a permitted transition between them, represents quantum-mechanically a virtual dipole oscillator of the frequency of the transition. Although the atom may be said to be in the ground state, any perturbation such as the approach of another atom, brings the other quantum states into play so that even an atom in its ground state acts on the perturbing atom as an oscillating dipole. The field of this dipole polarises the approaching atom, producing in it a dipole whose moment is proportional to $1/r^3$. The potential energy of interaction between the two dipoles is then proportional to $1/r^6$. Since the strength of the virtual dipole is approximately proportional to the polarisability of the atom, the coefficient C in the relation

$$V = hC/r^6 \tag{16}$$

is proportional to the product of the polarisabilities of the two atoms. Their values are large for states with strong combinations with levels of only slightly higher energy. The constant C tends to be of larger absolute value for excited states and is generally negative. The Franck–Condon principle leads, according to fig. VII, 8, to a displacement towards lower frequencies, given by

$$\Delta\nu_0 = (C' - C'')/r^6.$$

Equation (16) with $n = 6$, gives the distribution function[32]

$$I(\Delta\nu) \sim \Delta\nu^{-\frac{3}{2}} \tag{17}$$

which can be expected to hold in the long-wave wing of pressure broadened lines.

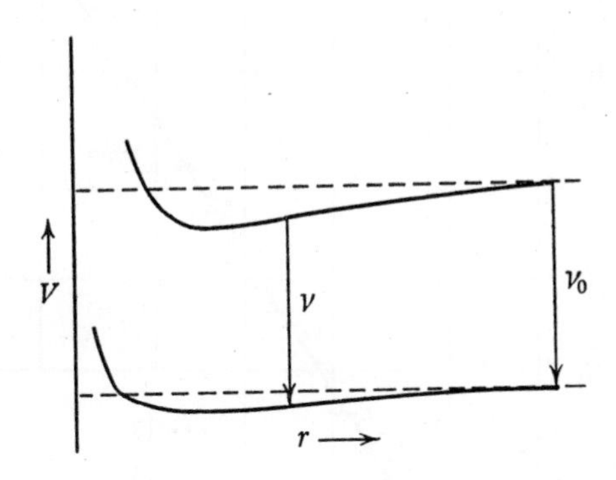

FIG. VII, 8. Franck–Condon principle applied to Van der Waals forces.

Measurements on broadening of the resonance lines of Na by the addition of argon[36] fully confirmed the prediction of the theory. Figure VII, 9 shows a logarithmic plot of the intensity distribution in the long-wave wing. In the range from 0·8 to 13 cm^{-1} the measured points lie closely on the theoretical curve. The measurements were carried out in absorption, so that $I(\nu)$ is, in this case, the extinction coefficient.

From the absolute values of the absorption in the wing, Minkowski could determine the value of C and thus the Weisskopf collision radius. The intensity distribution near the peak was found to be a Lorentz distribution ($I \sim 1/\Delta\nu^2$) with a collision radius in good agreement with that derived from the long wavelength wing.

Figure VII, 10 shows a similar plot for the line 2537 A of Hg[37] broadened by argon. The value of C is considerably smaller as a result of which the Lorentz distribution extends quite far from the maximum (dotted line) and rather abruptly passes into the statistical $\Delta\nu^{-\frac{3}{2}}$ distribution (solid line).

The value of C from the long wavelength wing was found to be much smaller than that from the Lorentz distribution near the peak. This appears to be due to the splitting of the excited energy level on approach of the perturbing atom. These results, and also the anomaly mentioned and its interpretation have been confirmed and extended by other observers.[38, 39]

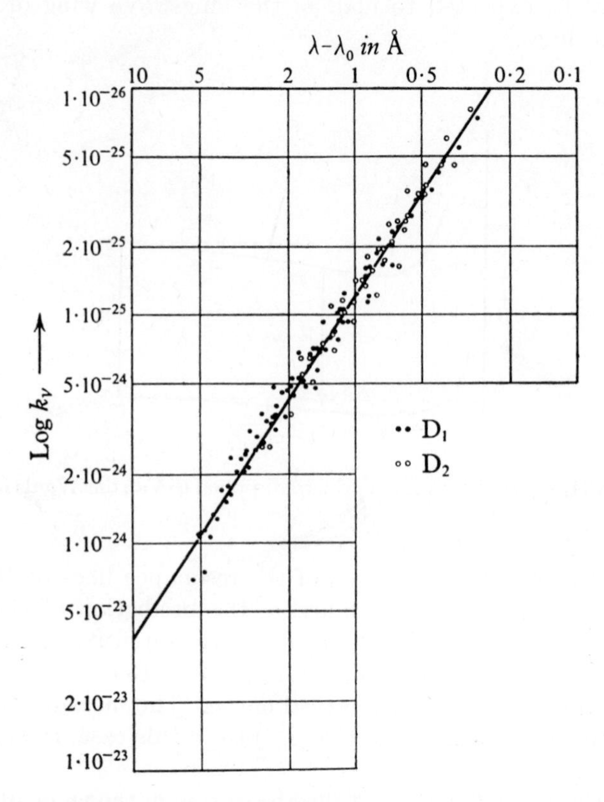

FIG. VII, 9. Statistical $(\Delta\nu)^{-3/2}$ distribution the long wave-length wings of the sodium resonance lines, broadened by argon (R. Minkowski, ref. 36).

In particular, the relation (17) has been found to be valid for both resonance lines 2537 A and 1850 A of mercury, broadened by various gases and by its own vapour[40, 41]. The value $n = 6$ of the exponent in (16) has thus been verified experimentally for several atoms to an accuracy of about $\pm\frac{1}{2}$. The validity of (17) for self-broadening confirms the theoretical result that the resonance forces largely cancel one another for higher densities.

A generalised calculation of the statistical intensity distribution including multiple encounters was carried out by Margenau[42] by a mathematical procedure similar to that used by Holtsmark for the Stark effect. The resulting formula

$$I(\Delta\nu) \sim \Delta\nu^{-\frac{3}{2}} \exp[-(\pi/4)(\overline{\Delta\nu}/\Delta\nu)] \tag{18}$$

passes into (17) for large $\Delta\nu$. The constant $\overline{\Delta\nu}$ is defined by

$$\overline{\Delta\nu} = C\left(\frac{4\pi}{3}N\right)^2$$

and is the frequency change produced by a single atom at about the average molecular distance in the gas. This generalised formula extends the range of validity of (17) towards smaller $\Delta\nu$ but only to a very limited extent: with decreasing $\Delta\nu$, the statistical theory loses its validity and becomes quite meaningless near the maximum.

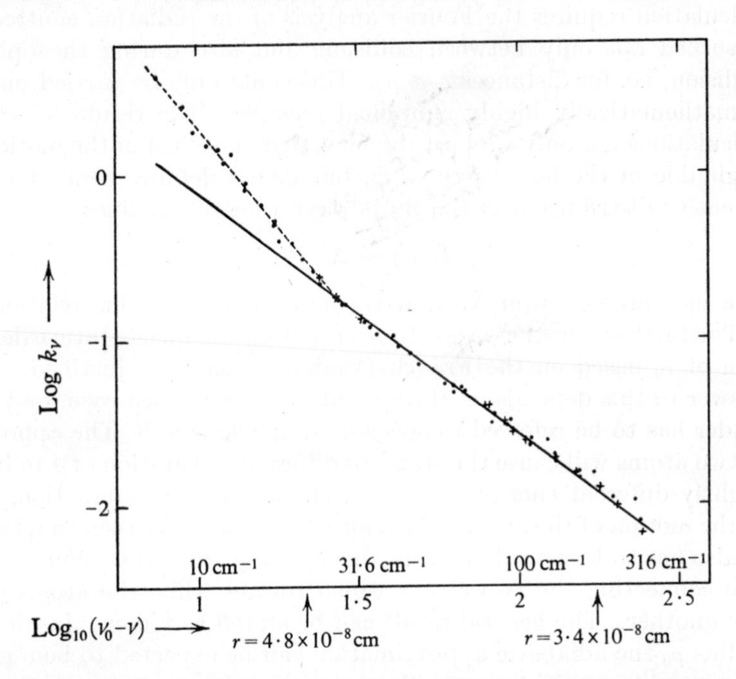

FIG. VII, 10. Intensity distribution (extinction coefficient) in the long wavelength wing of Hg resonance line 2536 A, broadened by argon.

4. Generalised theories

Both the collision- and the statistical-theory contain a number of assumptions whose validity requires closer investigation.

Accepting, at first, the classical view that the characteristic frequency ν_0 is a function of the interatomic distances alone, we have to ask to what extent the statistical intensity distribution agrees with the correct one which would result from a Fourier analysis of the time-dependent frequencies. To what extent, in other words, does the statistical distribution require corrections on account of the motion of the atoms.

Rough theoretical estimates and the good agreement with experiments lead one to expect that in the long-wave wings of the line, for fairly close encounters, the statistical theory is a good approximation. More recent theoretical work(43–47) has shown that for approaches within the Weisskopf radius ρ_0 the intensity distribution differs, in fact, very little from that given by the statistical theory. A rigorous calculation requires the Fourier analysis of the radiation emitted or absorbed not only between collisions but also during the optical collision, i.e. for distances $r < \rho_0$. This could only be carried out in a mathematically highly simplified form.(27) The results of these calculations not only support the view that the effect of the motion is negligible in the long-wave wing, but give a definite result for the intensity distribution in the short-wave wing of the line:

$$I(\Delta\nu) \sim \Delta\nu^{-7/2}. \qquad (19)$$

The measurements on Na and Hg agree well with this relation.

The further question arises to what extent the quasi-static calculation of ν_0 based on the Franck–Condon principle, is justified. The answer to this depends on the special features of each case, and the reader has to be referred to specialised articles.(34, 47) The approach of two atoms will cause the states of different orientation of **J** to have slightly different energy, and rapid changes of the interaction, due to the motion of the atoms, may cause transitions between the states. It also has to be considered that this space quantisation refers to the line connecting the two atoms which rotates while the atoms pass one another. The general result can be stated as follows: inside the radius ρ_0 the adiabatic approximation can be expected to be a good one, outside ρ_0 differences will arise in some cases, but even in this range the quasi-static assumption is a useful approximation.

From the various experimental and theoretical investigations, the following overall picture emerges. In the long-wave wings of pressure broadened lines the statistical theory of single encounters gives an adequate description of the intensity distribution. With decreasing $\Delta\nu$ multiple encounters become more important but also the statistical assumption less valid. The limit of validity is approximately given by the value of $\Delta\nu$ corresponding to the effect of a neighbouring atom at the distance of the Weisskopf radius. The half-value width and the shift of the maximum are fairly well described by the Lorentz collision theory in the form given to it by Lindholm. For finer details, such as the absolute values of the constant C, effects due to space quantisation and the influence of other atomic levels have to be considered.

Electrons, owing to their great velocity, act mainly according to the collision theory. It appears to be possible to estimate not only ion densities but also electron densities by means of broadening effects.(33) Broadening by ions and electrons has recently been subjected to a more thorough theoretical analysis(48, 49) and to experimental tests.(50)

5. States of high quantum numbers

There is one type of phenomena in which the broadening and shift of lines is due entirely to the co-operative action of a very large number of atoms. In the absorption spectra of alkali vapours one can observe lines whose upper state has very high values of the quantum number n. In these experiments, a buffer gas such as N_2 of pressure of several cm Hg is always present, and one easily estimates that under these conditions thousands of nitrogen molecules are present within the volume occupied by the ψ function of the outer electron. The foreign gas then acts like an almost continuous medium of slightly fluctuating density. Fermi(51) has derived the pressure shifts from two effects: the gas acts like a dielectric whose polarisation reduces the force of attraction between orbital electron and core, and it causes scattering of the electron. The results of these and later calculations(52) agree well with the observed widths and shifts of high members of the absorption series of alkali vapours. Especially striking is the fact that the shifts and widths due to foreign gases depend on the nature of the latter, but hardly at all on the alkali atom; the characteristic differences between the alkali atoms vanish for these high orbits.

APPENDIX TABLE 1a

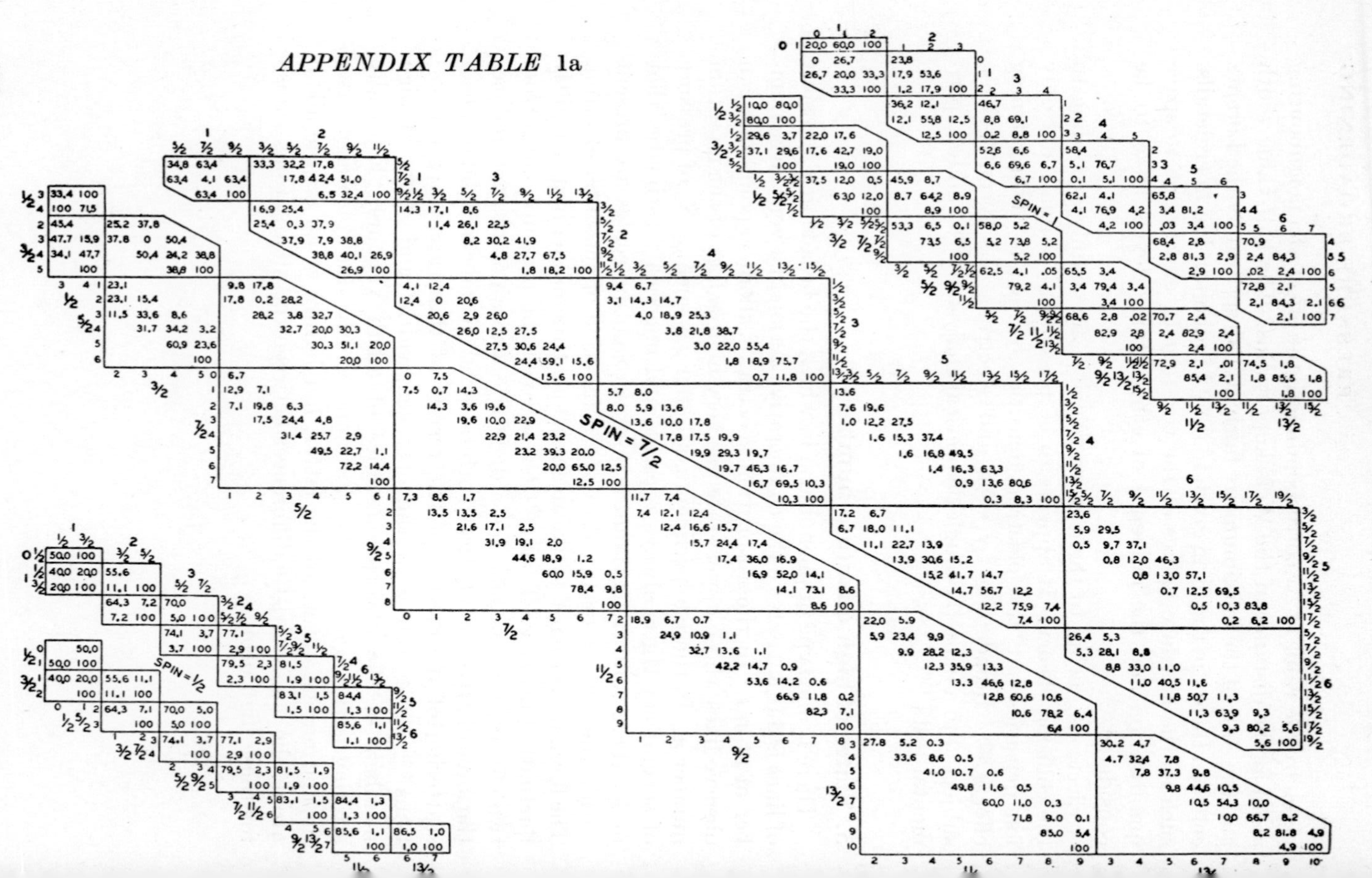

APPENDIX TABLE 1b

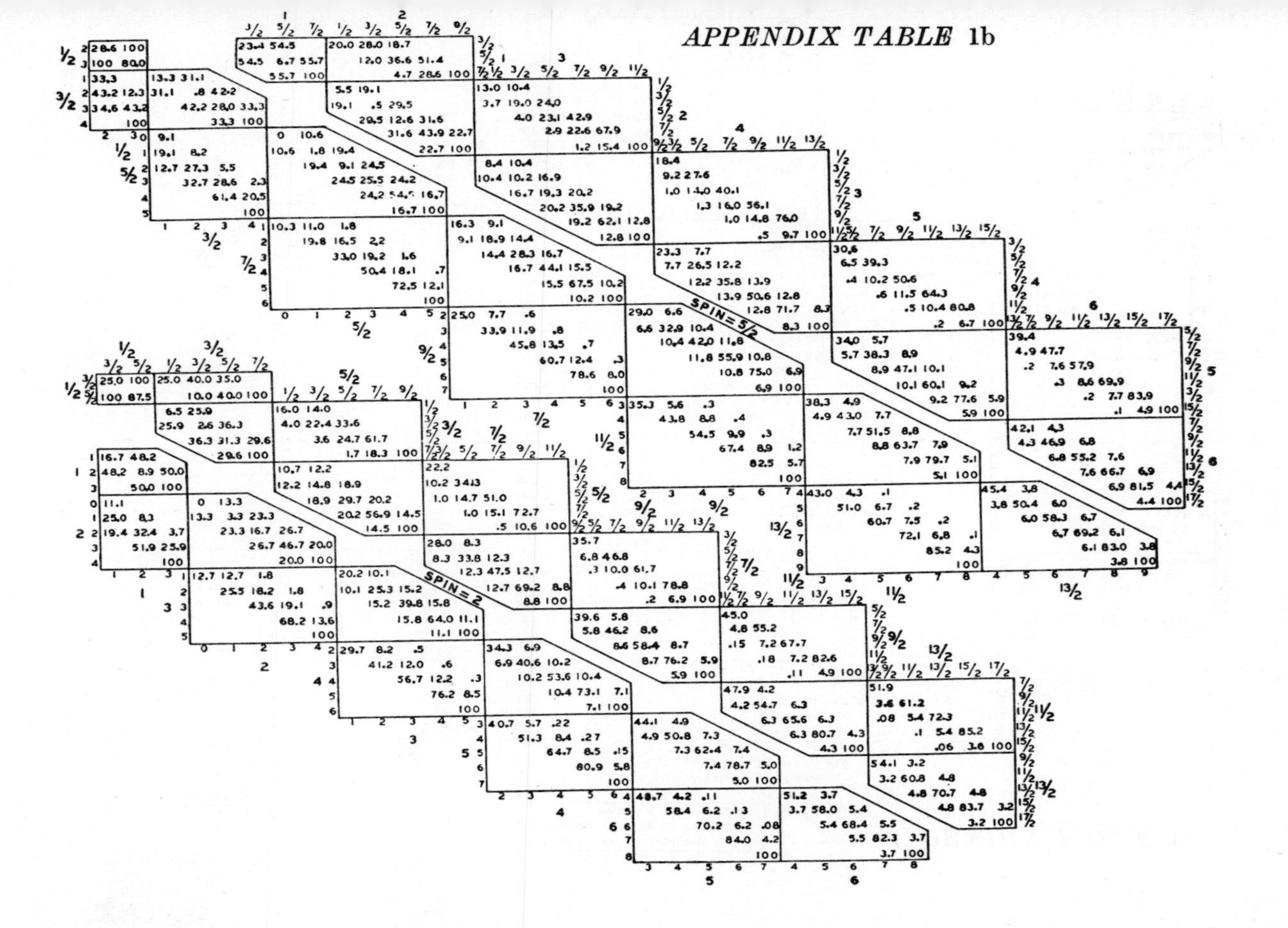

APPENDIX TABLE 1c

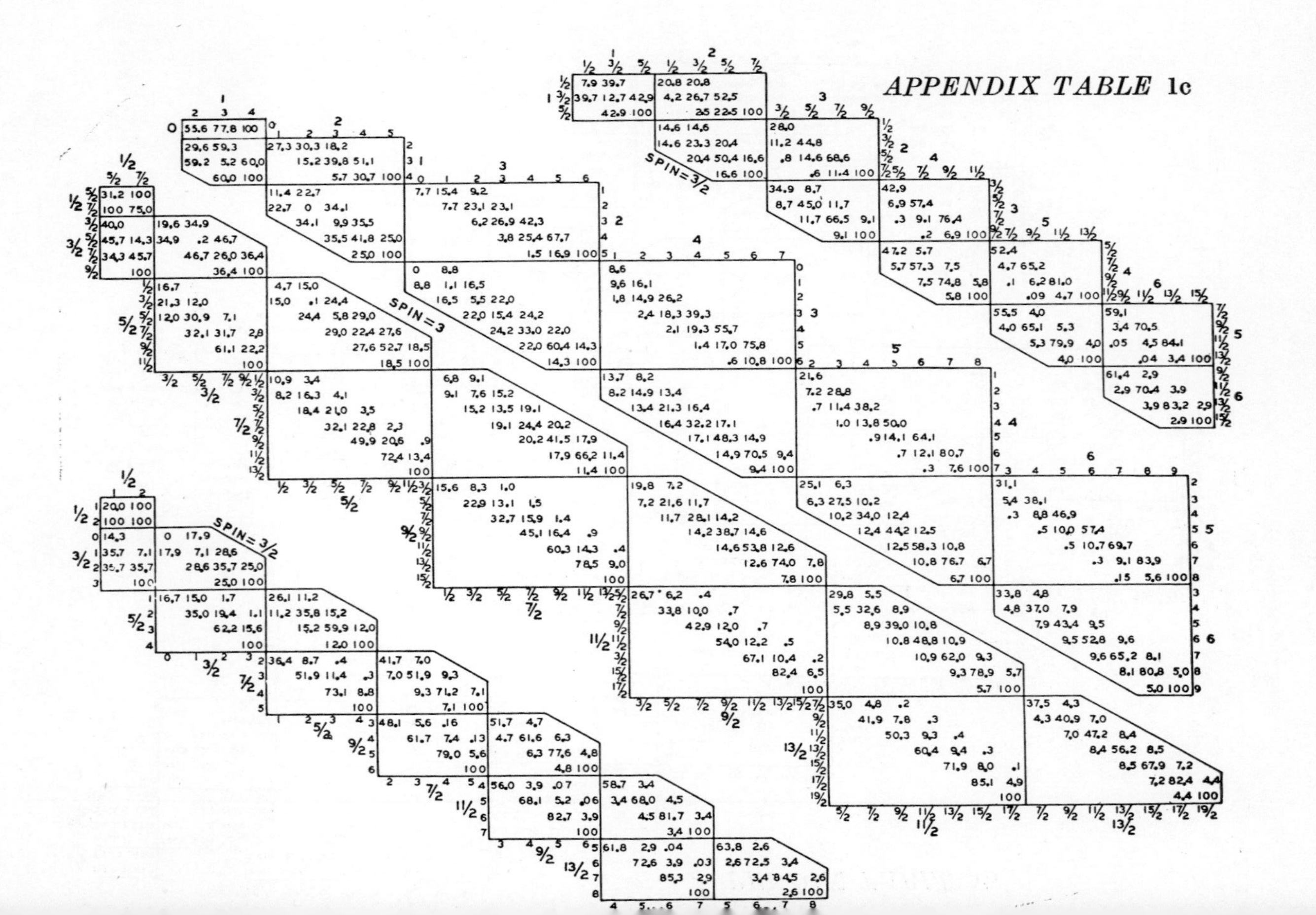

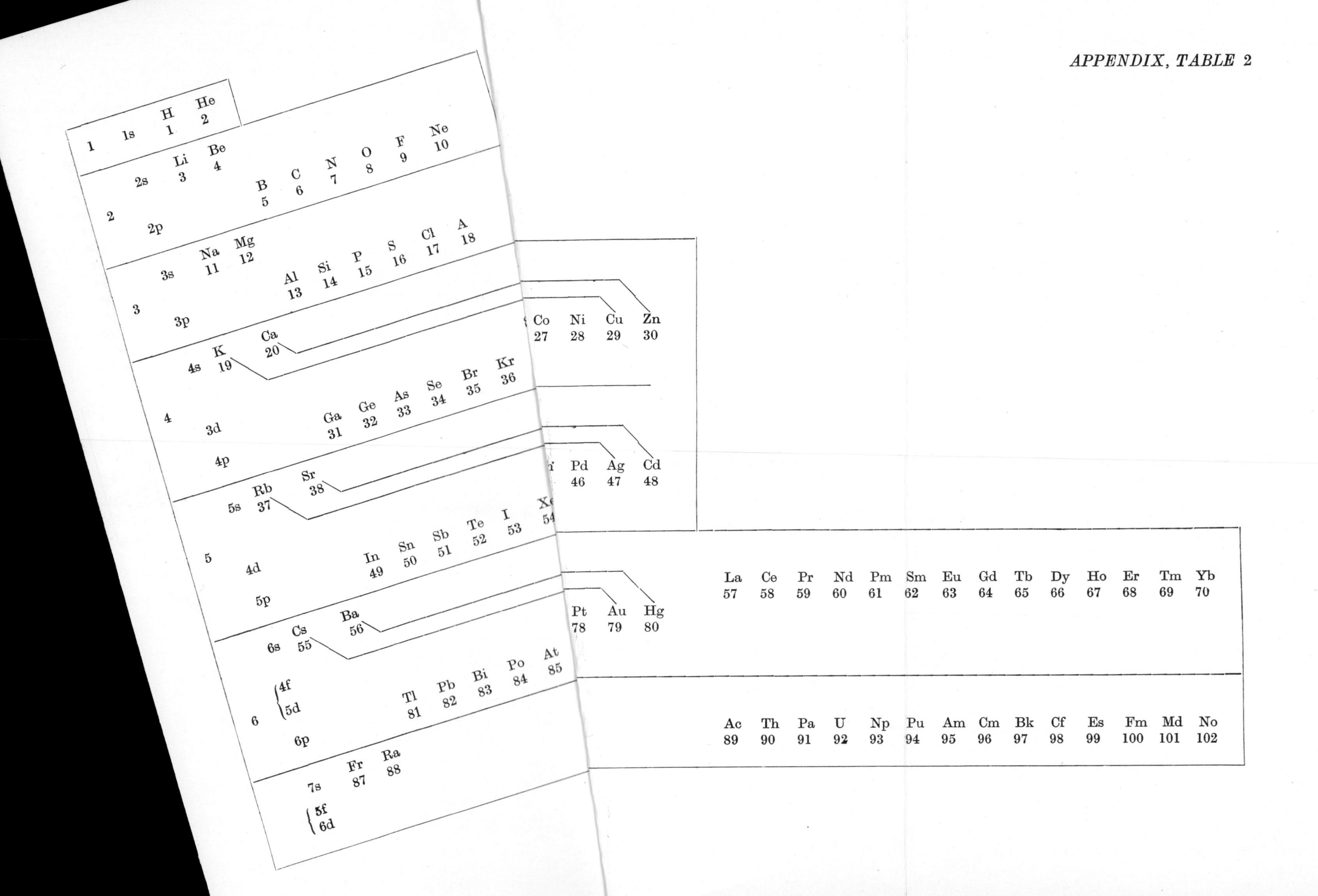

1 1s H 1 He 2
2 2s Li 3 Be 4
2p B 5 C 6 N 7 O 8 F 9 Ne 10
3 3s Na 11 Mg 12
3p Al 13 Si 14 P 15 S 16 Cl 17 A 18
4 4s K 19 Ca 20
3d Co 27 Ni 28 Cu 29 Zn 30
4p Ga 31 Ge 32 As 33 Se 34 Br 35 Kr 36
5 5s Rb 37 Sr 38
4d Pd 46 Ag 47 Cd 48
5p In 49 Sn 50 Sb 51 Te 52 I 53
6 6s Cs 55 Ba 56
4f La 57 Ce 58 Pr 59 Nd 60 Pm 61 Sm 62 Eu 63 Gd 64 Tb 65 Dy 66 Ho 67 Er 68 Tm 69 Yb 70
5d Pt 78 Au 79 Hg 80
6p Tl 81 Pb 82 Bi 83 Po 84 At 85
7s Fr 87 Ra 88
5f Ac 89 Th 90 Pa 91 U 92 Np 93 Pu 94 Am 95 Cm 96 Bk 97 Cf 98 Es 99 Fm 100 Md 101 No 102
6d

TABLE 4

Mass numbers (A), abundances and nuclear moments of natural isotopes

(Magnetic moments μ' in nuclear magnetons, quadrupole moments Q in units of 10^{-24} cm^2)

	Z	*A* (abund. in %)	*I*	μ'	*Q*
H	1	2(0·015)	1	0·86	0·0027
		1(99·985) 3(0)	1/2, 1/2	2·79; 2·98	0, —
He	2	4(100)	0	0	0
		3(10^{-4})	1/2	−2·13	—
Li	3	6(7·4)	1	0·82	—
		7(92·6)	3/2	3·26	
Be	4	9(100)	3/2	−1·18	±0·02
B	5	10(19)	3	1·80	0·07
		11(81)	3/2	2·69	0·036
C	6	12(98·9)	0	—	—
		13(1·1)	1/2	0·70	
N	7	14(99·6)	1	0·40	0·02
		15(0·38)	1/2	−0·28	
O	8	16(99·8) 18(0·20)	0, 0		
		17(0·037)	5/2	−1·89	−0·026
F	9	19(100)	1/2	2·63	
Ne	10	20(90·9) 22(8·8)	0, —	—	—
		21(0·26)	3/2	−0·66	
Na	11	23(100)	3/2	2·22	0·10
Mg	12	24(78·6) 26(11·3)			
		25(10·1)	5/2	−0·85	
		27(100)	5/2	3·64	0·15

TABLE 4 (*continued*)

Z		A (abund. in %)	I	μ'	Q
Si	14	28(92·2) 30(3·1)			
		29(4·7)	1/2	±0·55	
P	15	31(100)	1/2	1·13	
S	16	32(95) 34(4·2) 36(0·02)			
		33(0·75)	3/2	0·64	−0·06
Cl	17	35(75·5) 37(24·5)	3/2; 3/2	0·82; 0·68	−0·08; −0·06
A	18	36(0·34) 38(0·06) 40(99·6)			
K	19	40(0·01)	4	−1·30	
		39(93·1) 41(6·9)	3/2; 3/2	0·39; 0·22	0·1; ±0·1
Ca	20	40(97) 42(0·6) 44(2·0) 46(0·003) 48(0·2)			
		43(0·15)	7/2	−1·32	
Sc	21	45(100)	7/2	4·75	
Ti	22	46(7·9) 48(74) 50(5·3)			
		47(7·3) 49(5·5)	5/2; 7/2	−0·79; −1·10	
V	23	50(0·25)	6	3·34	
		51(99·75)	7/2	5·14	0·3
Cr	24	50(4·3) 52(83·7) 54(2·4)			
		53(9·5)	3/2	−0·47	
Mn	25	55(100)	5/2	3·46	≈0·4
Fe	26	54(5·8) 56(91·7) 58(0·31)			
		57(2·17)	1/2	<0·05	
Co	27	59(100)	7/2	4·64	0·5
Ni	28	58(67·7) 60(26·2) 62(3·7) 64(1·2)			
		61(1·25)		small	
Cu	29	63(69) 65(31)	3/2; 3/2	2·22; 2·38	−0·16; −0·15

Zn	30	64(49) 66(28) 68(18·5) 70(0·6)			
		67(4·1)	5/2	0·87	0·18
Ga	31	69(60·2) 71(39·8)	3/2; 3/2	2·01; 2·55	0·19; 0·12
Ge	32	70(20·5) 72(27·4) 74(36·7) 76(7·7)			
		73(7·7)	9/2	−0·88	−0·2
As	33	75(100)	3/2	1·43	0·3
Se	34	74(1) 76(9) 78(24) 80(50) 82(9)			
		77(7·5)	1/2	0·53	
Br	35	79(50·5) 81(49·5)	3/2; 3/2	2·10; 2·26	0·33; 0·28
Kr	36	78(0·3) 80(2) 82(11·5) 84(57) 86(17·5)			
		83(11·5)	9/2	0·97	0·2
Rb	37	85(72) 87(28)	5/2; 3/2	1·35; 2·74	0·3; 0·15
Sr	38	84(0·56) 86(9·9) 88(82·6)			
		87(7·0)	9/2	−1·09	
Y	39	89(100)	1/2	−0·14	
Zr	40	90(51·5) 92(17) 94(17·4) 96(2·8)			
		91(11·2)	5/2	−1·30	
Nb	41	93(100)	9/2	6·14	≈ −0·3
Mo	42	92(16) 94(9·1) 96(16·5) 98(24) 100(9·3)			
		95(16) 97(9·6)	5/2; 5/2	−0·93; −0·95	
Tc	43	[97] [99]	—; 9/2	—; 5·66	
Ru	44	96(5·5) 98(2·2) 100(12·7) 102(31·3) 104(18·3)			
		99(12·7) 101(17·0)	5/2; 5/2	−0·6; −0·7	
Rh	45	103(100)	1/2	−0·088	
Pd	46	102(1) 104(11) 106(27) 108(27) 110(13)			
		105(22)	5/2	−0·57	
Ag	47	107(51) 109(49)	1/2; 1/2	−0·11; −0·13	
Cd	48	106(1·2) 108(1) 110(12) 112(24) 114(28) 116(7·3)			
		111(12·7) 113(12·3)	1/2; 1/2	−0·59; −0·62	
In	49	113(4·3) 115(95·7)	9/2; 9/2	5·50; 5·51	≈ 1; ≈ 1

TABLE 4 (*continued*)

Z		A (abund. in %)	I	μ′	Q
Sn	50	112(1) 114(1) 116(14) 118(24) 120(33) 122(4·7) 124(6·0) 115(0·4) 117(8) 119(9)	1/2, 1/2, 1/2	−0·91; −0·99; −1·04	
Sb	51	121(57) 123(43)	5/2; 7/2	3·34; 2·53	≈ −0·6; ≈ −0·8
Te	52	120(0·1) 122(2·4) 124(4·6) 126(18·7) 128(32) 130(34·5) 123(0·85) 125(7·0)	1/2; 1/2	−0·73; −0·88	
I	53	127(100)	5/2	2·79	−0·7
Xe	54	124(0·1) 126(0·1) 128(1·9) 130(4·1) 132(27) 134(10·5) 136(9) 129(26) 131(21)	1/2, 3/2	−0·77; +0·67	—; −0·1
Cs	55	133(100)	7/2	2·56	−0·003
Ba	56	130(0·1) 132(0·1) 134(2·4) 136(7·8) 138(71·6) 135(6·6) 137(11·3)	3/2; 3/2	0·82; 0·93	
La	57	138(0·089) 139(99·91)	5 7/2	3·68 2·76	±0·9 0·27
Ce	58	136(0·2) 138(0·25) 140(89) 142(11)			
Pr	59	141(100)	5/2	3·8	−0·054
Nd	60	142(27) 144(24) 146(17) 148(5·7) 150(5·6) 143(12·2) 145(8·2)	7/2; 7/2	1·0; 0·6	± ≈ 1; ± ≈ 1
Pm	61	[145]	—	—	—
Sm	62	144(3) — 148(11) 150(7·4) 152(26·7) 154(23) 147(15) 149(14)	7/2; 7/2	−0·83; −0·68	
Eu	63	151(48) 153(52)	5/2; 5/2	3·42; 1·50	1·2; 2·5
Gd	64	152(0·2) 154(2) 156(20·5) 158(25) 160(22) 155(15) 157(16)	3/2; 3/2	−0·3; −0·4	1·1; 1·0
Tb	65	159(100)	3/2	±1·5	

Dy 66	156(0·05) 158(0·1) 160(2·3) 162(25) 164(28)			
	161(19) 163(25)	5/2; 5/2	−0·37; +0·51	1·1, 1·3
Ho 67	165(100)	7/2	±3·3	±2
Er 68	162(0·1) 164(1·5) 166(33·4) 168(27) 170(15)			
	167(23)	7/2	±0·5	±10
Tm 69	169(100)	1/2	−0·21	
Yb 70	168(0·14) 170(3) 172(22) 174(32) 176(13)			
	171(14·4) 173(16)	1/2; 5/2	0·45; −0·7	—; ≈ 3
Lu 71	176(2·6)	5–7	2·8	8
	175(97·4)	7/2	2	6
Hf 72	174(0·16) 176(5) 178(27) 180(35)			
	177(18·5) 179(13·7)	7/2; 9/2	0·61; −0·47	3; 3
Ta 73	180(0·012)			
	181(99·988)	7/2	2·1	3
W 74	180(0·14) 182(26·4) 184(30·6) 186(28·4)			
	183(14·4)	1/2	0·115	
Re 75	185(37) 187(63)	5/2; 5/2	3·14; 3·18	2·8; 2·6
Os 76	184(0·02) 186(1·6) 188(13) 190(26·4) 192(41)			
	187(1·6) 189(16)	1/2; 3/2	0·12; 0·65	—; 0·6
Ir 77	191(38·5) 193(61·5)	3/2; 3/2	0·2; 0·2	1; 1
Pt 78	190(0·01) 192(0·8) 194(33) 196(25) 198(7·2)			
	195(34)	1/2	0·60	
Au 79	197(100)	3/2	0·14	0·56
Hg 80	196(0·15) 198(10) 200(23) 202(30) 204(6·8)			
	199(17) 201(13)	1/2; 3/2	0·50; −0·61	—; 0·5
Tl 81	203(29·5) 205(70·5)	1/2; 1/2	1·60; 1·61	
Pb 82	204(1·4) 206(25) 208(52) 210(≡ RaD)			
	207(22)	1/2	0·58	
Bi 83	209(100)	9/2	4·04	−0·4
Ac 89	227	3/2	1·1	−1·7

TABLE 4 (*continued*)

Z	A (abund. in %)	I	μ'	Q
Pa 91	231	3/2		
U 92	234(0·006) 238(99·28)			
	[233] 235(0·72)	5/2; 7/2	±0·5; ±0·34	±3·4; ±4·0
Np 93	[237]	5/2	±6	
Pu 94	[239] [241]	1/2; 5/2	±0·4; ±1·4	
Am 95	[241] [243]	5/2; 5/2	1·4; 1·4	4·9; 4·9

TABLE 5

Symbols, definitions, constants*

(c.g.s. units are used throughout, in the Gaussian form: electric quantities in e.s.u., magnetic quantities in e.m.u.).

Symbols for angular momenta and (in brackets) their associated quantum numbers:

		vector	square of abs. value	z-component
electrons	orbital	$\mathbf{L}$	$\mathbf{L}^2 (l, L)$	L_z or L_z (m_l, M_L)
	spin	$\mathbf{S}$	$\mathbf{S}^2 (s, S)$	S_z or S_z (m_s, M_S)
	total	$\mathbf{J}$	$\mathbf{J}^2 (j, J)$	J_z or $J_z (m, M)$
nuclear spin		$\mathbf{I}$	$\mathbf{I}^2 (I)$	I_z or I_z (m_I)
overall total		$\mathbf{F}$	$\mathbf{F}^2 (F)$	F_z or F_z (m_F or m)

Mass number: A, charge number: Z, neutron number: N.
velocity of light: $c = 2{\cdot}99793 \times 10^{10}$ cm/sec.

Planck's constant: $h = 6{\cdot}6253 \times 10^{-27}$ erg.sec.
$\hbar = h/2\pi = 1{\cdot}05445 \times 10^{-27}$ erg.sec.
electron mass: m or $m_0 = 9{\cdot}1085 \times 10^{-28}$ g.
proton mass: $M_p = 1{\cdot}6724 \times 10^{-24}$ g.
ratio: $M_p/m = 1836{\cdot}12$.

Magnetic and electric field strength: $\mathscr{H}$ and $\mathbf{F}$ resp.
elementary charge: $e = 4{\cdot}8029 \times 10^{-10}$ esu.
(electron charge $= -e$)
specific charge of electron: $e/m = 5{\cdot}2730 \times 10^{17}$ esu.
$e/mc = 1{\cdot}75889 \times 10^{7}$ esu.

Bohr magneton: $\mu_0 = \hbar e/2mc = \gamma_0 \hbar = 0{\cdot}92732 \times 10^{-20}$ emu.

Magn. moment of electron: $-\mu_0 \times 1{\cdot}001145$.

Nuclear magneton: $\mu_m = \mu_0/1836{\cdot}12$.
$\mu' = \mu/\mu_n = g_I I$.

Radius of first Bohr orbit (for infinite mass of nucleus):
$a_0 = \hbar^2/me^2 = 0{\cdot}529173 \times 10^{-8}$ cm.

Electric quadrupole moment: $e \times Q$ in units of charge (esu) $\times \mathrm{cm}^2$.

Rydberg constant for infinite nuclear mass:
$R = R_\infty = 2\pi^2 me^4/ch^3 = 109{,}737{\cdot}31\ \mathrm{cm}^{-1}$.

Fine structure constant: $\alpha = e^2/(\hbar c) = 7{\cdot}2973 \times 10^{-3} = 1/137{\cdot}037$.

Avogadro's number: $N_0 = 6{\cdot}0248 \times 10^{23}$ per mole.

Boltzmann constant: $k = 1{\cdot}3805 \times 10^{-16}$ erg/degree.
1 electron volt (e.v.) $= 1{\cdot}6021 \times 10^{-12}$ erg,
corresp. to $8066{\cdot}0\ \mathrm{cm}^{-1}$ or 12398 A.U.
1 Rydberg $= hcR_\infty = 2{\cdot}1796 \times 10^{-11}$ erg $= 13{\cdot}605$ e.v.

* J. W. M. Du Mond and E. R. Cohen, *Rev. Mod. Phys.* **25**, 691, 1953; **27**, 363, 1955. J. A. Bearden and J. S. Thomsen, *Nuov. Cim.* **5** (Suppl.) 267, 1957.

APPENDIX

Magnetic splitting factors in multiplet-and h.f. structure:

a and A in ergs, a' and A' in cm^{-1}.

Quadrupole splitting factors: b and B in ergs, b' and B' in cm^{-1}.

Atomic units:†

atomic quantities are expressed in units of universal constants as follows: mass in units of m_0, charge in units of e, length in units of a_0, time in units of $\tau_0 = \hbar^3/me^4$ = time of $1/2\pi$ revol. in Bohr's first orbit $= 2{\cdot}419 \times 10^{-17}$ sec, angular momentum in units of $\hbar$, energy in units of $2Rhc$ (2 Rydberg).

† D. R. Hartree, *Proc. Cambr. Phil. Soc.* **24**, 89, 1926.

Bibliography

I.

1. S. Goudsmit and G. E. Uhlenbeck, *Naturwiss.* **13**, 953, 1925.
2. G. D. Liveing and J. Dewar, *Phil. Trans. Roy. Soc.*, **174**, 187, 1883.
3. J. R. Rydberg, *Astrophys. J.* **4**, 91, 1896.
4. W. Ritz, *Astrohpys. J.* **28**, 237, 1908.
5. W. Grotrian, *Graphische Darstellungen d. Spektren*, Vols. I and II, J. Springer, Berlin 1928.
6. J. Franck and G. Hertz, *Verhandl. Phys. Ges. Berlin*, **15**, 34, 1913.
7. B. Edlén, "Trans. Joint Com. Spectr.", *J. Opt. Soc. Amer.* **43**, 412, 1953.
8. G. R. Harrison, Mass. Inst. Techn. *Wavelength Tables*, New York 1939.
9. H. Kayser, *Tabelle d. Hauptlinien aller Elemente*, Springer 1926.
10. *Trans. Intern. Astr. Union*, **6**, 79, 1938.
11. A. G. Shenstone, *J. Opt. Soc. Amer.* **45**, 868, 1955.
12. P. G. Wilkinson, *J. Opt. Soc. Amer.* **45**, 862, 1955.
13. G. Herzberg, *Proc. Roy. Soc.* A **234**, 516, 1956.
14. R. D. Van Held and K. W. Meissner, *J. Opt. Soc. Amer.* **46**, 598, 1956.
15. C. E. Moore, *Atomic Energy Levels*, Vols. I, II and III, *Nat. Bur. Stand. Circular* 467, 1949–1958.
16. R. F. Bacher and S. Goudsmit, *Atomic Energy States*, McGraw-Hill 1932.
17. A. G. Shenstone, *Rep. Progr. Phys.* **5**, 210, 1938.
18. W. F. Meggers, *J. Opt. Soc. Amer.* **36**, 431, 1946.
19. W. F. Meggers, *J. Opt. Soc. Amer.* **43**, 415, 1953.
20. P. Brix and H. Kopfermann, *Landolt-Börnstein, Tabellen*, 6 ed. Vol. I, p. 1–69, Springer 1951.

II.

1. F. Reiche and W. Thomas, *Naturwiss.* **13**, 627, 1925.
2. W. Kuhn, *Z. Phys.* **33**, 408, 1925.
3. E. Wigner, *Phys. Z.* **32**, 450, 1931.
4. J. G. Kirkwood, *Phys. Z.* **33**, 521, 1932.
5. O. Laporte, *Z. Phys.* **23**, 135, 1924.
6. A. Rubinowitz and J. Blaton, *Ergebn. exakt. Naturw.* **11**, 176, 1932.
7. S. Mrozowsky, *Rev. Mod. Phys.* **16**, 153, 1944.
8. H. C. Brinkman, *Quantnmech. d. Mulstipolstrahlung*, Thesis Groningen, Noordhoff 1932.
9. A. Rubinowitz, *Rep. Progr. Phys.* **12**, 233, 1949.
10. I. I. Rabi, S. Millman, P. Kusch and J. R. Zacharias, *Phys. Rev.* **55**, 526, 1939.
11. F. Bloch, *Phys. Rev.* **70**, 460, 1946.
12. J. H. Gardner and E. M. Purcell, *Phys. Rev.* **76**, 1262, 1949.

13. M. Weingeroff, *Z. Phys.* **67**, 679, 1931.
14. A. S. King, *Astrophys. J.* **56**, 318, 1922 and following years.
15. J. S. Bowen, *Proc. Nat. Acad. Sci. Wash.* **14**, 30, 1928.
16. F. Becker and W. Grotrian, *Ergebn. exakt. Naturw.* **7**, 8, 1928.
17. B. Edlén, *Mon. Not. R. Astr. Soc.* **105**, 323, 1945.
18. H. C. Burger and P. C. van Cittert, *Z. Phys.* **51**, 638, 1928.

III.

1. G. Herzberg, *Ann. Phys. Lpz.* **84**, 565, 1927.
2. N. Bohr. *Phil. Mag.* **26**, 1, 476, 1913.
3. E. R. Cohen, *Phys. Rev.* **88**, 353, 1952.
4. J. W. M. Dumond, *Rev. Mod. Phys.* **25**, 691, 1953.
5. W. V. Houston, *Phys. Rev.* **30**, 608, 1927.
6. R. C. Williams, *Phys. Rev.* **54**, 568, 1938.
7. J. W. Drinkwater, O. Richardson and W. E. Williams, *Proc. Roy. Soc.* A**174**, 164, 1940.
8. B. Edlén, *Nova Acta Soc. Sci. Upsal.* IV, **9**, 6, 1933.
9. E. Tyrén, *Nova Acta Soc. Sci. Upsal.* IV, **12**, 24, 1940.
10. W. Wilson, *Phil. Mag.* **29**, 795, 1915.
11. A. Sommerfeld, *Ann. Phys. Lpz.* **51**, 1, 1916.
12. J. Stark, *Akad. Wiss. Berlin*, **40**, 932, 1913.
13. A. Lo Surdo, *Accad. Linc. Atti.* **22**, 665, 1913.
14. P. S. Epstein, *Ann. Phys. Lpz.* **50**, 489, 1916, and *Phys. Z.* **17**, 148, 1916.
15. K. Schwarzschild, *Akad. Wiss. Berlin* 1916, 548.
16. E. Schrödinger, *Ann. Phys. Lpz.* **80**, 437, 1926.
17. P. S. Epstein, *Phys. Rev.* **28**, 695, 1926.
18. L. Kassner, *Z. Phys.* **81**, 346, 1933.
19. M. Kiuti, *Z. Phys.* **57**, 658, 1929.
20. H. R. v. Traubenberg and R. Gebauer, *Z. Phys.* **54**, 307; **56**, 254, 1929, **62**, 289, 1930.
21. H. Mark and R. Wierl, *Z. Phys.* **55**, 156, 1929.
22. J. R. Oppenheimer, *Phys. Rev.* **31**, 66, 1928.
23. H. R. v. Traubenberg, R. Gebauer and G. Lewin, *Naturwiss.* **18**, 418, 1930.
24. C. Lanczos, *Z. Phys.* **65**, 431, 1930.
25. L. H. Thomas, *Nature*, **117**, 514, 1926.
26. J. Frenkel, *Z. Phys.* **37**, 243, 1926.
27. W. Heisenberg and P. Jordan, *Z. Phys.* **37**, 263, 1926.
28. P. A. M. Dirac, *Proc. Roy. Soc.* A **117**, 610, 1928.
29. R. C. Williams, *Phys. Rev.* **54**, 558, 1938.
30. S. Pasternack, *Phys. Rev.* **54**, 1113, 1938.
31. W. E. Lamb Jr. and R. C. Retherford, *Phys. Rev.* **72**, 241, 1947.
32. H. G. Kuhn and G. W. Series, *Nature* **162**, 373, 1948, *Proc. Roy. Soc.* A **202**, 127, 1950.
33. H. A. Bethe, *Phys. Rev.* **72**, 339, 1947.
34. E. E. Salpeter, *Phys. Rev.* **89**, 92, 1953.
35. W. E. Lamb, *Phys. Soc. London, Yearbook* 1958, p. 1.
36. T. A. Welton, *Phys. Rev.* **74**, 1157, 1948.
37. G. W. Series, *Proc. Roy. Soc.* A **208**, 277, 1951.

38. E. Lipworth and R. Novik, *Phys. Rev.* **108**, 1434, 1957.
39. G. Herzberg, *Proc. Roy. Soc.* A **234**, 516, 1956.
40. F. Paschen, *Ann. Phys. Lpz.* **82**, 689, 1927.
41. H. Kopfermann, H. Krüger and H. Öhlmann, *Z. Phys.* **126**, 760, 1949.
42. J. G. Hirschberg and J. E. Mack, *Phys. Rev.* **77**, 745, 1950.
43. K. Murakawa and S. Suwa, *J. Phys. Soc. Jap.* **7**, 467, 1952.
44. G. W. Series, *Proc. Roy. Soc.* A **226**, 377, 1954.
45. G. Herzberg, *Z. Phys.* **146**, 267, 1956.
46. R. Schlapp, *Proc. Roy. Soc.* A **119**, 313, 1928.
47. V. Rojansky, *Phys. Rev.* **33**, 1, 1929.
48. G. Lüders, *Ann. Phys. Lpz.* **8**, 301, 1951.
49. G. W. Series and K. W. H. Stevens, *Proc. Roy. Soc.* A **226**, 393, 1954.
50. J. S. Foster, *Rep. Progr. Phys.* **5**, 233, 1938.
51. B. Kullenberg, Dissertation. Lund 1938.
52. W. Gordon, *Ann. Phys. Lpz.* **2**, 1031, 1929.
53. T. E. Hull and L. Infeld, *Phys. Rev.* **74**, 905, 1948.
54. H. Bethe, *Handb. d. Phys.* **24**, 1, Springer 1933.
55. T. Takamine, L. S. Ornstein and J. M. W. Milatz, *Z. Phys.* **78**, 169, 1932.
56. A. Carst and R. Ladenburg, *Z. Phys.* **48**, 192, 1928.
57. L. S. Ornstein and H. C. Burger, *Z. Phys.* **83**, 177, 1933.
58. L. R. Maxwell, *Phys. Rev.* **38**, 1664, 1931.
59. J. Brochard, R. Chabbal, H. Chantrel and P. Jacquinot, *J. Phys. Radium* **13**, 433, 1952.
60. B. Edlén, *Ark. f. Fys.*, **4**, 441, 1952.
61. E. A. Hylleraas, *Z. Phys.* **54**, 347, 1929; **65**, 209, 1930.
62. C. L. Pekeris, *Phys. Rev.* **112**, 1649, 1958.
63. H. M. James and A. S. Coolidge, *Phys. Rev.* **51**, 857, 1937.
64. T. Kinoshita, *Phys. Rev.* **115**, 366, 1959.
65. G. Herzberg, *Proc. Roy. Soc.* A **248**, 309, 1958.
66. E. A. Hylleraas and J. Midtdal, *Phys. Rev.* **103**, 829, 1956.
67. L. C. Green et al., *Phys. Rev.* **104**, 1593, 1956.
68. E. Holoien, *Phys. Rev.* **104**, 1301, 1956
69. W. Heisenberg, *Z. Phys.* **39**, 499, 1926.
70. G. Breit, *Phys. Rev.* **36**, 383, 1929; **39**, 616, 1932.
71. D. R. Inglis, *Phys. Rev.* **61**, 297, 1942.
72. G. Araki, et al. *Phys. Rev.* **116**, 651, 1959.
73. I. Wieder and W. E. Lamb, Jr. *Phys. Rev.* **107**, 125, 1957.
74. D. R. Bates and A. Damgaard, *Phil. Trans.* **242**, 101, 1949.
75. Ta-You-Wu, *Phys. Rev.* **58**, 1114, 1940.
76. P. Lee and P. L. Weissler, *Phys. Rev.* **99**, 540, 1955.
77. L. R. Henrich, *Astrophys. J.* **99**, 59, 1944.
78. D. Chalonge, L. Divan and V. Kourganoff, *Ann. Astrophys.* **13**, 347, 1950.
79. H. S. W. Massey, *Negative Ions*, Camb. Univ. Press 1950.
80. W. Lochte-Holtgreven, *Naturwiss.* **38**, 258, 1951.
81. L. M. Branscomb and S. J. Smith, *Phys. Rev.* **98**, 1028, 1955.
82. S. Chandrasekhar, *Astophys. J.* **102**, 223, 395, 1945.
83. S. Geltman, *Phys. Rev.* **104**, 346, 1956.
84. R. W. Wood and R. Fortrat, *Astrophys. J.* **43**, 73, 1916.
85. E. R. Thackeray, *Phys. Rev.* **75**, 1840, 1949.
86. H. R. Kratz, *Phys. Rev.* **75**, 1844, 1949.

87. K. W. Meissner, L. G. Mundie and P. H. Stelson, *Phys. Rev.* **74**, **932, 1948.**
88. M. Born and W. Heisenberg, *Z. Phys.* **4**, 347, 1921.
89. I. Waller, *Z. Phys.* **38**, 635, 1926.
90. A. T. van Urk, *Z. Phys.* **13**, 268, 1923.
91. D. R. Hartree, *Proc. Cambr. Phil. Soc.* **24**, 426, 1928.
92. R. Jastrow, *Phys. Rev.* **73**, 60, 1948.
93. R. M. Sternheimer, *Phys. Rev.* **96**, 951, 1954.
94. K. Bockasten, *Phys. Rev.* **102**, 729, 1956.
95. H. M. James and A. S. Coolidge, *Phys. Rev.* **49**, 688, 1936.
96. K. Huang, *Phys. Rev.* **70**, 197, 1946.
97. D. R. Hartree, *Rep. Progr. Phys.* **11**, 113, 1948.
98. P. Gombas, *Z. Phys.* **118**, 164, 1941.
99. L. Biermann, *Z. Astrophys.* **22**, 157, 1943; *Nachr. Wiss. Gött.* **2**, **116, 1946.**
100. A. Landé, *Z. Phys.* **25**, 46, 1924.
101. H. B. G. Casimir, *On the interaction between atomic nuclei and electrons,* De Erven F. Bohn, N. V., Haarlem 1936.
102. A. G. Barnes and W. V. Smith, *Phys. Rev.* **93**, 95, 1954.
103. E. David, *Z. Phys.* **91**, 289, 1934.
104. B. Kockel and H. Wagenbreth, *Ann. Phys. Lpz.* **9**, 1, 1951.
105. H. E. White, *Phys. Rev.* **40**, 316, 1932.
106. R. F. Bacher, *Phys. Rev.* **43**, 269, 1933.
107. M. Philips, *Phys. Rev.* **44**, 644, 1933.
108. G. Araki, *Proc. Phys. Math. Soc. Jap.* **21**, 508, 1939.
109. T. Y. Wu, *J. Phys. Soc. Jap.* **4**, 343, 1949.
110. H. B. Dorgelo, *Phys. Z.* **26**, 756, 1925.
111. L. S. Ornstein and H. C. Burger, *Z. Phys.* **40**, 403, 1926.
112. S. Sambursky, *Z. Phys.* **49**, 731, 1928.
113. C. Füchtbauer and H. W. Wolff, *Ann. Phys. Lpz.* (**5**), **3**, 359, 1929.
114. E. Fermi, *Z. Phys.* **59**, 680, 1930.
115. B. Trumpy, *Z. Phys.* **57**, 787, 1929; **61**, 54; **66**, 720, 1930.
116. W. K. Prokofjew, *Z. Phys.* **58**, 255, 1929.
117. J. Hargreaves, *Proc. Camb. Phil. Soc.* **25**, 75, 1929.
118. L. Biermann and K. Lübeck, *Z. Astrophys.* **26**, 43, 1949.
119. J. N. Nanda, *Ind. J. Phys.* **19**, 1, 1945.
120. D. S. Villars, *J. Opt. Soc. Amer.* **42**, 552, 1952.
121. E. A. Hylleraas, *Arch. Math. Naturw.* B **48**, No. 4, 1945.
122. R. Ladenburg and E. Thiele, *Z. Phys.* **72**, 697, 1931.
123. R. Minkowski, *Z. Phys.* **36**, 839, 1926.
124. S. A. Korff, *Astrophys. J.* **76**, 124, 1932.
125. J. F. Heard, *Mon. Not. R. Astr. Soc.* **94**, 458, 1933.
126. M. Weingeroff, *Z. Phys.* **67**, 679, 1931.
127. M. L. Perl, I. I. Rabi and B. Senitzky, *Phys. Rev.* **98**, 611, 1955.
128. A. Filippow and W. Prokofjew, *Z. Phys.* **56**, 458, 1929.
129. A. Filippow, *Z. Phys.* **69**, 526, 1931.
130. R. W. Ditchburn, J. J. Jutsum and G. V. Marr, *Proc. Roy. Soc.* A **219**, 89, 1953.
131. D. R. Bates and H. S. W. Massey, *Proc. Roy. Soc.* A **177**, 329, 1941.
132. M. J. Seaton, *Proc. Roy. Soc.* A **208**, 408, 418, 1951.
133. F. L. Mohler, *Bur. Stand. J. Res. Wash.* **17**, 849, 1936: **19**, 447; **21**, 697, 873, 1938.

134. L. W. Davies, *Proc. Phys. Soc.* B **66**, 33, 1953.
135. S. Datta, *Proc. Roy. Soc.* **101**, 539, 1922.
136. A. L. M. Sowerby and S. Barratt, *Proc. Roy. Soc.* **110**, 192, 1926.
137. C. Shang-Yi and C. Chih-San, *Phys. Rev.* **75**, 81, 1949.
138. E. Segré, *Z. Phys.* **66**, 827, 1931.
139. H. G. Kuhn, *Z. Phys.* **61**, 805, 1930.
140. C. J. B. Bakker, *Proc. Amst.* **36**, 589, 1933.
141. E. Amaldi, *Rend. R. Accad. Nat. Lincei* **19**, 588, 1934.
142. F. A. Jenkins and E. Segré, *Phys. Rev.* **55**, 545, 1939.
143. D. R. Hartree and W. Hartree, *Proc. Roy. Soc.* A **883**, 588, 1936.
144. A. Landé, *Z. Phys.* **15**, 189; **19**, 112, 1923.
145. P. J. Rubenstein, *Phys. Rev.* **58**, 1007, 1940.
146. B. R. Judd, *Proc. Phys. Soc.* A **69**, 157, 1956.
147. R. de L. Kronig, *Z. Phys.* **33**, 261, 1925.
148. A. Sommerfeld and H. Hönl, *Akad. Wiss. Berlin* **9**, 141, 1925.
149. H. N. Russell, *Nature*, **115**, 835, 1925.
150. P. Dirac, *Proc. Roy. Soc.* A **111**, 281, 1926.
151. H. E. White and A. Y. Eliason, *Phys. Rev.* **44**, 753, 1933.
152. G. W. King and J. H. Van Vleck, *Phys. Rev.* **56**, 464, 1939; P. J. Rubenstein, *Phys. Rev.* **58**, 1007, 1940.
153. A. Landé, *Z. Phys.* **5**, 231, 1921.
154. F. Paschen and E. Back, *Physica* **1**, 261, 1921.
155. W. Heisenberg and P. Jordan, *Z. Phys.* **37**, 263, 1926.
156. C. G. Darwin, *Proc. Roy. Soc.* A 115, 1927.
157. L. S. Ornstein and H. C. Burger, *Z. Phys.* **28**, 135; **29**, 241, 1924.
158. K. Darwin, *Proc. Roy. Soc.* A **118**, 264, 1928.
159. N. A. Kent, *Astrophys. J.* **40**, 339, 1914.
160. P. Kapitza, P. G. Strelkow and E. Laurman, *Proc. Roy. Soc.* A **167**, 1, 1938.
161. P. Jacquinot, Thesis, Masson, Paris 1937.
162. F. A. Jenkins and E. Segré, *Phys. Rev.* **55**, 52, 1939.
163. D. Harting and P. F. A. Klinkenberg, *Physica* **14**, 669, 1949.
164. L. I. Schiff and H. Snyder, *Phys. Rev.* **55**, 59, 1939.
165. I. Rabi, S. Millman, P. Kusch and J. Zacharias, *Phys. Rev.* **55**, 526, 1939.
166. P. Kusch and H. M. Foley, *Phys. Rev.* **74**, 251, 1948.
167. J. Schwinger, *Phys. Rev.* **73**, 416, 1948.
168. P. Jacquinot and J. Brochard, *J. Phys. Radium VIII*, **10**, 27, 1949.
169. A. Abragam and J. H. Van Vleck, *Phys. Rev.* **92**, 1448, 1953.
170. R. Ladenburg, *Phys. Z.* **30**, 369, 1929.
171. J. S. Foster, *Rep. Progr. Phys.* **5**, 233, 1938.
172. Y. T. Yao, *Z. Phys.* **77**, 303, 1932.
173. H. Kopfermann and W. Paul, *Z. Phys.* **120**, 545, 1943.
174. E. Segré, *Rend. Acc. Lincei* **19**, 591, 596, 1934.
175. N. Tsi-Ze and C. Shin Piaw, *C. R. Acad. Sci. Paris* **198**, 2156, 1934. *J. Phys. Rad. VII*, **6**, 147, 1935.
176. J. S. Foster, *Proc. Roy. Soc.* A **114**, 47; **117**, 137, 1927.
177. R. Becker, *Z. Phys.* **9**, 332, 1922.
178. W. Thomas, *Z. Phys.* **34**, 586, 1925.
179. A. Unsoeld, *Ann. Phys. Lpz.* **82**, 390, 1927.
180. W. Lochte-Holtgreven, *Z. Phys.* **109**, 359, 1938.

181. B. Kullenberg, Dissert. Lund 1938.
182. G. Lüders, *Ann. Phys. Lpz.* **8**, 301, 1951.
183. G. W. Series and K. W. H. Stevens, *Proc. Roy. Soc.* A **226**, 393, 1954.

IV.

1. W. Kossel, *Verh. D. Phys. Ges.* **16**, 898, 953, 1914; **18**, 339, 1916; *Ann. Phys. Lpz.* **49**, 229, 1916.
2. R. Ladenburg, *Z. Elektrochem.* **26**, 262, 1920.
3. N. Bohr, *Z. Phys.* **9**, 1, 1922.
4. E. C. Stoner, *Phil. Mag.* **48**, 719, 1924.
5. D. R. Hartree, *Rep. Progr. Phys.* **11**, 113, 1947.
6. D. R. Hartree, *The Calculation of Atomic Structures*, Wiley 1957.
7. Landolt-Börnstein, *Zahlenwerte u. Funktionen*, **1**, 276, 1950.
8. D. R. Hartree, *Proc. Cambr. Phil. Soc.* **24**, 89, 1928, *Proc. Roy. Soc.* A **141**, 282, 1933.
9. J. C. Slater, *Phys. Rev.* **34**, 1293, 1929.
10. V. Fock, *Z. Phys.* **61**, 126, 1930.
11. A. J. Freeman, *Phys. Rev.* **91**, 1410, 1953.
12. D. R. Hartree and W. Hartree, *Proc. Roy. Soc.* A **164**, 167, 1938.
13. J. C. Slater, *Phys. Rev.* **81**, 385, 1951; **91**, 528, 1953.
14. C. Zener, *Phys. Rev.* **36**, 51, 1930.
15. J. C. Slater, *Phys. Rev.* **36**, 57, 1930.
16. P. M. Morse, L. A. Young and E. S. Haurwitz, *Phys. Rev.* **48**, 948, 1935.
17. H. Yilmaz, *Phys. Rev.* **100**, 1148, 1955.
18. A. Tubis, *Phys. Rev.* **102**, 1049, 1956.
19. P. O. Löwdin, *Phys. Rev.* **90**, 120, 1953; **94**, 1600, 1954; P. O. Löwdin and K. Appel, *Phys. Rev.* **103**, 1746, 1956.
20. L. H. Thomas, *Proc. Cambr. Phil. Soc.* **23**, 542, 1927.
21. E. Fermi, *Z. Phys.* **48**, 73; **49**, 550, 1928.
22. P. A. M. Dirac, *Proc. Cambr. Phil. Soc.* **26**, 376, 1930.
23. R. Latter, *Phys. Rev.* **99**, 510, 1955.
24. H. G. J. Moseley, *Phil. Mag.* **26**, 1024, 1913; **27**, 703, 1914.
25. W. Kossel, *Z. Phys.* **1**, 119, 1920.
26. N. Bohr and D. Coster, *Z. Phys.* **12**, 342, 1923.
27. L. Pauling and S. G. Goudsmit, *Structure of Line Spectra*, McGraw-Hill 1930, p. 180 ff.
28. W. A. Thatcher, *Proc. Roy. Soc.* A **172**, 242, 1939.
29. M. N. Lewis et al. *Phys. Rev.* **98**, 1020, 1955.
30. R. de L. Kronig, *Z. Phys.* **70**, 317; **75**, 191, 468, 1932.
31. J. D. Hanawalt, *Z. Phys.* **70**, 293, 1931; *Phys. Rev.* **37**, 715, 1931.
32. D. Coster and J. Veldkamp, *Z. Phys.* **70**, 306, 1931.
33. J. A. Bearden and H. Friedman, *Phys. Rev.* **58**, 387, 1940.
34. H. W. B. Skinner, *Rep. Progr. Phys.* **5**, 257, 1939; *Phil. Trans.* A **239**, 95 1940.
35. D. H. Tamboulian and D. E. Bedo, *Phys. Rev.* **104**, 590, 1956.
36. F. K. Richtmyer, S. W. Barnes and E. Ramberg, *Phys. Rev.* **46**, 843, 1934.
37. L. G. Parratt, *Phys. Rev.* **56**, 295, 1939.
38. C. H. Shaw, *Phys. Rev.* **57**, 877, 1940.

39. S. Idei, *Nature*, **123**, 643, 1929.
40. H. Bethe, *Z. Phys.* **60**, 603, 1930.
41. E. Segré, *Linc. Rend.* **14**, 501, 1931.
42. S. Kaufmann, *Phys. Rev.* **40**, 116, 1932.
43. G. Wentzel, *Ann. Phys. Lpz.* **66**, 437, 1921.
44. M. J. Druyvesteyn, *Z. Phys.* **43**, 707, 1927.
45. F. K. Richtmyer, *Rev. Mod. Phys.* **9**, 391, 1937.
46. P. Auger, Thesis, Paris, 1926.

V.

1. H. N. Russell and F. A. Saunders, *Astrophys. J.* **61**, 38, 1925.
2. F. Hund, *Z. Phys.* **33**, 345, 1925.
3. E. Wigner, *Z. Phys.* **40**, 492; **43**, 624; **45**, 601, 1927.
4. J. v. Neumann and E. Wigner, *Z. Phys.* **47**, 203; **49**, 73, 1928.
5. J. G. Slater, *Phys. Rev.* **34**, 1293, 1929.
6. G. Racah, *Phys. Rev.* **61**, 186; **62**, 438, 1942; **63**, 367, 1943; **76**, 1352, 1949.
7. G. Racah, *Phys. Rev.* **62**, 523, 1942.
8. H. A. Skinner and H. O. Pritchard, *Trans. Far. Soc.* **49**, 1254, 1953.
9. R. E. Trees, *Phys. Rev.* **83**, 756; **84**, 1089, 1951; **85**, 382, 1952.
10. G. Racah, *Phys. Rev.* **85**, 381, 1952.
11. G. Racah, *Lunds Univ. Arssk.* N.F. Avd. 2, **50**, No. 21 (*Proc. Rydberg Conf.* 1954).
12. N. Sack, *Phys. Rev.* **102**, 1302, 1956.
13. W. Bingel, *Z. Naturf.* **9a**, 675, 1954.
14. R. F. Bacher and S. Goudsmit, *Phys. Rev.* **46**, 948, 1934.
15. S. Meshkow and C. W. Ufford, *Phys. Rev.* **94**, 75, 1954.
16. F. Rohrlich, *Phys. Rev.* **101**, 69, 1956.
17. S. Goudsmit, *Phys. Rev.* **31**, 946, 1928.
18. S. Goudsmit and C. J. Humphreys, *Phys. Rev.* **31**, 960, 1928.
19. R. E. Trees, *Phys. Rev.* **82**, 683, 1951.
20. G. Araki, *Prog. Theor. Phys.* **3**, 152, 262, 1948.
21. R. de L. Kronig, *Z. Phys.* **31**, 885, 1925.
22. A. Sommerfeld and H. Hönl, *Ber. Akad. Berlin* **9**, 141, 1925.
23. H. N. Russell, *Proc. Nat. Acad. Wash.* **11**, 314, 1925.
24. P. Dirac, *Proc. Roy. Soc.* A **111**, 302, 1926.
25. W. V. Houston, *Phys. Rev.* **33**, 297, 1929.
26. E. U. Condon and G. H. Shortley, *Phys. Rev.* **35**, 1342, 1930.
27. O. Laporte and D. R. Inglis, *Phys. Rev.* **33**, 1337, 1930.
28. S. Goudsmit, *Phys. Rev.* **35**, 1325, 1930.
29. A. G. Shenstone, *Phil. Trans.* A **235**, 195, 1936.
30. G. H. Shortley and B. Fried, *Phys. Rev.* **54**, 749, 1938.
31. G. Racah, *Phys. Rev.* **61**, 537, 1942.
32. K. B. S. Eriksson, *Phys. Rev.* **102**, 102, 1956.
33. G. R. Harrison and M. H. Johnson, *Phys. Rev.* **38**. 757, 1931.
34. G. H. Shortley, *Phys. Rev.* **47**, 295, 1935.
35. W. Pauli, *Z. Phys.* **16**, 155, 1923.
36. J. Brochard and P. Jacquinot, *C. R. Acad. Sci. Paris* **223**, 507, 1946.
37. J. Brochard, *Ann. Phys. Paris* **6**, 108, 1951.
38. P. Pluvinage, *J. Phys. Radium* **13**, 405, 1952.

39. A. G. Shenstone and N. H. Russell, *Phys. Rev.* **39**, 415, 1932.
40. R. M. Langer, *Phys. Rev.* **35**, 649, 1930.
41. N. G. Whitelaw and J. H. Van Vleck, *Phys. Rev.* **41**, 389, 1932.
42. L. Pincherle, *Nuov. Cim.* **10**, 37, 1933.
43. G. O. Langstroth, *Proc. Roy. Soc.* A **142**, 286, 1933.
44. A. G. Shenstone, *Phys. Rev.* **37**, 1701; **38**, 873, 1931.
45. E. Fues, *Z. Phys.* **43**, 762, 1927.
46. R. de L. Kronig, *Z. Phys.* **50**, 347, 1928; **62**, 300, 1930.
47. E. Majorana, *Nuov. Cim.* **8**, 107, 1931.
48. W. R. S. Garton and A. Rajaratuam, *Proc. Phys. Soc.* A **68**, 1107, 1955.
49. B. H. Bransden and A. Dalgarno, *Proc. Phys. Soc.* A **66**, 904, 1953.
50. G. Wentzel, *Phys. Z.* **25**, 182, 1924.
51. H. N. Russell and F. A. Saunders, *Astrophys. J.* **61**, 38, 1925.
52. B. H. Bransden and A. Dalgarno, *Proc. Phys. Soc.* A **66**, 911, 1953.
53. B. Edlén, *Nova Acta Soc. Sci. Upsal.* IV, **9**, No. 6, 1934.
54. H. Beutler et al., *Z. Phys.* **86**, 495, 710; **87**, 19, 176, 188, 1933; **88**, 25; **91**, 131, 202, 218, 1934.
55. A. G. Shenstone, *Phys. Rev.* **72**, 411, 1947.
56. G. Racah, *Phys. Rev.* **61**, 194, 1942.
57. M. A. Catalan, F. Rohrlich and A. G. Shenstone, *Proc. Roy. Soc.* A **221**, 421, 1954.
58. J. H. Van Vleck, *Theory of electric and magnetic susceptibilities*, Oxford Univ. Press 1932.
59. W. F. Meggers, *Science* **105**, 514, 1947.
60. P. F. A. Klinkenberg, *Physica* **8**, 1, 1947.
61. F. Hund, *Z. Phys.* **33**, 855, 1925.
62. M. Fred and F. S. Tomkins, *J. Opt. Soc. Amer.* **47**, 1076, 1957.

VI.

1. W. Pauli, *Naturwiss.* **12**, 741, 1924.
2. N. A. Bogros, *C.R. Acad. Sci., Paris* **183**, 124, 1926.
3. D. A. Jackson and H. G. Kuhn, *Proc. Roy. Soc.* A **148**, 335, 1935.
4. R. Minkowski, *Z. Phys.* **95**, 274, 1935.
5. K. W. Meissner and K. L. Luft, *Ann. Phys. Lpz.* **28**, 667, 1937.
6. K. W. Meissner, *Rev. Mod. Phys.* **14**, 68, 1942.
7. J. Brossel and A. Kastler, *C.R. Acad. Sci. Paris* **229**, 1213, 1949.
8. J. Brossel and F. Bitter, *Phys. Rev.* **86**, 308, 1952.
9. G. W. Series, *Rep. Progr. Phys.* **22**, 280, 1959.
10. D. A. Jackson and H. G. Kuhn, *Proc. Roy. Soc.* A **165**, 303, 1938.
11. K. W. Meissner and K. F. Luft, *Z. Phys.* **106**, 362, 1937.
12. H. Schüler and H. Gollnow, *Naturwiss.* **22**, 511, 1934.
13. M. Van den Berg, P. F. A. Klinkenberg and P. Regnaut, *Physica* **20**, 37, 461, 1954.
14. G. Breit, *Phys. Rev.* **35**, 1447, 1930.
15. E. Fermi, *Z. Phys.* **60**, 320, 1930.
16. G. H. Wannier, *Phys. Rev.* **72**, 304, 1947.
17. J. E. Nafe, E. B. Nelson and I. I. Rabi, *Phys. Rev.* **71**, 914, 1947.
18. D. A. Jackson, *Proc. Roy. Soc.* **121**, 432, 1928.
19. M. H. Cohen, D. A. Goodings and V. Heine, *Proc. Phys. Soc.* **73**, 811, 1959.

20. S. Goudsmit, *Phys. Rev.* **43**, 636, 1933.
21. E. Fermi and E. Segré, *Z. Phys.* **82**, 729, 1933.
22. L. L. Foldy, *Phys. Rev.* **111**, 1093, 1958.
23. G. Racah, *Z. Phys.* **76**, 431, 1931.
24. J. E. Rosenthal and G. Breit, *Phys. Rev.* **41**, 495, 1932.
25. M. F. Crawford and A. L. Schawlow, *Phys. Rev.* **76**, 1310, 1949.
26. A. Bohr and V. Weisskopf, *Phys. Rev.* **77**, 94, 1950.
27. A. Bohr, *Phys. Rev.* **81**, 331, 1951.
28. F. Bitter, *Phys. Rev.* **76**, 150, 1949.
29. D. A. Jackson and H. G. Kuhn, *Proc. Roy. Soc.* A **167**, 205, 1938.
30. D. A. Jackson, *Proc. Roy. Soc.* A **147**, 500, 1934.
31. H. Schüler and T. Schmidt, *Z. Phys.* **94**, 457, 1935.
32. M. L. Perl, I. I. Rabi and B. Senitzki, *Phys. Rev.* **98**, 611, 1955.
33. J. J. Ritter and G. W. Series, *Proc. Roy. Soc.* **238**, 473, 1957.
34. U. Meyer-Bekhout, *Z. Phys.* **141**, 185, 1955.
35. K. Althoff, *Z. Phys.* **141**, 33, 1955.
36. D. Strominger, J. M. Hollander and G. T. Seaborg, *Rev. Mod. Phys.* **30**, 585 1958.
37. S. Goudsmit and E. Back. *Z. Phys.* **43**, 321, 1927.
38. G. Breit and I. I. Rabi, *Phys. Rev.* **38**, 2002, 1931.
39. S. Goudsmit and R. Bacher, *Phys. Rev.* **34**, 1499, 1501, 1929.
40. F. M. Kelly, H. G. Kuhn and Anne Pery, *Proc. Phys. Soc.* A **67**, 450, 1954.
41. S. Goudsmit, *Phys. Rev.* **37**, 663, 1931.
42. G. Breit and L. A. Wills, *Phys. Rev.* **44**, 470, 1933.
43. T. Schmidt, *Z. Phys.* **121**, 63, 1943.
44. R. E. Trees, *Phys. Rev.* **92**, 308, 1953.
45. C. Schwartz, *Phys. Rev.* **97**, 380, 1955.
46. H. Schüler and E. Jones, *Z, Phys.* **77**, 802, 1932.
47. S. Mrozowski, *Phys. Rev.* **67**, 161, 1945.
48. S. Goudsmit and R. Bacher, *Phys. Rev.* **43**, 894, 1933.
49. F. Les and H. Niewodnicsanski, *Bull. Acad. Pol. Sci.* (III) V, 299, 1957.
50. G. F. Koster, *Phys. Rev.* **86**, 148, 1952.
51. G. K. Woodgate and J. S. Martin, *Proc. Phys. Soc.* A**70**, 485, 1957.
52. V. Heine, *Phys. Rev.* **107**, 1002, 1957.
53. P. Güttinger and W. Pauli, *Z. Phys.* **67**, 743, 1931.
54. S. Suwa, *Phys. Rev.* **78**, 810, 1950; *J. Phys. Soc. Jap.* **6**, 231, 1951.
55. M. Fred, F. S. Tomkins, et al., *Phys. Rev.* **82**, 406, 1951.
56. D. S. Hughes and C. Eckart, *Phys. Rev.* **36**, 694, 1930; J. H. Bartlett and J. J. Gibbons, *Phys. Rev.* **44**, 538, 1930.
57. H. Schüler, *Z. Phys.* **42**, 487, 1927.
58. D. A. Jackson and H. G. Kuhn, *Proc. Roy. Soc.* **173**, 278, 1939.
59. L. C. Bradley and H. G. Kuhn, *Proc. Roy. Soc.* A **209**, 325, 1951.
60. A. P. Stone, *Proc. Phys. Soc.* A 68, 1152, 1955.
61. G. Araki, K. Mano and M. Ohta, *Phys. Rev.* **115**, 1222, 1959.
62. S. Wagner, *Z. Phys.* **141**, 122, 1955.
63. W. Pauli and R. Peierls, *Phys. Z.* **32**, 670, 1931.
64. J. E. Rosenthal and G. Breit, *Phys. Rev.* **41**, 459; **42**, 384, 1932.
65. P. Brix and H. Kopfermann, *Z. Phys.* **126**, 344, 1944.
66. H. G. Kuhn and S. A. Ramsden, *Proc. Roy. Soc.* A **237**, 485, 1956.
67. H. Schüler and H. Westmeyer, *Z. Phys.* **82**, 685, 1933.

68. E. K. Broch, *Arch. f. Math. Nat.* **48**, 25, 1945.
69. W. Humbach, *Z. Phys.* **133**, 589, 1952.
70. A. R. Bodmer, *Proc. Phys. Soc.* A **66**, 1041, 1953.
71. W. Humbach, *Z. Phys.* **141**, 59, 1955.
72. G. Breit, *Rev. Mod. Phys.* **30**, 507, 1958.
73. P. Brix and H. Kopfermann, *Rev. Mod. Phys.* **30**, 517, 1958.
74. J. Blaise, Thesis, Paris 1958.
75. P. F. A. Klinkenberg, *Physica* **11**, 327, 1945.
76. H. Arroe, *Phys. Rev.* **93**, 94, 1954.
77. P. Brix, H. v. Butlar, F. Houtermans and H. Kopfermann, *Z. Phys.* **79**, 836, 1950.
78. W. R. Hindmarsh and H. G. Kuhn, *Proc. Phys. Soc.* A **68**, 433, 1955.
79. H. G. Kuhn and A. G. Warner, *Proc. Roy. Soc.* A **245**, 330, 1958.
80. H. G. Kuhn and R. Turner, *Proc. Roy. Soc.* A., in the press.
81. R. H. Hughes, *Bull. Amer. Phys. Soc.* II **4**, 262, 1959.
82. W. H. King and H. G. Kuhn, unpublished work.
83. See G. M. Temmer, *Rev. Mod. Phys.* **30**, 498, 1958.
84. L. Wilets, L. Hill & K. Ford, *Phys. Rev.* **91**, 1488, 1953.
85. P. H. Stelson and F. K. McGowan, *Phys. Rev.* **110**, 489, 1958.
86. H. Arroe, *Phys. Rev.* **79**, 836, 1950.

VII.

1. P. A. M. Dirac, *Proc. Roy. Soc.* A **114**, 243, 710, 1927.
2. V. Weisskopf and E. Wigner, *Z. Phys.* **63**, 54; **65**, 18, 1930.
3. V. Weisskopf, *Ann. Phys. Lpz.* (5) **9**, 23, 1931.
4. M. Born, *Optik*, Springer, Berlin 1933, p. 470.
5. R. Minkowski, *Z. Phys.* **95**, 274, 1935.
6. A. C. G. Mitchell and M. W. Zemansky, *Resonance Radiation and Excited Atoms* Cambr. Univ. Press 1934.
7. R. Minkowski, *Z. Phys.* **36**, 839, 1926.
8. D. Roschdestwensky, *Ann. Phys. Lpz.* **39**, 307, 1912, *Trans. Opt. Inst. Leningrad* **2**, No. 13, 1921.
9. R. Ladenburg and G. Wolfsohn, *Z. Phys.* **63**, 616, 1930.
10. R. Ladenburg and E. Thiele, *Z. Phys.* **72**, 697, 1931.
11. R. Ladenburg and G. Wolfsohn, *Z. Phys.* **65**, 207, 1930.
12. W. Schütz, *Z. Phys.* **64**, 682, 1930.
13. W. Schütz, *Naturwiss.* **20**, 64, 1932.
14. L. C. Bradley and H. G. Kuhn, unpublished.
15. H. Hupfield, *Z. Phys.* **54**, 484, 1929.
16. S. Heron, R. W. P. Whirter and E. H. Rhoderick, *Proc. Roy. Soc.* A **234**, 565, 1956.
17. P. H. Garrett and H. W. Webb, *Phys. Rev.* **37**, 1686, 1931; **40**, 778, 1932.
18. H. D. Koenig and A. Ellett, *Phys. Rev.* **39**, 576, 1932.
19. P. Soleillet, *J. Phys. Radium* **3**, 59, 1932.
20. H. W. Webb and H. A. Messenger, *Phys. Rev.* **66**, 77, 1944.
21. H. C. Van de Hulst and J. J. M. Reesink, *Astrophys. J.* **106**, 121, 1947.
22. C. W. Allen, *Astrophysical quantities*, Athlone Press, 1955, p. 83.
23. H. A. Lorentz, *Amsterdam Proc.* **8**, 591, 1906.

24. J. Holtsmark, *Ann. Phys. Lpz.* **58**, 577, 1919; *Phys. Z.* **20**, 162, 1919; **25**, 73, 1924.
25. V. Weisskopf, *Phys. Z.* **34**, 1, 1933.
26. W. Lenz, *Z. Phys.* **80**, 423; **83**, 139, 1933.
27. E. Lindholm, *Ark. Mat. Ast. Phys.* **28B**, 1, 1941; **32A**, 1, 1945.
28. A. Jablonski, *Phys. Rev.* **68**, 78, 1945.
29. H. M. Foley, *Phys. Rev.* **69**, 616, 1946.
30. A. Jablonski, *Z. Phys.* **70**, 723, 1931.
31. F. London, *Z. Phys.* **63**, 245, 1930; *Z. Phys. Chem.* B **11**, 222, 1930.
32. H. G. Kuhn, *Phil. Mag.* (7) **18**, 987, 1934.
33. See W. Lochte-Holtgreven, *Rep. Progr. Phys.* **21**, 312, 1958.
34. A. Unsöld, *Physik d. Sternatmosphaeren*, 2nd ed., Springer 1955.
35. V. Weisskopf, *Z. Phys.* **75**, 287, 1932.
36. R. Minkowski, *Z. Phys.* **93**, 731, 1935.
37. H. G. Kuhn, *Proc. Roy. Soc.* A **158**, 212, 1937.
38. H. A. Rühmkorf, *Ann. Phys. Lpz.* **33**, 21, 1938.
39. L. Huldt and E. Knall, *Z. Naturf.* **9a**, 663, 1954.
40. H. G. Kuhn, *Proc. Roy. Soc.* A **158**, 230, 1937.
41. S. Robin and S. Robin, *C.R. Acad. Sci. Paris* **244**, 2375; **245**, 1056, 1957.
42. H. Margenau, *Phys. Rev.* **48**, 755, 1935.
43. G. Burkhardt, *Z. Phys.* **115**, 592, 1940.
44. L. Spitzer, *Phys. Rev.* **55**, 699; **56**, 39, 1939; **58**, 348, 1940.
45. P. W. Anderson, *Phys. Rev.* **86**, 809, 1952.
46. M. Baranger, *Phys. Rev.* **111**, 489, 494; **112**, 855, 1958.
47. A. C. Kolb and H. Green, *Phys. Rev.* **111**, 514, 1958.
48. H. Margenau and M. Lewis, *Rev. Mod. Phys.* **31**, 569, 1959.
49. L. A. Vainshtein and I. I. Sobel'man, *Optica i. Spectrosc.* **6** (4), 440, 1959,
50. L. A. Vainshtein, V. G. Koloshnikow, M. A. Mazing, S. L. Mandel'stam, and I. I. Sobel'man, *Isv. Ak. Nauk. S.S.S.R.*, Ser. Fiz. **22**, 718, 1958.
51. E. Fermi, *Nuov. Cim.* **11**, 157, 1934.
52. C. Reinsberg, *Z. Phys.* **105**, 460, 1937; **111**, 95, 1938.

Author Index

(The numbers in brackets refer to the bibliography)

Subject Index